RESEARCH AND PERSPECTIVES IN ALZHEIMER'S DISEASE
Fondation Ipsen

K.S. Kosik Y. Christen
D.J. Selkoe (Eds.)

Alzheimer's Disease: Lessons from Cell Biology

With 60 Figures and 4 Tables

Springer-Verlag
Berlin Heidelberg New York London Paris
Tokyo Hong Kong Barcelona Budapest

Kosik, K.S., M.D.
Associate Professor of Neurology and Neuroscience
Harvard Medical School
Brigham and Women's Hospital
221 Longwood Avenue
Boston, MA 02115
USA

Christen, Y., Ph.D.
Fondation IPSEN
24, rue Erlanger
75781 Paris
France

Selkoe, D.J., M.D.
Professor of Neurology and Neuroscience
Harvard Medical School
Brigham and Women's Hospital
221 Longwood Avenue
Boston, MA 02115
USA

ISBN 3-540-58744-6 Springer-Verlag Berlin Heidelberg New York

Library of Congress Cataloging-in-Publication Data. Alzheimer's disease: lessons from cell biology/K.S. Kosik. Y. Christen, D.J. Selkoe (eds.). p. cm. – (Research and perspectives in Alzheimer's disease) Proceedings of the Tenth Colloque médecine et recherche held in Paris on April 25, 1994. Includes bibliographical references and index. ISBN 3-540-58744-6. – ISBN 0-387-58744-6 1. Alzheimer's disease – Pathophysiology – Congresses. 2. Alzheimer's disease – Molecular aspects – Congresses. I. Kosik, K.S. (Kenneth S.), 1950– . II. Christen, Yves. III. Selkoe, Dennis J. IV. Colloque médecine et recherche (10th: 1994: Paris, France) V. Series. [DNLM: 1. Alzheimer's Disease – physiopathology – congresses. WM 220 A4753 1995] RC523.A3793 1995 616.8′31 – dC20 DNLM/DLC for Library of Congress 94-41048

Printed in Germany

Typesetting: Best-set Typesetter Ltd., Hong Kong

SPIN: 10484379 27/3130/SPS – 5 4 3 2 1 0 – Printed on acid-free paper

Preface

Like the unflinching gaze of Captain Ahab walking the deck of the Pequod, Alzheimer researchers have had their sights fixed firmly on the disease for many years. Now, as this volume amply demonstrates, accomplished researchers from other fields, who have thought deeply about cell biological problems are applying their insights to Alzheimer's disease. The contributions here represent the text versions of the proceedings from the tenth "Colloque médecine et recherche" of the Fondation IPSEN devoted to research on Alzheimer's disease. The symposium, entitled "Alzheimer's Disease: Lessons from Cell Biology" was held in Paris on April 25, 1994. As is apparent from the varied backgrounds of the contributors, the scientific pursuit of Alzheimer's disease has begun to meld with more basic disciplines, particularly cell biology. While on the one hand, new areas of specialization are continuously emerging, the boundaries of older disciplines are increasingly blurred. Perhaps for most of the years since the first descriptions of the disease in 1907, the science of Alzheimer's disease was descriptive, and lay in the province of pathologists. This time period, during which a great deal was learned about the topography of senile plaques and neurofibrillary tangles, culminated with an ultrastructural description of these hallmark structures.

The modern era of Alzheimer's disease research opened with the identification of the component proteins in plaques and tangles. From the direct peptide sequence data obtained on these proteins, it was possible to clone their genes; thus began the molecular foundation upon which subsequent developments relied. During this time period, which continues up to the present, molecular techniques have provided major insights, including the discovery of a mutation in the gene encoding the amyloid precursor protein. This mutation although exceedingly rare is the first identified cause of Alzheimer's disease and has catalyzed a rich area of research which seeks to understand, in cellular terms, the series of steps, beginning with a mutant APP gene, that leads to the Alzheimer phenotype. Now we must learn how the cell as a total entity orchestrates its myriad gene products in a way that can sustain brain function well into senescence or fail, and undergo an insidious degeneration. To that end, the editors have brought together both basic cell biologists and Alzheimer's disease researchers to achieve a timely synthesis of these disciplines. Conceptual problems over which cell biologists have pondered for many years, such as how cells organize their cytoplasm or regulate their phosphorylation state, are ripe for application to the specific cells affected by the Alzheimer disease process. Likewise the level of understanding achieved by Alzheimer scientists in identifying specific proteins involved in the disease, has brought the field to the doorstep of the cell biologist and has introduced novel concepts in how cells function.

It has been the aim of this series to look just beyond the present at the direction toward which future research is moving. There is no doubt that cell biology is that crucial intermediate between where we stand now, with the several identified proteins, such as tau and the amyloid precursor protein, to the cellular basis for intellectual function.

Ken Kosik
Yves Christen
Dennis Selkoe

Acknowledgements. The editors wish to thank Mrs Mary Lynn Gage for editorial assistance, Mrs Jacqueline Mervaillie for the organization of the meeting in Paris and Daniel Louvard, Jacques Nunez, Jacques Mallet and Alain Prochiantz for their collaboration as chairmen of the meeting.

Contents

Contributors

Allinquant, B.
CNRS URA 1414, Ecole Normale Supérieure, 46 rue d'Ulm,
75230 Paris Cedex 05, France

Baumann, K.
Max-Planck-Unit for Structural Molecular Biology, c/o DESY,
Notkestr. 85, 22603 Hamburg, Germany

Behl, C.
The Salk Institute for Biological Studies, 10010 N. Torrey Pines Rd.,
La Jolla, CA 92037, USA

Biernat, J.
Max-Planck-Unit for Structural Molecular Biology, c/o DESY,
Notkestr. 85, 22603 Hamburg, Germany

Bonzelius, F.
Department of Biochemistry and Biophysics, University of California,
San Francisco, CA 94143, USA

Bouillot, C.
CNRS URA 1414, Ecole Normale Supérieure, 46 rue d'Ulm,
75230 Paris Cedex 05, France

Bouras, C.
Department of Psychiatry, IUPG Bel-Air, University of Geneva,
School of Medicine, Geneva, Switzerland

Brion, J.P.
Laboratory of Pathology and Electron Microscopy,
Université Libre de Bruxelles, 808 route de Lennik, Bldg C-10,
1070 Brussels, Belgium

Brugg, B.
Université Pierre et Marie Curie, Institut des Neurosciences, (URA CNRS 1488) Lab. de Neurobiologie du Développement, 75005 Paris, France

Buée, L.
INSERM U422, Place de Verdun, 59045 Lille Cédex, France

Buée-Scherrer, V.
INSERM U422, Place de Verdun, 59045 Lille Cédex, France

Cheng, M.
The University of Texas Southwestern Medical Center at Dallas, Department of Pharmacology, 5323 Harry Hines Blvd., Dallas, TX 75235-9041, USA

Cheng, S.
Howard Hughes Medical Institute, The Rockefeller University, 1230 York Avenue, New York, NY 10021, USA

Cleveland, D.W.
Departments of Biological Chemistry and Neuroscience, 725 North Wolfe Street, The Johns Hopkins University School of Medicine, Baltimore, MD 21205, USA

Cobb, M.H.
The University of Texas Southwestern Medical Center at Dallas, Department of Pharmacology, 5323 Harry Hines Blvd., Dallas, TX 75235-9041, USA

Conreur, J.L.
Laboratory of Pathology and Electron Microscopy, Université Libre de Bruxelles, 808 route de Lennik, Bldg c-10, 1070 Brussels, Belgium

Couck, A.M.
Laboratory of Pathology and Electron Microscopy, Université Libre de Bruxelles, 808 route de Lennik, Bldg c-10, 1070 Brussels, Belgium

Couratier, P.
Unité de Neurobiologie Cellulaire, Laboratoire d'Histologie, Faculté de Médecine, 2 rue du Docteur Marcland, 87025 Limoges Cédex, France

Cras, P.
Laboratory of Neurogenetics, Born Bunge Foundation, Department of Biochemistry, University of Antwerp (UIA), Universiteits pleni 1, 2610 Antwerp, Belgium

Dang, A.
The University of Texas Southwestern Medical Center at Dallas, Department of Pharmacology, 5323 Harry Hines Blvd., Dallas, TX 75235-9041, USA

Davis, J.B.
The Salk Institute for Biological Studies, 10010 N. Torrey Pines Rd., La Jolla, CA 92037, USA. Present adress: Dept of Molecular Neuropathology, SKB, Coldharbour Rd, Harlow CM 19 SAD, UK

De Jager, P.L.
Howard Hughes Medical Institute, The Rockefeller University, 1230 York Avenue, New York, NY 10021, USA

De Strooper, B.
Experimental Genetics Group, Center for Human Genetics, K.U. Leuven, Campus Gasthuisberg O&N 6, 3000 Leuven, Belgium

Delacourte, A.
INSERM U422, Place de Verdun, 59045 Lille Cédex, France

Delhaye-Bouchaud, N.
Université Pierre et Marie Curie, Institut des Neurosciences, (URA CNRS 1488) Lab. de Neurobiologie du Développement, 75005 Paris, France

Dewachter, I.
Experimental Genetics Group, Center for Human Genetics, K.U. Leuven, Campus Gasthuisberg O&N 6, 3000 Leuven, Belgium

Drewes, G.
Max-Planck-Unit for Structural Molecular Biology, c/o DESY, Notkestr. 85, 22603 Hamburg, Germany

Ebert, D.
The University of Texas Southwestern Medical Center at Dallas, Department of Pharmacology, 5323 Harry Hines Blvd., Dallas, TX 75235-9041, USA

Esclaire, F.
Unité de Neurobiologie Cellulaire, Laboratoire d'Histologie, Faculté de Médecine, 2 rue du Docteur Marcland, 87025 Limoges Cédex, France

Felsenstein, K.
CNS Drug Discovery Unit, Bristol Myers Squibb, 5 Research Parkway, Wallingford, CT 06471, USA

Feng, L.
Howard Hughes Medical Institute, The Rockefeller University,
1230 York Avenue, New York, NY 10021, USA

Ferreira, A.
Harvard Medical School and Center for Neurologic Diseases, Department of Medicine (Division of Neurology), Brigham and Women's Hospital,
Boston, MA 02115, USA

Green, S.
Department of Biochemistry and Biophysics, University of California,
San Francisco, CA 94143, USA

Greenberg, S.M.
Harvard Medical School and Center for Neurologic Diseases, Department of Medicine (Division of Neurology), Brigham and Women's Hospital,
Boston, MA 02115, USA

Grote, E.
Department of Biochemistry and Biophysics, University of California,
San Francisco, CA 94143, USA

Gubbay, J.
Howard Hughes Medical Institute, The Rockefeller University,
1230 York Avenue, New York, NY 10021, USA

Gustke, N.
Max-Planck-Unit for Structural Molecular Biology, c/o DESY,
Notkestr. 85, 22603 Hamburg, Germany

Hantraye, P.
CNRS URA 1285, SHFJ-CEA, 4 Place du Général Leclerc, Orsay,
France

Heintz, N.
Howard Hughes Medical Institute, The Rockefeller University,
1230 York Avenue, New York, NY 10021, USA

Hendriks, L.
Laboratory of Neurogenetics, Born Bunge Foundation, Department of Biochemistry, University of Antwerp (UIA), Universiteits pleni 1,
2610 Antwerp, Belgium

Hepler, J.E.
The University of Texas Southwestern Medical Center at Dallas, Department of Pharmacology, 5323 Harry Hines Blvd., Dallas, TX 75235-9041, USA

Herman, G.
Department of Biochemistry and Biophysics, University of California, San Francisco, CA 94143, USA

Hof, P.R.
Departments of Neurobiology, Pathology, Geriatrics and Psychiatry, Mount Sinai Medical Center, New York, NY 10029, USA

Huber, G.
F. Hoffmann-La Roche, Basel, Switzerland

Hugon, J.
Unité de Neurobiologie Cellulaire, Laboratoire d'Histologie, Faculté de Médecine, 2 rue du Docteur Marcland, 87025 Limoges Cédex, France

Ihara, Y.
Department of Neuropathology, Institute for Brain Research, Faculty of Medicine, University of Tokyo, 7-3-1 Hongo, Bunkyoku, Tokyo 113, Japan

Kelly, R.B.
Department of Biochemistry and Biophysics, University of California, San Francisco, CA 94143, USA

Kincaid, R.L.
Department of Cell Biology, Human Genome Sciences, Inc., 9620 Medical Center Drive, Rockville, MD 20850, USA

Klier, F.G.
Scripps Research Institute, 10666 N. Torrey Pines Rd., La Jolla, CA 92037, USA

Knowles, R.
Harvard Medical School and Center for Neurologic Diseases, Department of Medicine (Division of Neurology), Brigham and Women's Hospital, Boston, MA 02115, USA

Kopmels, B.
Université Pierre et Marie Curie, Institut des Neurosciences, (URA CNRS 1488) Lab. de Neurobiologie du Développement, 75005 Paris, France

Kosik, K.S.
Harvard Medical School and Center for Neurologic Diseases, Department of Medicine (Division of Neurology), Brigham and Women's Hospital, Boston, MA 02115, USA

Leclerc, N.
Harvard Medical School and Center for Neurologic Diseases, Department of Medicine (Division of Neurology), Brigham and Women's Hospital, Boston, MA 02115, USA

Lemaigre-Dubreuil, Y.
Université Pierre et Marie Curie, Institut des Neurosciences, (URA CNRS 1488) Lab. de Neurobiologie du Développement, 75005 Paris, France

Lesort, M.
Unité de Neurobiologie Cellulaire, Laboratoire d'Histologie, Faculté de Médecine, 2 rue du Docteur Marcland, 87025 Limoges Cédex, France

Leveugle, B.
Departments of Neurobiology, Pathology, Geriatrics and Psychiatry, Mount Sinai Medical Center, New York, NY 10029, USA

Lichtenberg-Kraag, B.
Max-Planck-Unit for Structural Molecular Biology, c/o DESY, Notkestr. 85, 22603 Hamburg, Germany

Lorent, K.
Experimental Genetics Group, Center for Human Genetics, K.U. Leuven, Campus Gasthuisberg O&N 6, 3000 Leuven, Belgium

Mandelkow, E.
Max-Planck-Unit for Structural Molecular Biology, c/o DESY, Notkestr. 85, 22603 Hamburg, Germany

Mandelkow, E.-M.
Max-Planck-Unit for Structural Molecular Biology, c/o DESY, Notkestr. 85, 22603 Hamburg, Germany

Mariani, J.
Université Pierre et Marie Curie, Institut des Neurosciences, (URA CNRS 1488) Lab. de Neurobiologie du Développement, 75005 Paris, France

Martin, J.-J.
Laboratory of Neurogenetics, Born Bunge Foundation, Department of Biochemistry, University of Antwerp (UIA), Universiteits pleni 1, 2610 Antwerp, Belgium

Matter, K.
Department of Cell Biology, Yale University School of Medicine, 333 Cedar Street, PO Box 3333, New Haven, CT 06520, USA

Mellman, I.
Department of Cell Biology, Yale University School of Medicine, 333 Cedar Street, PO Box 3333, New Haven, CT 06520, USA

Moechars, D.
Experimental Genetics Group, Center for Human Genetics, K.U. Leuven, Campus Gasthuisberg O&N 6, 3000 Leuven, Belgium

Morishima-Kawashima, M.
Department of Neuropathology, Institute for Brain Research, Faculty of Medicine, University of Tokyo, 7-3-1 Hongo, Bunkyoku, Tokyo 113, Japan

Moya, K.L.
CNRS URA 1285 and INSERM U334, SHFJ-CEA, 4 Place du Général Leclerc, Orsay, France

Norman, D.J.
Howard Hughes Medical Institute, The Rockefeller University, 1230 York Avenue, New York, NY 10021, USA

Octave, J.N.
Laboratory of Neurochemistry, Université Catholique de Louvain, Clos Chapelle aux Champs, 1200 Brussels, Belgium

Perl, D.P.
Departments of Neurobiology, Pathology, Geriatrics and Psychiatry, Mount Sinai Medical Center, New York, NY 10029, USA

Pfeffer, S.R.
Department of Biochemistry, Stanford University School of Medicine, Stanford, CA 94305-5307, USA

Pollack, N.
Department of Cell Biology, Yale University School of Medicine, 333 Cedar Street, PO Box 3333, New Haven, CT 06520, USA

Prochiantz, A.
CNRS URA 1414, Ecole Normale Supérieure, 46 rue d'Ulm, 75230 Paris Cedex 05, France

Ramaswami, M.
Department of Biochemistry and Biophysics, University of California, San Francisco, CA 94143, USA

Robbins, D.
The University of Texas Southwestern Medical Center at Dallas, Department of Pharmacology, 5323 Harry Hines Blvd., Dallas, TX 75235-9041, USA

Roberts, S.
CNS Drug Discovery Unit, Bristol Myers Squibb, 5 Research Parkway, Wallingford, CT 06471, USA

Roome, J.
CNS Drug Discovery Unit, Bristol Myers Squibb, 5 Research Parkway, Wallingford, CT 06471, USA

Schubert, D.
The Salk Institute for Biological Studies, 10010 N. Torrey Pines Rd., La Jolla, CA 92037, USA

Selkoe, D.J.
Harvard Medical School, Brigham and Women's Hospital, Boston, MA 02115, USA

Sindou, P.
Unité de Neurobiologie Cellulaire, Laboratoire d'Histologie, Faculté de Médecine, 2 rue du Docteur Marcland, 87025 Limoges Cédex, France

Van Broeckhoven, C.
Laboratory of Neurogenetics, Born Bunge Foundation, Department of Biochemistry, University of Antwerp (UIA), Universiteits pleni 1, 2610 Antwerp, Belgium

van de Goor, J.
Department of Biochemistry and Biophysics, University of California, San Francisco, CA 94143, USA

Van Leuven, F.
Experimental Genetics Group, Center for Human Genetics, K.U. Leuven, Campus Gasthuisberg O&N 6, 3000 Leuven, Belgium

Vermersch, P.
INSERM U422, Place de Verdun, 59045 Lille Cédex, France

Wattez, A.
INSERM U422, Place de Verdun, 59045 Lille Cédex, France

Wille, H.
Max-Planck-Unit for Structural Molecular Biology, c/o DESY, Notkestr. 85, 22603 Hamburg, Germany

Wollman, E.E.
Hôpital Cochin INSERM 283, Paris, France

Xu, Z.
Department of Biological Chemistry, 725 North Wolfe Street, The Johns Hopkins University School of Medicine, Baltimore, MD 21205, USA

Yamamoto, E.
Department of Cell Biology, Yale University School of Medicine, 333 Cedar Street, PO Box 3333, New Haven, CT 06520, USA

Yardin, C.
Unité de Neurobiologie Cellulaire, Laboratoire d'Histologie, Faculté de Médecine, 2 rue du Docteur Marcland, 87025 Limoges Cédex, France

Zhen, E.
The University of Texas Southwestern Medical Center at Dallas, Department of Pharmacology, 5323 Harry Hines Blvd., Dallas, TX 75235-9041, USA

Zuo, J.
Howard Hughes Medical Institute, The Rockefeller University, 1230 York Avenue, New York, NY 10021, USA

Endocytotic Pathways in Neurons

E. Grote, F. Bonzelius, G. Herman, M. Ramaswami,*
J. van de Goor, S. Green, and *R.B. Kelly*

Summary

The amyloid precursor protein is a cell surface protein of neurons that undergoes rapid endocytosis and degradation in lysosomes. In diseased brains the proteolysis of the protein in the lysosomes is aberrant, giving rise to novel proteolytic products, including the ß-amyloid fragment. We have examined the processes of endocytosis and protein degradation in neurons and the neuroendocrine PC12 cell line. We find evidence for two different classes of endosome that internalize different surface proteins. One of these classes of endosomes is enriched in cell bodies, where it carries out housekeeping functions. The other is in axonal processes, where its functions appear to include synaptic vesicle biogenesis. We have followed the biogenesis of synaptic vesicles in PC12 cells using a synaptic vesicle protein, VAMP or synaptobrevin, that is epitope tagged on its lumenal domain. No neural-specific protein is required for the endocytosis of VAMP since it is endocytosed rapidly in transfected CHO cells. Mutational analysis of VAMP showed that the signal for endocytosis was constrained to a small sequence which showed no homologies with other known endocytotic signals. Domains required for sorting to synaptic vesicles were distributed throughout the length of the molecule. The sequence required for endocytosis, which was contained within a region predicted to be a coiled coil, was also required for targeting to synaptic vesicles.

Internalization of synaptic vesicle proteins into the nerve terminal can also be studied using the neuromuscular junction of Drosophila larvae. The *shibire* locus controls endocytosis. Mutations in this locus arrest the synaptic vesicle life cycle at the internalization step, leading to loss of synaptic vesicle content and paralysis. We show that the *shibire* arrest precedes a calcium-independent step.

To study what targets proteins from the endosome to the lysosome, we studied a surface protein that is rapidly degraded in cells, P-selectin. The rapid degradation in lysosomes of this protein could be attributed to an exon

* Department of Biochemistry and Biophysics, University of California, San Francisco, CA 94143, USA

K.S. Kosik et al. (Eds.)
Alzheimer's Disease: Lessons from Cell Biology

that codes for a 10 amino acid region of the cytoplasmic tail. The amyloid precursor, which is also rapidly degraded, may have an equivalent lysosomal targeting domain.

Introduction

Often it is important that a membrane protein carry out its function only when it is in a defined location in the cell. If a membrane protein has a defined intracellular address that guides it through the biosynthetic organelles to its final destination, the address must be encoded in some fashion in its protein sequence. An address or sorting domain may be recognized by specialized sorting machinery, such as an adaptin and its associated clathrin coat, that removes the protein from its donor compartment. This we shall call *direct* sorting. Alternatively, the protein sequence may have no sorting information itself but have a conformation that allows it to associate with another protein that has a direct sorting signal. Finally, a protein may interact with a second protein in such a way that it causes the protein to be retained in a donor compartment. We call the latter two mechanisms *associative* sorting, since it requires association with another protein. To understand sorting, then, one needs to identify alternative sorting pathways, sorting domains, and the sorting machinery.

Endocytosis has been a favorite pathway for the study of sorting because it is possible to label proteins on the cell surface and follow their fates. Most information has accrued for endocytosis into the receptor recycling pathway (Fig. 1) used to internalize nutrients, such as low density lipoprotein (LDL)

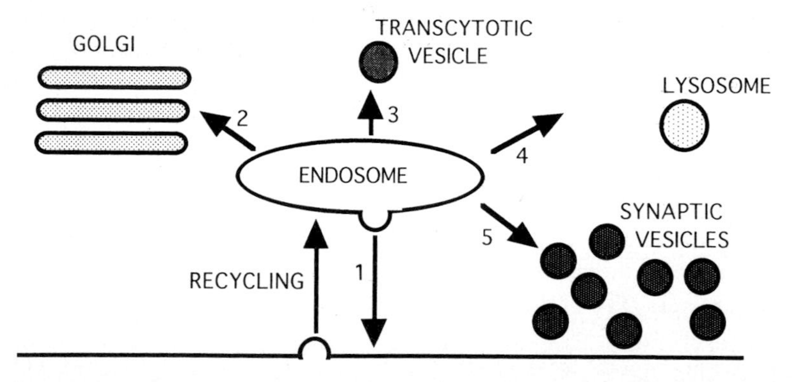

Fig. 1. Endosome-derived pathways in a cell. *1* Recycling pathway that carries nutrients such as iron or low density lipoprotein into the cell. *2* Return of membrane proteins from the cell surface to the Golgi complex, where they can be re-utilized. *3* Generation of transcytotic vesicles that carry membrane proteins from basolateral to apical membranes, and vice versa. The vesicles carrying membrane proteins to the axon may be analogous to transcytotic vesicles. *4* Targeting to the lysosome to give down-regulation of surface proteins. *5* Diversion of proteins derived from the plasma membrane into cytoplasmic storage vesicles such as synaptic vesicles

and transferrin (Trowbridge et al. 1993). A second function carried out by cellular endocytosis is the recycling of secretory vesicle membrane back to the Golgi complex to be re-utilized by the biosynthetic machinery. It is not known if this pathway is identical to the partially characterized pathway taken by cell surface proteins such as the LDL and transferrin receptors (Green and Kelly 1992). Some cell surface proteins are unusually short-lived, expressed briefly on the cell surface and then internalized, and sent to the lysosome for degradation. This is a small, but important class of surface proteins, including the EGF-EGF receptor complex, the CD4 receptor for HIV and the ß-amyloid precursor protein. Membrane proteins that are rapidly degraded appear, in at least one case, to have a special sorting domain that targets them to the lysosome. Finally, endocytosis is used to generate pools of small intracellular vesicles that can be mobilized to the cell surface on chemical stimulation of the cell. Examples include the small cytoplasmic vesicles containing the facilitative glucose transporter, GLUT 4, which fuse with the plasma membrane in the presence of insulin, and the synaptic vesicles which fuse at the nerve terminal releasing neurotransmitter.

One of the tasks of those studying endocytosis is to discover if the same well-characterized machinery that is used for receptor recycling is used in the other endocytotic pathways. Specifically, it is important to know if the same sorting domains are used and if they are recognized by the same sorting machinery. To address these issues we have studied neuroendocrine cells, which express all of the pathways described in Figure 1. Our data suggest that there are at least two different types of endosome in the pheochromocytoma cell line, PC12, that an unusual sorting signal regulates the endocytosis of synaptic vesicle proteins, and that a 10 amino acid domain selectively targets surface proteins to the lysosome for degradation. We have used two blocks to endocytosis, mutations in the dynamin protein and intoxication by black widow spider venom in the absence of calcium, to separate the steps of the endocytotic pathways.

Results

Two Classes of Endosome

In hippocampal neurons the transferrin receptor, a marker for housekeeping endosomes, is in the cell bodies but not in the axonal processes. Endocytosis can be detected in the axon using a bulk phase marker (Mundigl et al. 1993). From the similarity between sorting processes in epithelial cells and neurons (Pietrini et al. 1994), we predicted that the polymeric immunoglobulin receptor (pIgR), which undergoes basal to apical transcytosis in epithelia, would be targeted to axonal endosomes. To test this prediction, we expressed pIgR in PC12 cells. In the presence of NGF, PC12 cells extend processes that have some of the properties expected of axons. One such property is the accumulation of synaptic vesicle proteins at the tips of the processes. We found that the fraction of cells that accumulated the synaptic vesicle protein, synaptophysin, at the tips of their processes was much higher than the fraction that accumulated the transferring receptor, a housekeeping endosome marker, or the mannose phosphate receptor, a late endosome marker. In contrast, the distribution of pIgR was almost indistinguishable by immunofluorescence from that of synaptophysin (Bonzelius et al. 1994). We conclude that the pIgR allows us to detect an endosome that is distinct from the housekeeping endosome, since it is enriched in axons from which housekeeping endosomes are excluded. Although synaptophysin is probably also present in axonal endosomes, it is present in synaptic vesicles, from which pIgR is excluded.

To verify that the distribution of endosome markers truly represents a distribution of endosome functions, we compared the uptake of a ligand targeted to housekeeping endosomes (antibody to the LDL receptor) to the uptake of antibody to pIgR or the physiological ligand for the pIgR (dimeric IgA). Once again the distributions were quite different, with uptake of antibody to the LDL receptor restricted almost exclusively to the cell bodies, whereas the ligands for pIgR were also found at the tips (Bonzelius et al. 1994). The concordance between the data from neurons and epithelial cells leads us to conclude that the different tasks in a cell that involve endocytosis are not all executed by a single endocytotic mechanism; rather different endosomes are available for different tasks. For convenience we refer to the class of endosomes that exclude housekeeping markers as specialized endosomes.

Synaptic Vesicles and the Endocytotic Pathways

The synaptic vesicle proteins, synaptotagmin, SV2 and VAMP, are excluded from housekeeping endosomes but are associated with the specialized endosomes of the axon (Mundigl et al. 1993). Only synaptophysin is present in both housekeeping and specialized endosomes (for review, see Kelly 1993). The targeting of synaptic vesicle membrane proteins to endosomes has implicated an endosomal intermediate in synaptic vesicle biogenesis. Because endocytosis is best studied in cells in culture, much of the work on synaptic vesicle biogenesis has been done using PC12 cells. The synaptic vesicles present in these neuroendocrine cells have many, if not all, of the properties associated with the synaptic vesicles of brain (Clift-O'Grady et al. 1990). They appear to be required for acetylcholine release from depolarized PC12 cells (Bauerfeind et al. 1993). Their biogenesis was argued to involve a novel endocytotic sorting mechanism since they exclude markers of housekeeping endosomes (Cameron et al. 1991; Linstedt and Kelly 1991). If the explanation of these observations was that synaptic vesicle biogenesis involves specialized endosomes, then two possibilities could be considered:

either synaptic vesicles would contain markers of the specialized endosomes, or they would not. Our data indicating that pIgR is a marker for specialized endosomes in PC12 cells allowed this issue to be addressed in transfected PC12 cells. We found that pIgR is completely excluded from the synaptic vesicles but is recovered in a fraction with the properties of endosomes (Bonzelius et al. 1994).

We can combine the above information on the targeting of synaptic vesicle proteins into one speculative model (Fig. 2). A surprising feature is that the synaptic vesicle proteins fall into two classes: those that go to the housekeeping endosomes and to the axonal endosomes, and those that go directly to the axon. In this feature neurons again closely resemble epithelial cells, where some membrane proteins go directly to the apical surface and others go first to the basolateral ones, from which they are sorted into transcytotic vesicles for transport to the apical surface. Whether synaptophysin is first delivered to the surface of the neuronal cell body before it is axonally transported remains to be determined.

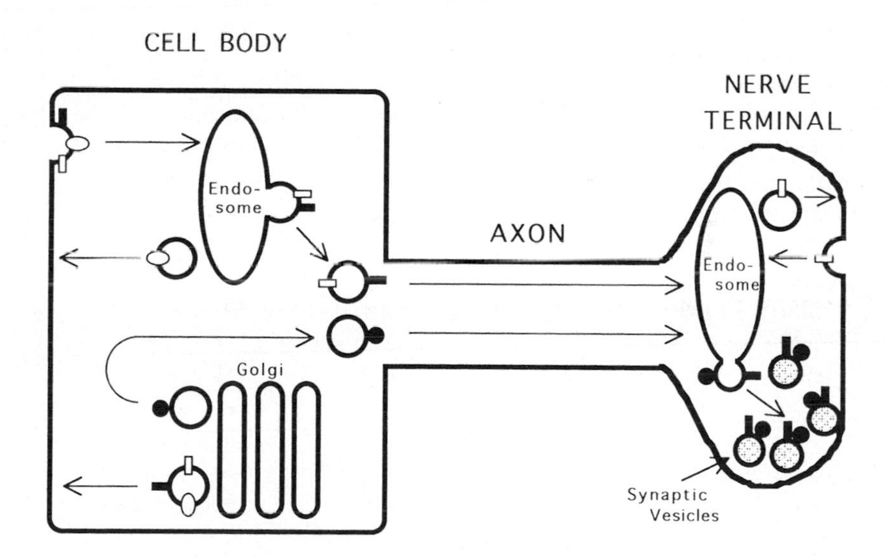

Fig. 2. Membrane pathways involved in synaptic vesicle biogenesis in the neuron. Some synaptic vesicle proteins (VAMP, SV2 and synaptotagmin; *filled circles*) go directly from the Golgi complex to the axonal endocytotic apparatus by axonal transport. Three other classes of membrane protein go to the plasma membrane and the "housekeeping" endosomes of the cell body. Some (*open ovals*) recycle between these endosomes and the cell surface, never leaving the endosome. The transferrin receptor is an example. Others (*rectangles*) are transported to the axon by a process similar to transcytosis in epithelial cells. Some proteins recycle between axonal endosomes and the axonal surface. Although no neuronal protein of this type has been identified, pIgR follows this route in transfected cells. Synaptic vesicle proteins (*filled circles and rectangles*) that follow either of the routes to the axon are sorted together from axonal endosomes into synaptic vesicles

Targeting Signals in VAMP

If the endocytosis of synaptic vesicle proteins involves a pathway different from that taken by a recycling receptor such as the LDL receptor, then it should involve a new class of sorting domains, different from the tyrosine-based and dileucine-based signals already identified for housekeeping endosomes (Trowbridge et al. 1993). To look for such domains we chose one of the simplest synaptic vesicle proteins, originally called VAMP and subsequently renamed synaptobrevin. VAMP plays a pivotal role in vesicle fusion, interacting with the plasma membrane proteins syntaxin and SNAP-25 (Rothman and Warren 1994). The lumenal domain is very small, only a couple of amino acids in rat VAMP 2 (Fig. 3). Sorting information, therefore, must be in the transmembrane or cytoplasmic domains. The cytoplasmic domain consists of two elements: a region close to the plasma membrane with a predicted α-helical coiled coil structure that is conserved between classes of VAMP and across species boundaries, and a variable amino terminal domain. To allow us to modify these regions without affecting the ability of antibodies to recognize VAMP, we appended a T-antigen (TAg) epitope tag to the lumenal carboxy-terminus of VAMP. Endocytosis of the epitope-tagged molecule and variants created by in vitro mutagenesis could be readily followed using a labelled antibody to TAg.

Initial constructions demonstrated that the amino-terminal 30 amino acids were not required for rapid endocytosis (Grote and Kelly, in preparation). Indeed their removal enhanced the initial endocytotic rate almost

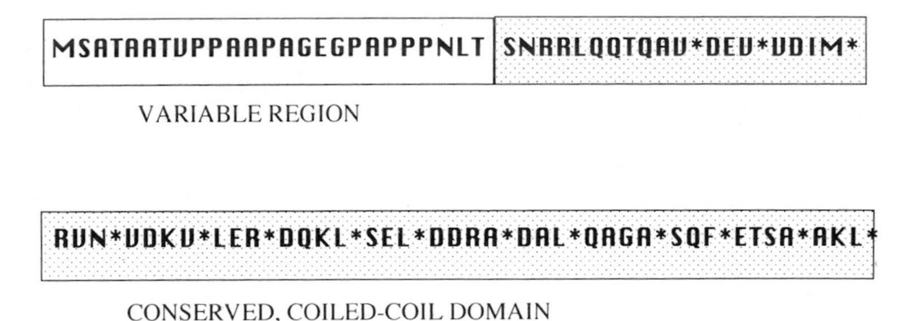

Fig. 3. The domains of VAMP (synaptobrevin) are a short carboxy-terminal, lumenal domain, a transmembrane region, a conserved region and a variable region. The conserved region has hydrophobic residues (*asterisks*) every three to four amino acids, characteristic of a protein with a coiled-coil domain

two-fold. Similarily, replacement of the transmembrane domain with that of another Type 2 membrane protein, the transferrin receptor, had no discernible effect on initial endocytotic rates. In contrast, removal of 60 amino acids from the amino terminus reduced endocytotic rates to background levels.

The signal for rapid endocytosis, therefore, seemed to be localized within the conserved domain of VAMP between amino acid 30 and the transmembrane region. To identify the internalization domain more precisely, a set of 10 amino acid deletions was generated covering this region. Only one deletion impaired endocytosis. The essential amino acids within that domain were identified by alanine-scanning mutagenesis. The domain identified in this way was very short, as are most endocytotic signals, but did not correspond to any known signal. We conclude that endocytosis of the synaptic vesicle protein VAMP is by a novel endocytotic mechanism. One such novel pathway could be involved in targeting surface proteins to specialized endosomes since, as we have argued here, specialized endosomes give rise to synaptic vesicles.

The sequence described above could be involved in direct sorting of VAMP into an endocytotic pathway, or it could permit associative sorting by allowing VAMP to interact or associate with another protein that has a direct sorting domain. If it must interact with another protein, then a protein with VAMP-binding capability is present in many cells, not just neuroendocrine cells, because expression of VAMP in non-neural cells such as Chinese hamster ovary (CHO) cells shows that it undergoes even more rapid endocytosis than in PC12 cells. Therefore, we can exclude the possibility that VAMP must interact with other synaptic vesicle proteins in order to be internalized.

When the mutant forms of VAMP that are poorly endocytosed in PC12 cells were expressed in CHO cells, their rates of endocytosis were almost identical to that of wild type VAMP. The reason that the mutations had a more dramatic effect in PC12 cells is not clear at this moment. What the result suggests, however, is that the endocytotic machinery which recognizes this new sorting domain is enhanced in PC12 cells. The recognition signal used in CHO cells has not yet been identified.

Sorting of VAMP into Synaptic Vesicles

Synaptic vesicles arise either directly from the plasma membrane or from a class of specialized endosomes. If there are two endocytotic routes operating in parallel, then mutations affecting rapid endocytosis might have no effect on synaptic vesicle biogenesis, and vice versa. Mutations eliminating endocytosis were found to be defective also in synaptic vesicle targeting, consistent with a direct link between endocytosis and vesicle biogenesis (Grote and Kelly, in preparation).

Whereas the signal for endocytosis is restricted to only a small, well-defined domain in VAMP, the domains required for correct vesicle targeting are distributed throughout the VAMP molecule. Thus, many of the constructs that are endocytosed quite normally are poorly targeted to vesicles. It appears as if the overall conformation of VAMP is required for targeting to synaptic vesicles much more stringently than it is required for endocytosis.

A Mutation Affecting Endocytosis in Drosophila

An alternative strategy we have taken to explore endocytosis in the neuron has been to study vesicle recycling at the neuromuscular junction of the fruitfly. The nerve terminals at Drosophila neuromuscular junctions are quite large, about 10 microns in diameter. To monitor vesicle recycling we used uptake of the dye FM1-43, whose fluorescence is markedly enhanced when it binds membranes (Betz and Bewick 1992). In the absence of stimulation no endocytosis is detected, presumably because basal endocytosis is slow (Ramaswami et al. 1994). Stimulation, either electrically or by potassium depolarization, causes strong labeling of the nerve terminal with the dye. The distribution of the label overlapped that of a synaptic vesicle marker, synaptotagmin. The dye can be readily released by washing the cells and then stimulating in the presence of calcium, or adding black widow spider venom. When these studies were repeated using the temperature sensitive *shibire* mutation, which is defective in most forms of cellular endocytosis, no dye uptake was observed at non-permissive temperatures, verifying the claims made from electron microcopy (Koenig and Ikeda 1989).

The *shibire* mutation allows us to arrest the synaptic vesicle cycle at a defined step, after exocytosis but before endocytosis. When the *shibire* function was blocked we found that synaptotagmin, a synaptic vesicle marker, was present on the cell surface, coincident in distribution with a surface membrane marker. When we examined another marker of nerve terminals, the cysteine string protein, its distribution paralleled that of synaptotagmin. These morphological characterizations are supported by cell fractionation studies. Synaptic vesicles, identified by their sedimentation velocity and antibodies to synaptotagmin, can be identified in homogenates of Drosophila heads (van de Goor et al., in preparation). In extracts from *shibire* heads, synaptic vesicles could only be isolated if the flies had been maintained at permissive temperatures prior to sacrifice. If they were maintained at non-permissive temperatures, the synaptotagmin was recovered in faster sedimenting membranes, presumably the presynaptic plasma membranes. We found that the cysteine string protein co-distributed with synaptotagmin during these manipulations. We conclude then that synaptic vesicle proteins are transferred to the plasma membrane in Drosophila, despite earlier confounding observations (Koenig and Ikeda 1989). We also conclude that the cysteine string protein is an authentic synaptic vesicle protein and that it stays on the synaptic vesicle throughout its cycle.

The *shibire* mutation allows us also to separate in time the processes of exocytosis and endocytosis that are normally coupled together during neurotransmitter release. When the temperature is lowered from non-permissive levels, endocytosis is detected by dye labelling of the terminal. Calcium is not required for the rapid endocytosis, indicating that the signal for rapid endocytosis is still functional in these terminals, long after cytoplasmic calcium levels have returned to normal. In contrast, endocytosis is not seen in terminals intoxicated with black widow spider venom in the absence of calcium. Calcium is clearly required for recovery from this block to endocytosis. We can thus resolve two independent steps in the vesicle recycling pathway, one requiring calcium and the other not.

Lysosomal Targeting

From earlier data (Green and Kelly 1992) one predicts that those proteins which are rapidly internalized and degraded should have a sorting signal targeting them directly to lysosomes, and removing that signal should increase its half-life. We have found this to be the case for P-selectin (Green et al. 1994).

P-selectin is an adhesion protein normally stored in secretory granules of platelets and endothelial cells. When the cells are appropriately stimulated, their secretory granules fuse with the plasma membrane and P-selectin is expressed on the cell surface. By this regulated secretory pathway, new protein can be rapidly expressed on the cell surface without the need for new protein synthesis. When on the cell surface P-selectin makes the cell adhesive; since enhanced adhesiveness is only a transient need for a cell, the P-selectin is rapidly removed and sent to the lysosome for degradation.

To understand how P-selectin is degraded, we expressed it in PC12 cells. As expected, it has a short half-life. When its cytoplasmic tail was used to replace that of the LDL receptor, the chimeric protein was rapidly degraded. When a 10 amino acid domain was deleted from the cytoplasmic tail of P-seletin, its internalization rate was unaffected, but it had the turnover time of the LDL receptor (Green and Kelly 1992). Since the 10 amino acid signal for lysosomal degradation did not resemble any sequence found in a lysosomal membrane protein, we believe a new sorting pathway may have been identified.

Discussion

The β-amyloid precursor protein is targeted to the cell surface of neurons, where it is transiently expressed prior to its degradation in lysosomes. If one wished to perturb that pathway, to alter the rate or the nature of this proteolysis in the brain, it would be helpful to have an understanding of the membrane protein trafficking pathways, particularly those involving endocytosis in the neuron. Our studies of endocytosis in neuronal cells, particularly PC12 cells, have provided some of the needed background. The identification of two classes of endosomes with different distributions in the neuron makes it necessary to decide which class of endosome is involved in β-amyloid protein precursor processing. Given its abundance in the cell body, the housekeeping endosome seems a more reasonable prediction.

Some of the alternative fates that face a membrane protein internalized into the early endosome are illustrated in Figure 4. They can be recycled back to the cell surface in recycling endosomes (filled circles) or retained in a large carrier vesicle. The membrane proteins that stay in the external membrane of the carrier vesicle (filled rectangles) are recovered, usually by recycling through the Golgi complex. Those membrane proteins destined for rapid degradation (filled circles) are sorted into vesicular structures that remain within the lumen of the carrier vesicle. Our work suggests that membrane proteins with short half-lives are targeted to the lumenal vesicles by a novel sorting signal. We have defined a short amino acid sequence that is responsible for this targeting. If a similar signal exists in the β-amyloid

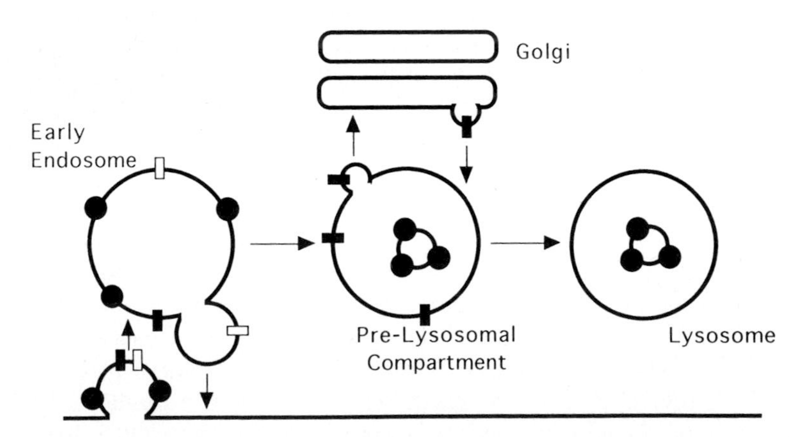

Fig. 4. Targeting of membrane proteins to lysosomes. Membrane proteins in the early endosome can recycle back to the cell surface (*open rectangles*) or be carried on to late endosomes. In late endosomes the proteins destined for rapid degradation are sequestered within internal vesicular structures, whereas those that remain on the external membrane are reutilized, perhaps after recycling through the Golgi. In vitro mutagenesis now identifies a domain in a rapidly turning over protein that targets it for lysosomal degradation

protein, its sequence could be determined using the same strategy we have employed for P-selectin. If such a signal could be found, then it may be possible to perturb the degradation of the β-amyloid protein, without a generally deleterious effect on the cell, by perturbing the interaction of the sorting machinery of this specialized pathway with its sorting domain.

We have used the strategy of identifying novel sorting signals to show that the endocytosis of a synaptic vesicle protein, VAMP or synaptobrevin, involves a novel class of sorting signal, presumably targeting it to a pathway that is not yet fully characterized. This pathway is more prominent in PC12 cells compared with CHO cells. One possibility is that it is a signal for targeting to the specialized endosome of axons, and that there are few specialized endosomes in CHO cells.

The domain specifying endocytosis of VAMP in PC12 cells is short and can be recognized even when the overall conformation of the protein has been perturbed by large deletions. In contrast, delivery of VAMP to the synaptic vesicle has a much more stringent requirement for overall protein conformation. Changes almost anywhere in the variable, conserved or transmembrane regions have some effect. A requirement for overall structure appears to us to be more consistent with an associative type of sorting. Consistent with this view is our discovery that single amino acid mutations in domains thought to be important for formation of the predicted coiled-coil structures (Lupas et al. 1991) are very effective in inhibiting vesicle biogenesis. Coiled-coil domains are frequently found to be involved in protein-protein associations. Indeed they are believed to be responsible for the interaction of VAMP with its plasma membrane target, syntaxin (Calakos et al. 1994). The data are consistent, therefore, with a model of synaptic vesicle biogenesis that involves two steps (Fig. 5). The first uses a novel endocytosis domain recognized by cellular sorting machinery to target synaptic vesicle proteins to a specialized endosome. In the endosome, association between vesicle proteins occurs, inhibiting re-cycling to the plasma membrane and facilitating targeting to the synaptic vesicle.

Although, in our opinion, the model presented in Figure 5 is the most plausible, there are other possibilities. To illustrate these, we can consider two extreme alternatives (Fig. 6). In one, all the targeting is provided by three different targeting domains, with two domains required for synaptic vesicle biogenesis. One narrow domain specifies endocytosis by a mechanism that is enhanced in PC12 cells; a larger domain containing additional features is required for targeting to synaptic vesicles. In the other model, VAMP has no targeting information at all, but reaches its intracellular destination by associating with other proteins that confer correct targeting. To explain the data we need to assume that the interacting proteins are not identical in PC12 and CHO cells.

If VAMP is interacting via its coiled-coil domain with another synaptic vesicle protein prior to being sorted into the synaptic vesicle, then it must dissociate from its partner to interact with syntaxin, since this interaction

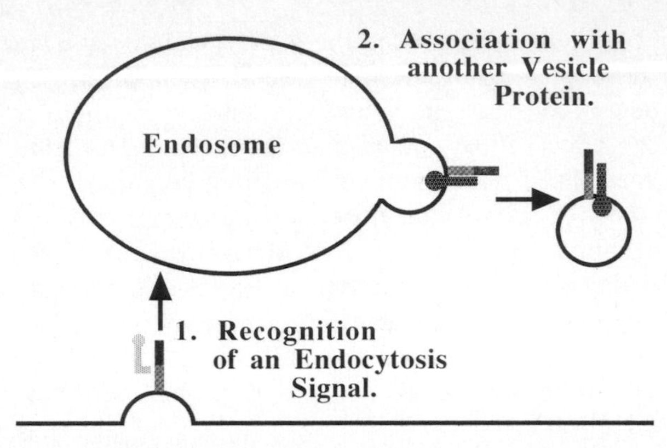

Fig. 5. A two-step model of synaptic vesicle biogenesis. The first step involves the recognition of a molecule such as VAMP by intracellular sorting machinery, responsible for targeting VAMP to specialized endosomes. In the second step, synaptic vesicle proteins associate via the coiled-coil domain, and this association leads to diversion into synaptic vesicles

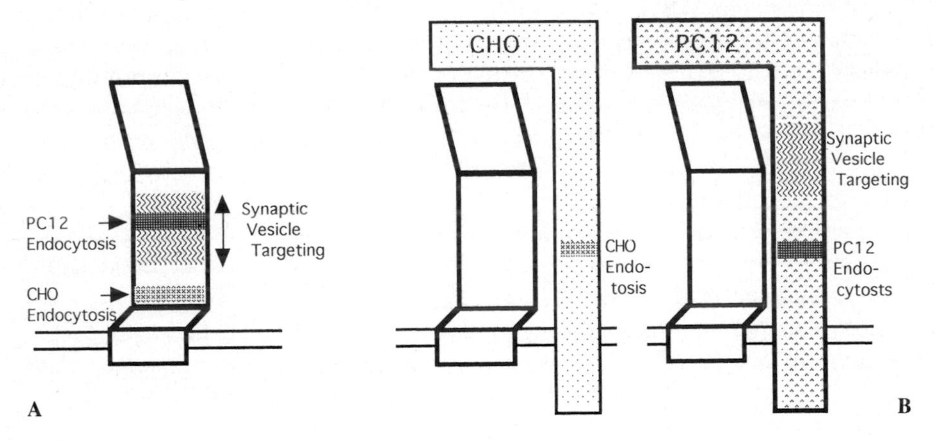

Fig. 6. Two models of sorting of synaptic vesicle proteins. These two models illustrate two extremes in synaptic vesicle membrane protein sorting. **A** All the sorting is via direct sorting domains. The synaptic vesicle sorting domain is highly extended and includes the domain specifying endocytosis. **B** The domains described here specify association with another protein, and both endocytosis and synaptic vesicle targeting of VAMP are by associative sorting. To explain the differences between PC12 and CHO cells, we postulate different carrier proteins. Excluded from these models is any attempt to explain how synaptic vesicle proteins are targeted to axons

uses the same coiled-coil domain implicated in sorting. Synaptic vesicle docking, therefore, would involve the exchange by VAMP of one binding partner for another.

Similarily, after exocytosis the interaction between VAMP and syntaxin must be reversed to allow vesicle recycling. If reversal of the docking interactions were to have an as yet unidentified calcium requirement, we would have identified a step in the recycling pathway that could be inhibited in black widow spider intoxicated terminals.

References

Bauerfeind R, Regnier-Vigouroux A, Flatmark T, Huttner WB (1993) Selective storage of acetylcholine, but not catecholamines, in neuroendocrine synaptic-like microvesicles of early endosomal origin. Neuron 11: 105–121

Betz WJ, Bewick GS (1992) Optical analysis of synaptic vesicle recycling at the frog neuromuscular junction. Science 255: 200–203

Bonzelius F, Herman GA, Cardone MH, Mostov KE, Kelly RB (1994) The polymeric immunoglobulin receptor accumulates in specialized endosomes but not synaptic vesicles within the neurites of transfected neuroendocrine PC12 cells. J Cell Biol (in press)

Calakos N, Bennett MK, Peterson KE, Scheller RH (1994) Protein-protein interactions contributing to the specificity of intracellular vesicular trafficking. Science 263: 1146–1149

Cameron PL, Sudhof TC, Jahn R, DeCamilli P (1991) Colocalization of synaptophysin with transferrin receptors: implications for synaptic vesicle biogenesis. J Cell Biol 115: 151–164

Clift-O'Grady L, Linstedt AD, Lowe AW, Grote E, Kelly RB (1990) Biogenesis of synaptic vesicle like structures in a pheochromocytoma cell line PC12. J Cell Biol 110: 1693–1703

Green SA, Kelly RB (1992) Low density lipoprotein receptor and cation-independent mannose phosphate receptor are transported from the cell surface to the Golgi apparatus at equal rates in PC12 cells. J Cell Biol 117: 47–55

Green SA, Setiadi H, McEver RP, Kelly RB (1994) The cytoplasmic domain of P-selectin contains a sorting determinant that mediates rapid degradation in lysosomes. J Cell Biol 124: 435–448

Kelly RB (1993) Storage and release of neurotransmitter. Cell 72: 43–55

Koenig J, Ikeda K (1989) Disappearance and reformation of synaptic vesicle membrane upon transmitter release observed under reversible blockage of membrane retrieval. J Neurosci 9: 3844–3860

Linstedt AD, Kelly RB (1991) Synaptophysin is sorted from endocytotic markers in neuroendocrine PC12 cells but not transfected fibroblasts. Neuron 7: 309–317

Lupas A, van Dyke M, Stock J (1991) Predicting coiled-coils from protein sequences. Science 252: 1162–1164

Mundigl O, Matteoli M, Daniell L, Thomas-Reetz A, Metcalf A, Jahn R, DeCamilli P (1993) Synaptic vesicle proteins and early endosomes in cultured hippocampal neurons: differential effects of brefeldin A in axons and dendrites. J Cell Biol 122: 1207–1221

Pietrini G, Suh YJ, Edelman L, Rudnick G, Kaplan MJ (1994) The axonal gamma-butyric acid transporter GAT-1 is sorted to the apical membranes of polarized epithelial cells. J Biol Chem 269: 4668–4674

Ramaswami M, Krishnan KS, Kelly RB (1994) Intermediates in synaptic vesicle recycling revealed by optical imaging of Drosophila neuromuscular junctions. Neuron 13: 363–375

Rothman JE, Warren G (1994) Implications of the SNARE hypothesis for intracellular membrane topology and dynamics. Curr Biol 4: 220–233

Trowbridge IS, Collawn JF, Hopkins CR (1993) Signal-dependent membrane protein trafficking in the endocytotic pathway. Ann Rev Cell Biol 9: 163–206

Mechanisms of Molecular Sorting in Polarized Cells: Relevance to Alzheimer's Disease

*I. Mellman**, *K. Matter*, *E. Yamamoto*, *N. Pollack*, *J. Roome*, *K. Felsenstein*, and *S. Roberts*

Summary

The plasma membranes of many cell types, such as epithelial cell and neurons, are polarized into distinct domains. In epithelial cells, we have found, that the targeting of newly synthesized membrane proteins is governed largely by a ubiquitously distributed set of cytoplasmic domain sorting signals. These determinants may be superficially related to well-characterized signals for localization at plasma membrane-coated pits. Their inactivation results in transport to the apical surface by a second targeting determinant found in the extracellular and/or membrane-anchoring domains. Importantly, these sorting signals are recognized in both be Golgi complex and in endosomes, indicating that a common mechanism is responsible for polarized sorting of both newly synthesized proteins and proteins internalized by endocytosis. The Alzheimer's amyloid precursor protein (APP) is expressed by polarized cell types such as endothelial and neuronal cells. Thus, we analyzed its intracellular transport in the MDCK kidney epithelial cell line and asked whether mutations in the APP gene that lead to familial early onset Alzheimer's disease have any effect on the polarity of APP transport or the release of amyloidogenic fragments. Wild type APP is transported to the basolateral surface due to a basolateral targeting signal in its cytoplasmic domain. While both the α-cleaved soluble APPs and amyloidogenic Aβ fragments are released into the basolateral medium, β-cleaved APPs is released apically. Removal of the APP cytoplasmic domain results in apical secretion of the Aβ fragment. Thus, Aβ release but not β-APPs release follows the polarity of the membrane-bound precursor, suggesting that β-cleavage occurs in the Golgi prior to polarized sorting of APP whereas the γ-cleavage responsible for Aβ release occurs after sorting at either the apical or basolateral domains. While the Swedish FAD mutation has no effect on polarity of APP transport, mutations affecting position 717 in the membrane-anchoring domain appear to alter the polarity of release of APP cleavage

* Department of Cell Biology, Yale University School of Medicine, 333 Cedar Street, PO Box 3333, New Haven, CT 06520, USA

K.S. Kosik et al. (Eds.)
Alzheimer's Disease: Lessons from Cell Biology

products. This finding suggests that polarity of Aβ release may be a contributing factor in the pathogenesis of Alzheimer's disease.

Introduction

The large majority of cells in all multi-cellular organisms exhibit distinctly polarized distributions of their plasma membrane components and/or intracellular organelles. In many instances, cell polarity is known to be a major determinant of cell or tissue function. Perhaps the best understood example is the case of the cells comprising the epithelial cell sheets that line the lumens of absorptive organs such as kidney and intestine (Mellman et al. 1993; Rodriguez-Boulan and Nelson 1989; Rodriguez-Boulan and Powell 1992; Wandinger-Ness and Simons 1990). As in other epithelia, the surfaces of these cells are differentiated into structurally and biochemically distinct domains: an apical domain, often characterized by abundant microvilli, that is enriched in proteins involved in nutrient transport as well as in glycophospholipid-anchored glycoproteins and sphingomyelin-containing glycolipids; and a basolateral domain containing most of the membrane proteins involved in the "housekeeping functions" of individual cell. Thus, the apical domain is often thought of as containing components specific to the function of a given epithelium while the basolateral domain shares most of its components in common with "non-specific" or non-polarized cells. The apical and basolateral domains are separted by junctional complexes which adhere adjacent cells to one another and also serve as a molecular fence to prevent mixing of asymmetrically arrayed proteins and lipids.

How epithelial or other polarized cells generate and maintain distinct apical and basolateral domains within a continuous lipid bilayer has emerged as a central problem in modern cell biology. It is a problem that pertains not only to epithelial cells but also to cells such as neurons, which maintain strikingly different axonal and dendritic surfaces without the benefit of junctional complexes to serve as a molecular fence to separate the two domains. Nevertheless, it now appears likely that the mechanisms underlying polarity in cell types as diverse as epithelia and neurons are more likely to be fundamentally similar than fundamentally different.

Our group has attempted to solve the problem of polarity by studying the transport of membrane proteins in a model epithelial cell line, Madin-Darby canine kidney (MDCK) cells. MDCK cells can be conveniently grown as epithelial sheets on permeable polycarbonate filter supports (e.g., Transwell® units, Corning-Costar, Cambridge, MA) that permit ready access to both the apical and basolateral surfaces. Our strategy has been to transfect cDNAs encoding various wild type and mutant membrane proteins to determine whether specific targeting signals could be identified, and ultimately to use this information to elucidate the cellular mechanisms that decode these signals. As summarized below, these efforts have already uncovered a surprisingly simple logic on which the entire system appears to be based.

Development of the Concept of Epithelial Cell Polarity

Early efforts to identify the molecular mechanisms responsible for the polarized distribution of membrane proteins in MDCK cells focussed on the analysis of viral glycoproteins, due to the pioneering work of Enrique Rodriguez-Boulan and David Sabatini (1978). These investigators demonstrated that when MDCK cells were infected with a range of different types of enveloped viruses, budding virions were relesed in a polarized fashion: e.g., influenza virus from the apical surface, vesicular stomatitis virus (VSV) from the basolateral surface. The polarity of budding was soon found to reflect the polarized insertion of the influenza and VSV spike glycoproteins, HA and G protein, respectively. These critical observations proved that individual membrane proteins contained sorting signals that specified their transport to the apical and/or basolateral surfaces. Work by Simons and colleagues next extended this concept to show that sorting occurred primarily upon exit of the newly synthesized proteins from a network of tubules continuous with the trans aspect of the Golgi complex, the "trans-Golgi network" or TGN (Griffiths and Simons 1986; Wandinger-Ness et al. 1990).

Although initial attempts to identify the actual sorting signals on HA and G protein were inconclusive, several considerations gave rise to the long-held prediction that targeting into the apical pathway represented the specific sorting event that underlay polarized transport from the TGN. Indeed, in absorptive epithelium, it is the apical side that is most cell type-specific in the sense that the apical surface contains most of the membrane proteins responsible for epithelial cell function. Enzymes involved in hydrolysis, amino acid, and sugar transport are all found on the apical plasma membrane, which, moreover, is characterized by highly differentiated microvillar outcroppings that aid in the absorption process. In contrast, the basolateral surface contained most of the membrane proteins responsible for the constitutive or "housekeeping" functions common to all cells, polarized and non-polarized (Fig. 1). Thus, the apical surface appeared to be a privileged site, transport to which whould appear most likely to require a specific sorting signal; in the absence of this signal, basolateral transport was viewed as occurring by default. The nature of the signal was obscure, but appeared to involve a determinant in a protein's extracellular and/or membrane-anchoring domain because, importantly, soluble forms of HA were secreted apically despite the absence of a membrane anchor or cytoplasmic tail. In addition, most GPI-anchored proteins (as well as sphingoglycolipids) were found at the apical surface, further indicating that cytoplasmic domain determinants could not be involved in apical transport (Rodriguez-Boulan and Powell 1992).

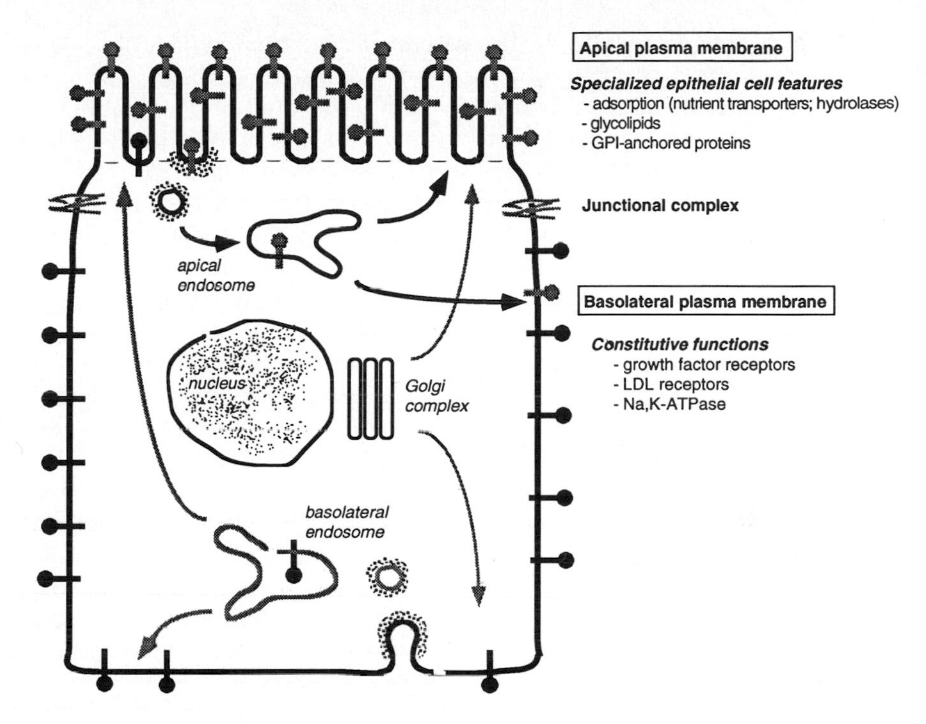

Fig. 1. General organization of polarized epithelial cells. (Taken from a figure to be published in Current Opinion in Cell Biology)

Polarized Sorting of Membrane Proteins in the Golgi Complex Is Controlled by Distinct Cytoplasmic Domain Determinants

Recent work, however, has fundamentally changed our views of polarized transport in epithelial cells. It is now clear that sorting is controlled, perhaps entirely, by the existence of novel signals present in the cytoplasmic domains of proteins that are targeted to the basolateral – and not the apical – plasma membrane. These signals were first discovered by expressing a family of IgG Fc receptors (FcRII) in MDCK cells and finding that, while the wild type receptor was efficiently transported to the basolateral surface, a receptor whose cytoplasmic domain was deleted was transported apically (Hunziker and Mellman 1989). Moreover, basolateral transport appeared to correlate with the presence of the receptor's clathrin coated pit localization domain, further suggesting a relationship between the mechanisms underlying basolateral sorting and endocytosis. These findings were rapidly generalized to a variety of other receptors (LDL receptor, NGF receptor, transferrin receptor) and other non-receptor membrane proteins including, interestingly enough, influenza HA containing an artifical cytoplasmic tail coated pit localization domain (Brewer and Roth 1991; Hunziker et al. 1991; Le Bivic et al. 1991). Interestingly, both basolateral targeting and endocytosis were often found to be dependent on critical tyrosine residues.

Due to this superficial relationship between coated pit localization and basolateral targeting, we next performed a detailed analysis of the amino acid sequences involved in polarized sorting using the human LDL receptor, as the receptor's coated pit localization domain had already been extensively characterized. In fact, we found that the LDL receptor cytoplasmic domain contains not one, but two basolateral sorting signals. The first was found in a region proximal to the membrane in a position that overlapped with the receptor's tyrosine-dependent coated pit localization domain. However, apart from a common dependency on this tyrosine at position 18 of the tail, the basolateral targeting activity of the proximal determinant was otherwise distinct from the coated pit localization signal (Fig. 2). In other words, none of the other residues required for coated pit localization was required for basolateral targeting (Matter el al. 1992, and manuscript in preparation). The second signal was in the distal portion of the LDL receptor's cytoplasmic tail. While in a region of the tail that has nothing to do with endocytosis, this second "distal" signal was also found to be critically dependent on a tyrosine (at position 35). The distal signal was also found to be significantly stronger than the proximal signal, in that it would direct efficient basolateral transport even at very high levels of receptor expression.

While differing in region and relationship to the receptor's coated pit signal, the proximal and distal basolateral determinants were related to each other by their tyrosine dependence. In addition, both signals required a single amino acid adjacent to each tyrosine (glutamine or glycine) as well as a cluster of acidic residues found several positions on the COOH-terminal side of each tyrosine (Matter et al., in preparation). These common features may not only relate the two signals to one another, but also may form the basis for a sequence motif that may ultimately be used to predict the

LDL receptor cytoplasmic tail contains two tyrosine-dependent signals for basolateral transport

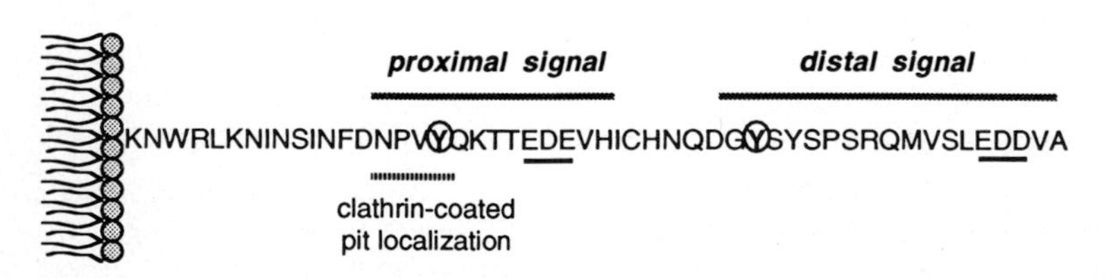

Fig. 2. Basolateral targeting motifs in the LDL receptor cytoplasmic domain. (Taken from Matter et al. 1993)

presence of basolateral signals in other membrane proteins (Matter et al., in preparation). At present, however, the motif is too degenerate and the number of examples are too few to determine if this is indeed the case.

In addition, it is clearly the case that some basolateral signals exist that are distinct from those defined in LDL receptor. For example, the polymeric immunoglobulin receptor, which is responsible for trans-epithelial transport of secretory IgA and IgM, contains an incompletely characterized signal which is nevertheless clearly tyrosine-independent. Even more interesting is the IgG Fc receptor. In this case, basolateral targeting is specified by a motif consisting of two adjacent leucine residues, a signal which is also completely responsible for the receptor's ability to localize at clathrin-coated pits (Matter et al., in preparation). This "di-leucine" type signal is turning up in a variety of leukocyte receptors and has been implicated in both endocytosis and targeting to lysosomes. (Letourneur and Klausner 1992). Unlike the tyrosine-dependent motifs, we have not been able to identify any differences in sequence requirements for basolateral targeting vs. endocytosis. Consequently, it is still conceivable that there is a structural and/or functional relationship between signals for coated pit localization and polarized sorting.

Finally, in all cases, it is clear that these signals specify transport of newly synthesized proteins directly to the basolateral surface. The removal or inactivation of these signals by deletion or point mutagenesis prevented basolateral delivery and instead resulted in vectorial transport to the apical surface. Consequently, the signals were decoded at the level of the TGN in MDCK cells (Matter et al. 1993).

Pathways of Polarized Sorting: The Simple Logic of Intracellular Transport

Taken togenther, these results suggest some important and rather simple conclusions concerning the mechanism of polarized sorting of newly synthesized plasma membrane proteins in MDCK cells. First, it appears likely that the vast majority of membrane proteins reach the basolateral surface due to the presence of one or more classes of novel cytoplasmic domain sorting signals. Second, these signals are widely distributed and not restricted to proteins expressed only in epithelial cells. Thus, the mechanism of polarized sorting revealed in MDCK cells may exist in other polarized cell types as well. Third, in the absence or inactivation of a basolateral signal, a membrane protein is generally targeted preferentially to the apical surface as opposed to reaching both the apical and basolateral domains "randomly." Thus, apical transport may not occur by default, as was originally and erroneously proposed for basolateral transport, but may instead make use of a second signal for apical transport whose activity is revealed upon inactivation of the basolateral signal. This situation predicts a simple hierarchical relationship between basolateral and apical transport, with basolateral transport being dominant.

These considerations also suggest some rather specific predictions concerning the types of pathways that emanate from the TGN in MDCK cells. As summarized in Figure 3, it now seems very likely that no more than three such routes are required. The first involves sequestration of basolateral membrane proteins in a distinct population of transport vesicles which are subsequently targeted to the basolateral surface. Presumably, accumulation in basolateral transport vesicles is accomplished due to the interaction of the cytoplasmic tail sorting signal with a cytoplasmic coat protein, at least functionally analogous to the clathrin coat. Indeed, a second pathway must exist that is clathrin-dependent, and mediates transport of proteins destined for the endocytic pathway (e.g., mannose-6-phosphate receptor) to endosomes via TGN-derived clathrin-coated vesicles. Segregation into this pathway is due to the presence of cytoplasmic domain coated pit localization signals that specify accumulation in TGN coated pits, in addition to plasma membrane coated pits. Membrane (and secretory) proteins that lack the signals to enter either the basolateral or endosomal pathways are sequestered into the third class of nascent vesicle, that which gives rise to transport to the apical surface. What specifies accumulation in apical vesicles is unknown. It may involve selective exclusion from both of the other two pathways. More likely, however, is the possibility that yet another general property of many glycoproteins allows their low affinity interaction with a sorting molecule that partitions into apical vesicles. Conceivably, this could be a lectin (that binds carbohydrate moieties on glycoproteins), the presence of a differentiated lipid microdomain (perhaps set up by lateral interactions of glycolipids), or protein such as the src-kinase substrate caveolin which is also

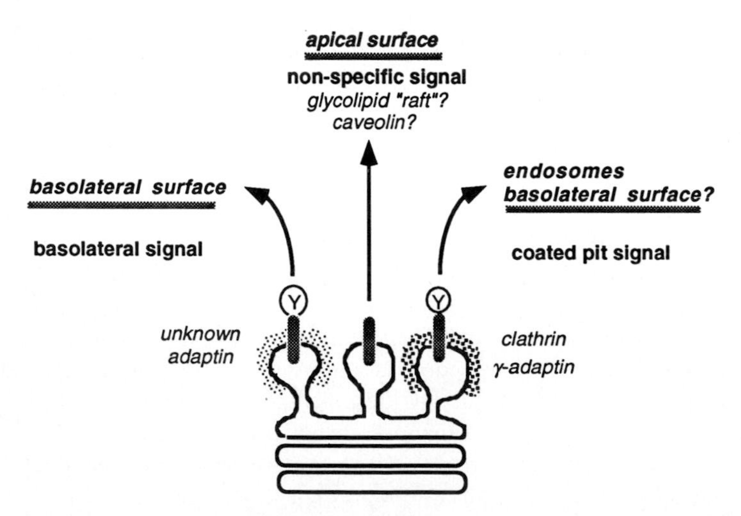

Fig. 3. Pathways of sorting and transport in the trans-Golgi network (TGN). (Taken from a figure to be published in Current Opinion in Cell Biology)

known to reach the apical surface (Wandinger-Ness and Simons 1990; Zurzolo et al. 1994).

Understanding the mechanism of sorting, the nature of the coat proteins that may be involved, and the features that distinguish apical and basolateral transport vesicles represents the major problems for future research. The identification of discrete sorting determinants that specify basolateral transport has provided the means to address these problems directly.

The Generality of Sorting Mechanisms: Common Problems, Common Solutions

While in MDCK cells and probably all kidney epithelial cells apical and basolateral proteins are sorted in the TGN and vectorially transported to their final destinations, other epithelia have different strategies. Hepatocytes, for example, transport all newly synthesized membrane proteins from the TGN to the basolateral (sinusoidal) surface where they are internalized and delivered to endosomes, and apical proteins are selectively sorted to the apical surface by the process of transcytosis. Basolateral proteins are returned by recycling back to the basolateral surface. Even in MDCK cells, the analogous sorting events must occur continuously during the normal course of endocytosis at the apical and basolateral domains. In other words, LDL receptors internalized at the basolateral surface must be correctly returned to their site of origin rather than being included in transcytotic vesicles and transported to the apical side. This behavior contrasts with that of receptors such as the polymeric immunoglobulin receptor, which when expressed in MDCK cells mediates the basolateral to apical transport of dimeric IgA (Apodaca et al. 1991; Deitcher et al. 1986).

Such considerations would predict that the same sorting signals recognized in the TGN might also be recognized in endosomes. Furthermore, apical proteins such as dipeptidyl peptidase IV are expressed apically in both hepatocytes and MDCK cells despite the differences in sorting sites (Bartles et al. 1987; Le Bivic et al. 1990; Matter et al. 1990). Our recent work has demonstrated directly that endosomes and the TGN do indeed decode the same sorting signals. For example, inactivation of the LDL receptor's distal basolateral sorting signal converts it into a transcytotic protein which is as efficient at basolateral to apical transport as is the polymeric immunoglobulin receptor (Matter et al. 1993). Thus, partial inactivation of a signal which decreases the efficiency of basolateral sorting in the TGN leads to a corresponding decrease in basolateral sorting efficiency in endosomes. This satisfying finding illustrates that multiple sorting organelles are very likely to use precisely the same sorting machinery: a common solution was developed in response to a common problem in epithelial cell sorting.

Even more remarkable is the suggestion of Dotti and Simons that neurons, a polarized cell type entirely distinct from epithelial cells, nevertheless seem to use the same basic strategy to distinguish between axonal and dendritic membrane protein components. Although the sample size remains quite small, it appears that proteins that reach the apical surface in MDCK cells (HA, GPI-anchored proteins) are generally targeted to the axonal surface, whereas basolateral proteins (VSV G protein) are restricted to the dendritic domain (Dotti and Simons 1990). Although different cell types may interpret them differently, all cells seem to use a common language of sorting signals that underlies all types of polarized sorting.

Sorting and Alzheimer's Disease

What does all this have to do with Alzheimer's disease? Alzheimer's precursor protein (APP) is a type I membrane protein expressed by a variety of polarized cell types, including neurons and cells of epithelial origin. Irrespective of which cell type is the most relevant for the production of amyloidogenic Aβ peptide, the polarity of peptide release is likely to be an important factor in the pathogenesis of the disease. Moreover, knowing the cellular distribution of APP itself may provide some insight into the actual function of this protein, even separate from its role in pathogenesis.

Extensive work over the past several years has shown that APP is subject to a series of endoproteolytic cleavage events that generates various fragments, including the Aβ peptide (Selkoe 1993). At least three general cleavage events can be distinguished due to the activity of three presumptive endoproteases termed "secretases." As summarized in Figure 4, these include: α-secretase, which cleaves in the middle of the 40 residue Aβ peptide, thus preventing its amyloidogenic potential; β-secretase, which defines the amino terminal end of the Aβ peptide; and γ-secretase, which clips at a site within the membrane-anchoring domain to yield (in conjunction with β-secretase) the potential amyloidogenic Aβ peptide. The nature of these endopeptidases or their intracellular locations remains unknown. In part, at least some cleavages can occur following dleivery of APP to one or more organelles of the endocytic pathway. It is also likely, however, that APP cleavage occurs during transit through biosynthetic organelles (Haass et al. 1992).

Several rare mutations are known that lead to early onset familial Alzheimer's disease (FAD; Fig. 4). In the best-studied case, members of a large Swedish kindred were found to have a mutation in the APP ectodomain near the site of α-secretase cleavage (codons 670/671; lys $\rightarrow$ met and asn $\rightarrow$ leu). Mutations at this site significantly increased the amount of Aβ released by increasing the amount of β-cleavage that occurred (Citron et al. 1992). A second well-studied mutation occurs at codon 717, a site within the membrane spanning domain, and induces a val to ile, phe, or gly substitution. Interestingly, this second mutation does not increase the overall amount of Aβ produced, but may change the site of the γ-cleavage (producing a peptide of 42 residues in length as opposed to 40 residues).

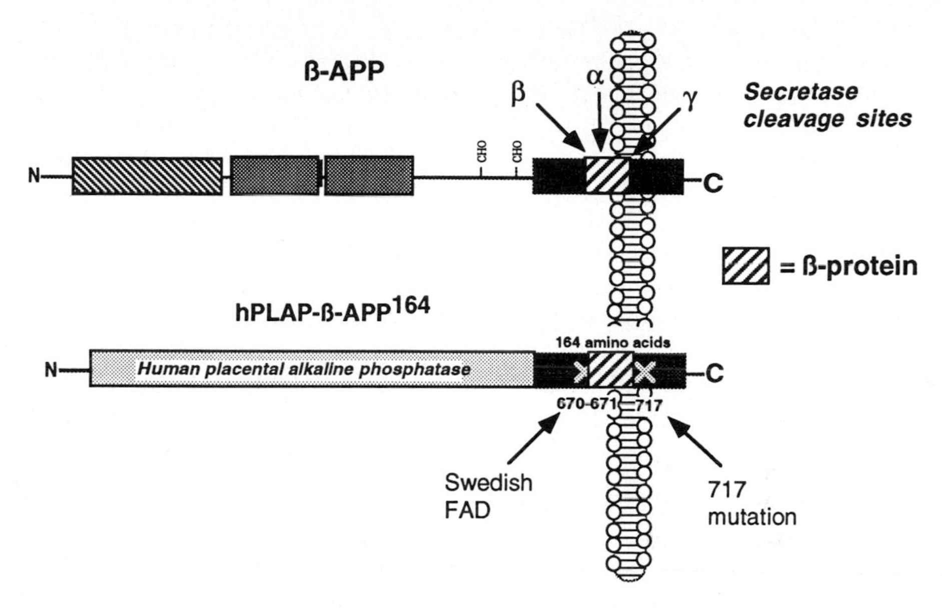

Fig. 4. Cleavage of the APP precursor and hPLAP-APP fusion protein

The sites and kinetics of APP cleavage would play a major role in determining the polarity of Aβ release if APP was expressed in a polarized fashion. This possibility recently motivated us, and others, to begin analyzing the behavior of APP when expressed in a model polarized cell, namely MDCK cells. Dennis Selkoe and his colleagues have already shown that APP expression is polarized: APP is clearly targeted to the basolateral plasma membrane where Aβ is released into the medium (Haass et al. 1994). Given our interest in the polarity problem in general, we have recently begun to examine whether any of the known APP mutations either directly or indirectly affect the polarity of Aβ release, perhaps by altering the timing or locations at which the various cleavages occur. Since this work is not yet published, it will be summarized only in general terms.

For most of our experiments, we made use of an APP reporter construct in which much of the extracellular domain of APP was replaced by human placental alkaline phosphatase (hPLAP). In previous work, we have shown that this reporter behaves in a fashion that is indistinguishable from authentic APP (Roberts et al. 1994). Moreover, an hPLAP-APP construct bearing the Swedish FAD mutation, when expressed in H4 human glioma cells, generated excess Aβ protein to the same extent as did the actual APP mutant. Interestingly, some Aβ could also be detected intracellularly prior to its appearance in the medium, suggesting that it was generated on the secretory pathway.

To determine whether APP used the same type of sorting signals as other proteins to reach the basolateral surface, we expressed the hPLAP construct bearing the wild type as well as several different mutant cytoplasmic tails in MDCK cells. Polarity was initially monitored by assaying the release of enzymatically active hPLAP into the medium (the result of either α- or β-secretase cleavage). As noted by Selkoe (this volume), wild type hPLAP-APP was targeted by >95% to the basolateral surface; basolateral targeting was also accomplished if the APP cytoplasmic tail was replaced by the cytoplasmic tail of another basolateral protein (VSV G protein). The presumptive coated pit localization domain of APP was not required, since it could be removed by internal deletion of the NPTY tetrapeptide and without affecting the polarity of hPLAP release. Thus, like LDL receptor, the APP cytoplasmic domain is very likely to have two basolateral sorting determinants. Truncation of the cytoplasmic tail leaving just the first two cytoplasmic lysine residues, however, resulted in the quantative transport of the APP to the apical surface and subsequent release of hPLAP into the apical medium. Expression of the Swedish FAD mutant had no effect on the overall polarity of release of APPs. However, expression of the 717 mutant – bearing the wild type APP cytoplasmic tail – completely altered the polarity of expression. In this case, ~50% of the soluble cleaved product was released into both the apical and basolateral medium. This is a most unexpected and intriguing result, since there has not yet been any suggestion that amino acid residues within membrane-anchoring domains can have effects on polarity.

We next characterized the secreted products more carefully. First, we found that the large ectodomain fragments of the APP constructs were released in a reproducibly polarized fashion. β-APPs, i.e., the large amino terminal fragment generated by β-secretase, was always released into the apical medium, irrespective of the type of APP cDNA expressed. Thus, we conclude that β-cleavage occurs prior to polarized sorting in the TGN: if β-APPs is removed from its membrane anchor and cytoplasmic domain prior to sorting, it would be targeted – along with other soluble secretory glycoproteins (for the reasons discussed in detail above) – to the apical surface since it would no longer be attached to the APP basolateral sorting signal. Thus, the relevant β-secretase activity probably occurred somewhere in the Golgi complex. The large α-cleaved product (α-APPs) was found in both the basolateral and/or apical media, suggesting that the α-secretase was located after sorting in the TGN and in both the apical and basolateral domains.

The behavior of Aβ release essentially mirrored the overall polarity of the APP expressed. Thus, wild type and Swedish FAD generated Aβ only in the basolateral medium, whereas the tail-minus mutant generated Aβ in the apical medium. The 717 mutant released Aβ equally at both surfaces. Again, these results strongly suggest that the γ-secretase must act after sorting occurs and also must be found in both the apical and basolateral domains. Whether the γ-secretase is found on the plasma membrane, in post-Golgi vesicles, or in endosomes remains an important unknown.

These results are interesting in several respects. First, although still somewhat preliminary, they put some constraints on the intracellular sites at which the various secretase enzymes are likely to be found: β-secretase in the cisternal Golgi, α- and γ-secretase in post-Golgi compartments that occur at both cytoplasmic/plasma membrane domains of MDCK cells. Second, they show that amyloidogenic Aβ peptide can be productively generated at either the apical or basolateral domains, suggesting that alterations in polarity of APP transport can result in alterations in the polarity of Aβ release. Third, they suggest that the polarity of Aβ release must be considered as another possible contributing factor in the pathogenesis of Alzheimer's disease, in addition to the amount and type of Aβ produced. The reasons for this are as follows. In cells expressing wild type APP, 95% of the very small amount of Aβ generated is released into the basolateral medium. While the Swedish FAD mutation does not alter this polarity, the absolute number of Aβ molecules released apically increases concomitant with the increase in total Aβ released. More importantly, the 717 mutant does not increase the amount of Aβ released but radically alters the polarity of release. Although the 717 mutant is thought to generate more of the 1–42 than 1–40 Aβ peptide, the absolute amount of Aβ released apically is increased by 5- to 10-fold relative to wild type. This is approximately the same amount of Aβ released apically by cells expressing the Swedish FAD. Consequently, it is impossible to distinguish whether the disease correlates with the overall amount of Aβ released, the type of Aβ released, or the polarity of Aβ released. Any one, or a combination, of these factors may provide the key to understanding the mechanism of plaque deposition and development of the disease.

References

Apodaca G, Bomsel M, Arden I, Breitfeld PP, Tang K, Mostov KE (1991) The polymeric immunoglobulin receptor. A model protein to study transcytosis. J Clin Invest 87: 1877–1882

Bartles JR, Feracci HM, Stieger B, Hubbard AL (1987) Biogenesis of the rat hepatocyte plasma membrane in vivo: Comparison of the pathways taken by apical and basolateral proteins using subcellular fractionation. J Cell Biol 105: 1241–1251

Brewer CB, Roth MC (1991) A single amino acid change in the cytoplasmic domain alters the polarized delivery of influenza virus hemagglutinin. J Cell Biol 114: 413–421

Citron M, Oltersdorf T, Haass C, McConlogue L, Hung AY, Seubert P, Vigo PC, Lieberburg I, Selkoe DJ (1992) Mutation of the β-amyloid precursor protein in familial Alzheimer's disease increases β-protein production. Nature 360: 672–674

Deitcher DL, Neutra MR, Mostov KE (1986) Functional expression of the polymeric immunoglobulin receptor from cloned cDNA in fibroblasts. J Cell Biol 102: 911–919

Dotti CG, Simons K (1990) Polarized sorting of viral glycoproteins to the axon and dendrites of hippocampal neurons in culture. Cell 62: 63–72

Griffiths G, Simons K (1986) The trans Golgi network: sorting at the exit site of the Golgi complex. Science 234: 438–443

Haass C, Koo EH, Mellon A, Hung AY, Selkoe DJ (1992) Targeting of cell-surface β-amyloid precursor protein to lysosomes: alternative processing into amyloid-bearing fragments. Nature 357: 500–503

Haass C, Koo EH, Teplow DB, Selkoe DJ (1994) Polarized secretion of beta-amyloid precursor protein and amyloid beta-peptide in MDCK cells. Proc Natl Acad Sci (USA) 91: 1564–1568

Hunziker W, Mellman I (1989) Expression of macrophage-lymphocyte Fc receptors in MDCK cells: polarity and transcytosis differ for isoforms with or without coated pit localization domains. J Cell Biol 109: 3291–3302

Hunziker W, Harter C, Matter K, Mellman I (1991) Basolateral sorting in MDCK cells requires a distinct cytoplasmic domain determinant. Cell 66: 907–920

Le Bivic A, Quaroni A, Nichols B, Rodriguez-Boulan E (1990) Biogenetic pathways of plasma membrane proteins in Caco-2, a human intestinal epithelial cell line. J Cell Biol 111: 1351–1361

Le Bivic A, Sambuy Y, Patzak A, Patil N, Chao M, Rodriguez-Boulan E (1991) An internal deletion in the cytoplasmic tail reverses the apical localization of human NGF receptor in transfected MDCK cells. J Cell Biol 115: 607–618

Letourneur F, Klausner RD (1992) A novel di-leucine motif and a tyrosine-based motif independently mediate lysosomal targeting and endocytosis of CD3 chains. Cell 69: 1143–1157

Matter K, Brauchbar M, Bucher K, Hauri H-P (1990) Sorting of endogenous plasma membrane proteins occurs from two sites in cultured human intestinal epithelial cells (Caco-2). Cell 60: 429–437

Matter K, Hunziker W, Mellman I (1992) Basolateral sorting of LDL receptor in MDCK cells: the cytoplasmic domain contains two tyrosine-dependent targeting determinants. Cell 71: 741–753

Matter K, Whitney JA, Yamamoto EM, Mellman I (1993) Common signals control LDL receptor sorting in endosomes and the Golgi complex of MDCK cells. Cell 74: 1053–1064

Mellman I, Yamamoto E, Whitney JA, Kim M, Hunziker W, Matter K (1993) Molecular sorting in polarized and non-polarized cells: common problems, common solutions. J Cell Sci Suppl 17: 1–7

Roberts SB, Ripellino JA, Ingalls KM, Robakis NK, Felsenstein KM (1994) Non-amyloidogenic cleavage of the β-amyloid precursor protein by an integral membrane metalloendopeptidase. J Biol Chem 269: 3111–3116

Rodriguez-Boulan E, Nelson WJ (1989) Morphogenesis of the polarized epithelial cell phenotype. Science 245: 718–725

Rodriguez-Boulan E, Powell SK (1992) Polarity of epithelial and neuronal cells. Ann Rev Cell Biol 8: 395–427

Rodriguez-Boulan E, Sabatini DD (1978) Asymmetric budding of viruses in epithelial monolayers: a model system for study of epithelial polarity. Proc Natl Acad Sci USA 75: 5071–5075

Selkoe DJ (1993) Physiological production of the β-amyloid protein and the mechanism of Alzheimer's disease. [Review]. Trends Neurosci 16: 403–409

Wandinger-Ness A, Simons K (1990) The polarized transport of surface proteins and lipids in epithelial cells. In: Hanover J, Steer S (eds) Intracellular trafficking of proteins. Cambridge, Cambridge University Press, pp 575–612

Wandinger-Ness A, Bennett MK, Antony C, Simons K (1990) Distinct transport vesicles mediate the delivery of plasma membrane proteins to the apical and basolateral domains of MDCK cells. J Cell Biol 111: 987–1000

Zurzolo C, van't Hof W, Rodriguez-Boulan E (1994) VIP21/caveolin, glycosphingolipid clusters and the sorting of glycophosphatidylinositol-anchored proteins in epithelial cells. EMBO J 13: 42–53

Regulation of Receptor Trafficking by ras-Like GTPases

*S.R. Pfeffer**

Summary

The trafficking of proteins, such as the amyloid precursor, between the Golgi complex, the cell surface, and the endocytic compartment is regulated by a family of ras-like GTPases termed "rab" proteins. We are trying to understand the mechanism by which rab proteins regulate receptor trafficking. We study rab9 and its role in facilitating the transport of proteins between late endosomes and the trans Golgi network. Stable expression of a dominant negative mutant form of rab9 protein, rab9 S21N, strongly inhibited the transport of mannose 6-phosphate receptors from late endosomes to the trans Golgi network in cultured cells. When this pathway is blocked, other aspects of endocytosis and receptor trafficking proceed normally. In addition, transport vesicle formation appears to require incorporation of a rab protein in its GTP-bound conformation, since receptors were not trapped in nonfunctional transport vesicles in cells expressing rab9 S21N protein. This is a surprise, since rab proteins are believed to facilitate transport vesicle targeting and/or fusion. Our findings suggest that transport vesicle formation is regulated by the availability of transport factors assembled in some fashion on the surface of newly forming transport vesicles.

A fraction of rab proteins are present in the cytosol, with GDP bound, complexed to a protein termed GDI. We have succeeded in reconstituting the selective targeting of prenylated rab9 protein onto late endosome membranes and have shown that this process is accompanied by endosome-triggered nucleotide exchange. Our results show that GDI presents functional rab9 protein to the transport machinery. GDI is then displaced by a protein on the surface of late endosomes. After nucleotide exchange, the activated rab protein becomes incorporated into a transport vesicle, which can bud off and be delivered to its appropriate target. After vesicle fusion, GTP is hydrolyzed by the rab protein, and unoccupied GDI retrieves the rab from the target membrane and delivers it to its membrane of origin.

* Department of Biochemistry, Stanford University School of Medicine, Stanford, CA 94305-5307, USA

K.S. Kosik et al. (Eds.)
Alzheimer's Disease: Lessons from Cell Biology

These events underlie the transport of receptors between membrane-bound compartments of the endocytic and exocytic pathways.

Introduction

The amyloid β-peptide (Aβ), encoded within the β-amyloid precursor protein (β-APP), is likely to play a significant role in the formation of extracellular amyloid plaques in the cerebral cortex and cerebral and meningeal blood vessels. Work in many laboratories has begun to elucidate the mechanisms by which the Aβ peptide is generated (for review, see Haass and Selkoe 1993; Ashall and Goate 1994). The amyloid precursor traverses the conventional secretory pathway, and probably travels from the Golgi complex directly to the plasma membrane. A consensus NPXY sequence in its cytoplasmic domain mediates the endocytosis and recycling of the protein from endosomes back to the plasma membrane. Agents that influence the rates of overall endocytosis and receptor recycling, such as phorbol esters, can influence the rate with which proteolytic products of β-APP are formed. These studies imply that an understanding of the intracellular trafficking of β-APP may provide important clues towards interfering with Aβ generation in living cells. Moreover, elucidation of the proteases which generate β-APP products, and characterization of their localizations in cells, will also provide useful information regarding the pathogenesis of Alzheimer's disease.

We study the transport of receptors within the endocytic pathway. We focus on a family of ras-like GTPases, termed rab proteins, that are key regulators of receptor-mediated endocytosis, endosome fusion, and receptor-recycling. Dominant negative mutations in these proteins can have severe consequences on the rates of endosome fusion, endocytosis, and receptor recycling, and can alter the steady state sizes of entire organelles. For this reason, rab proteins are thought to be master regulators of membrane trafficking.

Pathways of Intracellular Transport

Figure 1 depicts the transport routes taken by proteins from the Golgi complex to the cell surface and from the cell surface to endocytic compartments. It is most likely that the β-APP is delivered directly from the trans Golgi network (TGN) to the cell surface. Moreover, work from many labs suggests that β-APP undergoes a cycle of internalization and recycling, most likely through the early endosome and perhaps also the late endosomes. Some β-APP may be degraded within late endosomes. Only a small percentage of total membrane proteins is transported from late endosomes to the TGN. This group includes the mannose 6-phosphate receptor, which delivers newly synthesized lysosomal enzymes from the TGN to the endocytic pathway.

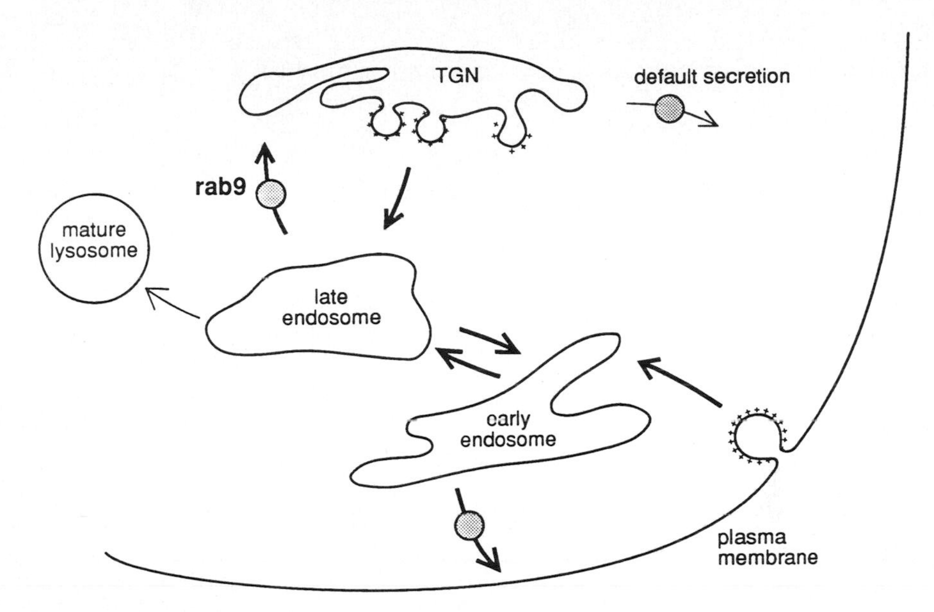

Fig. 1. Trafficking pathways taken by mannose 6-phosphate receptors and/or lysosomal enzymes. Bold arrows demarcate MPR pathways; thin arrows represent paths taken by lysosomal enzymes released from MPRs or secreted by default. Rab9S21N blocked only MPR recycling from late endosomes to the trans Golgi network (TGN); other pathways appeared to function at normal rates. The rab9S21N block led to increased default secretion, but did not alter significantly the overall subcellular distribution of MPRs at steady state

Rab GTPases

Rab GTPases represent a family of over 30 proteins that are localized to the surfaces of distinct membrane-bound organelles (Takai et al. 1992; Pfeffer 1992; Zerial and Stenmark 1993; Novick and Brennwald 1993; Ferro-Novick and Novick 1993). These proteins function in the processes by which transport vesicles dock and/or fuse with their cognate, target membranes. Like ras, they interconvert between GDP- and GTP-bound conformations, as part of a catalytic cycle of vesicle delivery and rab protein recycling. Current models suggest that rab proteins catalyze membrane fusion in their GTP-bound conformations (Fig. 2). They are then thought to be acted upon by rab-specific GTPase activating proteins (GAPs), converting them to their GDP-bound forms. A cytosolic protein, termed GDI, has the capacity to retrieve GDP-bound rab proteins from cellular membranes. GDI has been proposed to retrieve rab proteins from their fusion targets, and recycle them back to their membranes of origin.

GDI has now also been shown to deliver cytosolic rab proteins to the appropriate organelle, using permeabilized cells (Ullrich et al. 1994) or enriched membrane fractions (Soldati et al. 1994). Purified, prenylated rab5

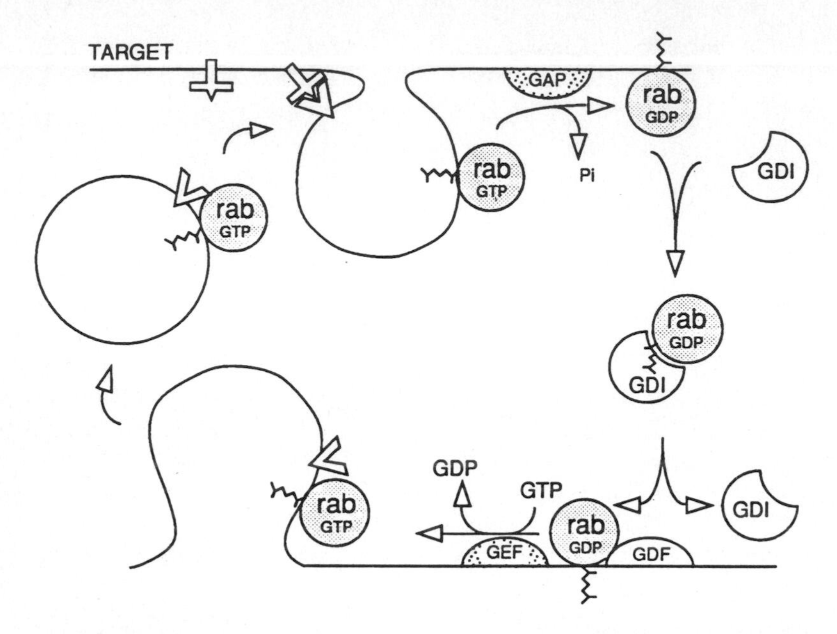

Fig. 2. Prenylated rab proteins occur in the cytosol as a complex with GDI (center, right). GDI presents the complexes to specific organelles; this may be catalyzed by a GDI-displacement factor (GDF) in conjunction with a nucleotide exchanger (GEF). Rab proteins, with GTP bound, are recruited into nascent transport vesicles and may catalyze association of V-SNARES (on the vesicle) with T-SNARES (on the target membrane) to accomplish transport vesicle delivery. After fusion, a GTPase-activating protein (GAP) increases the GTPase rate of the rab protein, converting it into its GDP conformation. Unoccupied GDI then retrieves the rab protein for another round of vesicular transport

or rab9 proteins were reconstituted with purified GDI and membrane transfer was then monitored. Targeting was accompanied by the displacement of GDI, followed by the exchange of bound GDP for GTP. Since membrane recruitment lagged the nucleotide exchange process, we proposed that a GDI-displacement factor, or GDF, catalyzes the recruitment of rab proteins onto specific cytoplasmic organelles (Soldati et al. 1994).

Direct proof that GDI presents functional rab proteins to specific membrane-bound organelles comes from immunodepletion experiments in which removal of GDI and rab proteins bound to GDI led to complete loss of the ability of cytosol to stimulate intracellular transport of proteins between late endosomes and the trans Golgi network in vitro (Dirac-Svejstrup et al. 1994). GDI also increased the efficiency with which rab proteins were utilized by the transport machinery, by increasing the selectivity with which they were delivered to their cognate membrane targets. Together, these experiments demonstrate that GDI functions to deliver functional rab proteins to their correct membrane targets. Perhaps by binding tightly to geranylgeranyl groups, GDI suppresses non-specific membrane association of hydrophobic rab proteins and thereby enhances the recognition of rab protein structural determinants by the organelle-specific rab recruitment machinery.

Rab9 Drives a Single Transport Step: Transport from Late Endosomes to the TGN

Newly synthesized lysosomal enzymes bind to mannose 6-phosphate receptors (MPRs) in the TGN and are carried to prelysosomes, where they are released. MPRs then return to the TGN for another round of transport. We showed recently that the rab9 protein is localized primarily to the surface of late endosomes and can facilitate the transport of MPRs from late endosomes to the TGN in vitro (Lombardi et al. 1993). To investigate the physiological significance of our findings, and to explore the significance of MPR recycling in living cells, we generated a dominant inhibitory form of rab9 protein based upon well-characterized mutations of Ras (Riederer et al. 1994).

We used site-directed mutagenesis to generate rab9 mutant proteins which might block intracellular transport upon expression in mammalian cells. The most informative mutant was rab9 S21N. This mutation is equivalent to Ras S17N, a dominant inhibitory mutation of the Ras protein (Feig and Cooper 1988; Farnsworth and Feig 1991). In Ras, serine 17 has been postulated to participate through its hydroxyl moiety in the coordination of a magnesium ion at the active site (Pai et al. 1989, 1990). Ras S17N binds GDP in strong preference over GTP and appears to act as a dominant inhibitor by sequestering a nucleotide exchange factor, thus increasing the concentration of Ras-GDP (Feig and Cooper 1988). This hypothesis was confirmed directly for an analogous mutation in rab3A (T36N) which was shown to bind with 10-fold greater affinity than wild type rab3A to a rab3A-specific, nucleotide exchange factor (Burstein et al. 1992).

A stable CHO cell line expressing rab9 S21N (N21) was generated by conventional means. These cells expressed rab9 S21N protein at approximately two-fold higher levels than the endogenous, wild type rab9 protein. We used an assay devised by Kornfeld and colleagues to detect the transport of proteins from the cell surface to the TGN in living cells (Duncan and Kornfeld 1988). Metabolically labeled surface MPRs were de-sialylated by incubation of cells with neuraminidase at 37°C. Under these conditions, a large proportion of MPRs cycle through the surface, where they can be acted upon by the neuraminidase. Cells are then washed and incubated further to permit endocytosis and intracellular transport. Unlike most receptors which remain in the endocytic pathway, MPRs return to the TGN and can be modified by the sialyltransferases located there. The re-acquisition of sialic acid by MPRs over time can then be determined by chromatography on a column of a sialic acid-specific lectin.

In control cells, a significant fraction of MPRs re-acquired sialic acid during a three-hour post-neuraminidase incubation. In contrast, MPR resialylation was inhibited dramatically in cells expressing rab9 S21N. Resialylation was not completely blocked, but displayed a lag of ≥3 hours. Thus, rab9 S21N displayed a dominant phenotype when expressed at two-fold higher levels than endogenous rab9 and interfered with the process by which MPRs are recycled to the TGN in living cells.

According to current models of rab function, a mutant rab protein should display a defect along a single route of intracellular transport. To verify the specificity of rab9 action, we investigated the capacity of cells expressing rab9 S21N to carry out both fluid phase and receptor-mediated endocytosis. Wild type cells and N21 cells displayed identical rates of horseradish peroxidase uptake. Thus, fluid phase endocytosis was unaltered by rab9 S21N overexpression.

Receptor-mediated endocytosis was tested by monitoring the ability of N21 cells to bind and internalize extracellularly administered β-glucuronidase via cell surface MPRs. When the rate of β-glucuronidase uptake in N21 cells was corrected for the difference in surface binding sites, the absolute rates of enzyme internalization per surface binding site were identical for N21 and control cells. These data confirm that the pathway of receptor-mediated endocytosis is not influenced by rab9 S21N protein expression. Moreover, they show that the 300 kD MPRs are functional in N21 cells in terms of their capacity for ligand binding and receptor-mediated endocytosis.

The specificity of the transport block induced by rab9 S21N expression was further confirmed by examining the rate of biosynthetic protein transport of glycoproteins through the secretory pathway. For this purpose, we examined the rate with which the 300 kD MPR was transported from the endoplasmic reticulum to the medial Golgi, as monitored by the rate with which it acquired endoglycosidase H-resistant oligo-saccharides. As expected, rab9 S21N had no influence on the rate of biosynthetic protein transport; MPRs in control and N21 cells acquired endo H-resistant oligo-saccharides with a half-time of ~60 min in CHO cells, consistent with previous reports.

Rab9 S21N Interferes with Lysosomal Enzyme Targeting

The recycling of MPRs from late endosomes to the TGN is believed to be an important step in the delivery of newly synthesized lysosomal enzymes from the TGN to prelysosomes. If rab9 S21N blocks this process selectively, expression of rab9 S21N should decrease the number of available MPRs within the TGN, and thus interfere with the targeting of newly synthesized lysosomal hydrolases. The expression of rab9 S21N had severe consequences in terms of the delivery of cathepsin to prelysosomes and/or lysosomes. The rate with which cathepsin D was processed, indicative of MPR-dependent targeting, was more than three-fold lower in N21 cells when compared with control cells.

If the receptor-mediated delivery of newly synthesized lysosomal enzymes from the TGN to lysosomes is disrupted, cells expressing rab9 S21N should display an increased level of lysosomal enzyme secretion via a bulk-flow, default pathway. This prediction was confirmed upon analysis of the secretion of newly synthesized hexosaminidase from control and N21 cell lines. Control cells and N21 cells synthesized hexosaminidase at identical rates (~14 units/mg/hr). However, N21 cells secreted more than twice as much of the newly synthesized hexosaminidase as control cells. Assays of enzyme secretion were carried out in the presence of excess mannose 6-phosphate (man6P) to block interaction with surface MPRs. Omission of man6P led to a decreased accumulation of hexosaminidase in the extracellular medium, indicating that the cells were able to re-internalize the secreted enzyme. The ability of man6P to increase the levels of detected secreted hydrolases confirmed that MPRs were fully functional in terms of their capacity to endocytose extracellular lysosomal hydrolases. In summary, N21 cells secreted a larger proportion of their lysosomal enzyme content than control cells and compensated for this by an endocytic recapture process.

These experiments confirm that the Ras-like GTPase, rab9, is essential for the recycling of 300 kD MPRs between endosomes and the TGN in living cells. Stable expression of a dominant negative form of rab9 blocked this transport step specifically, while the rates of biosynthetic protein transport and fluid phase and receptor-mediated endocytosis remained unchanged. Moreover, expression of rab9 S21N at only two-fold higher levels than wild type rab9 protein was sufficient to yield this phenotype.

Cells rendered defective in lysosomal enzyme sorting by expression of rab9 S21N compensated for this defect by increasing their rates of synthesis of a variety of lysosomal enzymes. Endogenous levels of cathepsin D, hexosaminidase and β-glucuronidase were all elevated in N21 cells. In addition, the level of the lysosomal membrane glycoprotein, lgpB, was increased 3.6-fold and MPRs levels increased 1.6-fold. Thus, cells seem to have the capacity to induce the coordinate synthesis of a group of organelle-specific proteins.

MPRs cycle between the TGN and late endosomes in what is referred to as the biosynthetic pathway, and between the cell surface and early endosomes along the conventional endocytic pathway (Fig. 1). These pathways are interconnected, in that the entire pool of MPRs cycles through the cell surface in less than an hour. However, it has not yet been established precisely how these pathways are connected. MPRs in the TGN might escape the biosynthetic pathway and be delivered to the surface by default. Alternatively, MPRs might shuttle directly from late endosomes to the cell surface, or from late endosomes to early endosomes, from which they could then recycle to the cell surface. Since a severe block in their transport back to the TGN did not alter their overall cycling properties or their steady state morphological distribution, our data provide the first indication that MPRs can be transported from late endosomes back to the cell surface, either directly, or via early endosomes. Such a pathway has also been documented for a lysosomal membrane glycoprotein in chick fibroblasts.

Nascent Transport Vesicle Formation Requires rab-GTP

The first rab protein family member was discovered as the product of a yeast gene (*sec4*) which, when mutated, led to the accumulation of secretory vesicles (Salminen and Novick 1987). It was therefore surmised that rab proteins function in membrane docking and/or fusion events. The ability of rab5 protein to regulate endosome fusion and the requirement for ypt1/rab1 in ER-to-Golgi transport vesicle fusion supported this notion (Pfeffer 1992; Zerial and Stenmark 1993). At first glance, these observations would lead to the prediction that inactive rab proteins would lead to the accumulation of transported proteins in non-functional transport vesicles. Surprisingly, for rab protein mutants which bind GDP in strong preference to GTP, this prediction turns out not to be correct.

The S21N mutation in rab9 (Riederer et al. 1994) and the comparable mutation in rab 1A (Nuoffer et al. 1994) proteins displayed biochemical properties similar to those described for ras S17N. However, elevated levels of prenylated rab1A S25N appeared to block protein *export* from the ER, rather than lead to the accumulation of non-functional transport vesicles (Nuoffer et al. 1994). Similarly, rab9 S21N blocked transport of proteins from late endosomes to the trans Golgi network, without sequestering mannose 6-phosphate receptors in non-functional transport intermediates. Rather, mannose 6-phosphate receptors retained full access to all of the other intracellular transport pathways along which they traversed (Riederer et al. 1994).

These experiments suggest, for the first time, that vesicle budding is tightly coupled to the proper assembly of transport constituents into nascent transport vesicles. In the absence of sufficient quantities of rab proteins in their GTP conformations, transport vesicles fail to form. This feature may not be true for all factors needed for membrane targeting, since ER-derived transport vesicles form with apparently normal efficiency in cells depleted of Bos1p, a membrane protein required for transport vesicle fusion (Lian and Ferro-Novick 1993).

Conclusion

Rabs are clearly needed for the processes of membrane vesicle docking and fusion, and it now appears that the appropriate rab protein, in its GTP-bound conformation, must be present for transport vesicle formation to occur. But what are rab proteins actually doing? Since hybrid rab proteins can function at different steps in intracellular transport without altering the routing of transported proteins (Brennwald and Novick 1993), rabs appear not to specify a transport vesicle's target. Rather, targeting seems to involve V- and T-SNARE proteins present on vesicles and target membranes (Rothman and Warren 1994). However, SNARE interactions must be tightly regulated (Novick and Brennwald 1993; Zerial and Stenmark 1993; Rothman and Warren 1994). V-SNAREs should only be active when present on a fully formed transport vesicle; otherwise, entire organelles would fuse. Moreover, T-SNAREs must be released from V-SNAREs after fusion of transport vesicles, to permit recycling of V-SNAREs and their re-incorporation into nascent transport vesicles. Rab GTPases may thus regulate membrane trafficking by only permitting SNARE interactions at the right time and at the right place. As such, they represent key points at which cells can regulate specific aspects of protein trafficking, including that of the beta amyloid protein precursor.

References

Ashall F, Goate AM (1994) Role of the β-amyloid precursor protein in Alzheimer's disease. Trends Biochem Sci 19: 42–46

Brennwald P, Novick P (1993) Interaction of three domains distinguishing the Ras-related GTP-binding proteins Ypt1 and Sec4. Nature 362: 560–563

Burstein ES, Brondyk WH, Macara IG (1992) Amino acid residues in the Ras-like GTPase rab3A that specify sensitivity to factors that regulate the GTP/GDP cycling of rab3A. J Biol Chem 267: 22715–22718

Dirac-Svejstrup AB, Soldati T, Shapiro AD, Pfeffer SR (1994) Rab-GDI presents functional rab9 to the intracellular transport machinery and contributes selectivity to rab9 membrane recruitment. J Biol Chem 269: 15427–15430

Duncan J, Kornfeld S (1988) Intracellular movement of two mannose 6-phosphate receptors: return to the Golgi apparatus. J Cell Biol 106: 617–628

Farnsworth CL, Feig LA (1991) Dominant inhibitory mutations in the Mg^{++}-binding site of RasH prevent its activation by GTP. Mol Cell Biol 11: 4822–4829

Feig LA, Cooper GM (1988) Inhibition of NIH 3T3 cell proliferation by mutated Res proteins with preferential affinity for GDP. Mol Cell Biol 8: 3235–3243

Ferro-Novick S, Novick P (1993) The role of GTP-binding proteins in transport along the exocytic pathway. Ann Rev Cell Biol 9: 575–599

Haass C, Selkoe DJ (1993) Cellular processing of β-amyloid precursor protein and the genesis of amyloid β-peptide. Cell 75: 1039–1042

Lian JP, Ferro-Novick S (1993) Bos1p, an integral membrane protein of the endoplasmic reticulum to golgi transport vesicles, is required for their fusion competence. Cell 73: 735–745

Lombardi D, Soldati T, Riederer MA, Goda Y, Zerial M, Pfeffer SR (1993) Rab9 functions in transport between late endosomes and the trans Golgi network. EMBO J 12: 677–682

Novick P, Brennwald B (1993) Friends and Family. The role of the Rab GTPases in vesicular traffic. Cell 75: 597–601

Nuoffer C, Davidson HW, Matteson J, Meinkoth J, Balch WE (1994) A GDP-bound form of Rab1 inhibits protein export from the endoplasmic reticulum and transport between Golgi compartments. J Cell Biol 125: 225–237

Pai EF, Kabsch W, Krengel U, Holmes KC, John J, Wittinghofer A (1989) Structure of the guanine-nucleotide binding domain of the Ha-Ras oncogene product p21 in the triphosphate conformation. Nature 1989: 209–214
Pai EF, Krengel U, Petsko GA, Goody RS, Kabsch W, Wittinghofer A (1990) Refined crystal structure of the triphosphate conformation of Ha-ras at 1.35 Å resolution: implications for the mechanism of GTP hydrolysis. EMBO J 9: 2351–2359
Pfeffer SR (1992) GTP-binding proteins in intracellular transport. Trends Cell Biol 2: 41–45
Riederer MA, Soldati T, Shapiro AD, Lin J, Pfeffer SR (1994) Lysosome biogenesis requires Rab9 function and receptor recycling from endosomes to the trans Golgi network. J Cell Biol 125: 573–582
Rothman JE, Warren G (1994) Implications of the SNARE hypothesis for intracellular membrane topology and dynamics. Curr Biol 4: 220–233
Salminen A, Novick PJ (1987) A Ras-like protein is required for a post-Golgi event in yeast secretion. Cell 49: 527–538
Soldati T, Shapiro AD, Dirac Svejstrup AB, Pfeffer SR (1994) Membrane targeting of the small GTPase rab9 is accompanied by nucleotide exchange. Nature 369: 76–78
Takai Y, Kaibuchi K, Kikuchi A, Kawata M (1992) Small GTP-binding proteins. Int Rev Cytol 133: 187–230
Ullrich O, Horiuchi H, Bucci C, Zerial M (1994) Membrane association of rab5 mediated by GDP-dissociation inhibitor and accompanied by GDP/GTP exchange. Nature 368: 157–160
Zerial M, Stenmark H (1993) Rab GTPases in vesicular transport. Curr Opin Cell Biol 5: 613–620

Alzheimer's Disease and Hemorrhagic Stroke: Their Relationship to βA4 Amyloid Deposition

*L. Hendriks**, *P. Cras*, *J.-J. Martin*, and *C. Van Broeckhoven*

Summary

Distinct mutations have been reported in approximately 5% of early-onset Alzheimer disease (AD) families in the gene coding for the amyloid precursor protein (APP) located at chromosome 21q21.2. Mutations in APP have also been found in families segregating hemorrhagic stroke due to congophilic βA4 amyloid angiopathy (CAA) both in the presence and absence of AD. These mutations are located close to known proteolytic cleavage sites in APP, either at the N-terminal or C-terminal site or within the sequence of βA4 amyloid, the proteolytic product found in AD and CAA brain lesions. cDNA transfection experiments have indicated that these APP mutations interfere with the normal processing of APP, causing either an overproduction of βA4 amyloid or a longer βA4 amyloid that is more prone to aggregation. We have initiated cDNA transfection experiments in COS-1 cells and have studied the production of βA4 amyloid. Our results confirm previous observations in other cell types, i.e., overproduction of βA4 amyloid is not a consistent finding in all known APP mutations.

Introduction

Alzheimer's disease (AD) is the major form of progressive dementia in the elderly. In a demented case, AD is diagnosed by post-mortem examination of the brain showing severe neuronal cell loss and particular brain lesions in the cortex and hippocampus named neurofibrillary tangles and senile plaques. The neurofibrillary tangles are intraneuronal inclusions of paired helical filaments composed of abnormally phosphorylated tau. The senile plaques result from deposition in the brain parenchyma and cerebral blood vessel walls of βA4 amyloid. The βA4 amyloid deposits are surrounded by dystrophic neurites. βA4 amyloid is a 4 kDa proteolysis product of a larger precursor protein, the amyloid precursor protein (APP). The gene encoding

* Laboratory of Neurogenetics, Born Bunge Foundation, Department of Biochemistry, University of Antwerp (UIA), Universiteitsplein 1, 2610 Antwerp, Belgium

K.S. Kosik et al. (Eds.)
Alzheimer's Disease: Lessons from Cell Biology

APP is localized at chromosome 21q21.2 (Kang et al. 1987; Tanzi et al. 1987) and contains 18 exons (Yoshikai et al. 1990). Three major splicing variants have been identified containing the βA4 amyloid sequence, i.e., the APP695, APP751 and APP770 isoforms, of which APP695 is the major isoform found in brain (Tanzi et al. 1988; Kitaguchi et al. 1988; Ponte et al. 1988). Some of the AD pathological lesions are also seen in patients suffering from hereditary cerebral hemorrhage with amyloidosis-Dutch type (HCHWA-D), a rare disease segregating in families in the Netherlands (*for review* see Haan et al. 1991). HCHWA-D is characterized by recurrent hemorrhagic strokes due to congophilic amyloid angiopathy (CAA). The amyloid that is deposited in the leptomeningeal and cortical blood vessel walls was identified as βA4 amyloid. Although small βA4 amyloid aggregates are also seen in the parenchyma, termed diffuse plaques, no senile plaques, neurofibrillary tangles or dystrophic neurites are observed. Approximately 50% of the HCHWA-D patients die at the first hemorrhagic stroke and those that survive develop a progressive dementia of the multi-infarct type.

We described a family with patients presenting with either probable AD according to NINCDS criteria or hemorrhagic stroke due to CAA (Hendriks et al. 1992). Immunohistochemical analysis of brain biopsy material obtained at surgery of a stroke patient showed extensive vascular βA4 amyloid deposition, diffuse βA4 amyloid plaques and dystrophic neurites. Recently, we obtained autopsy material of a demented case and immunohistochemical analysis confirmed that the brain pathology in this patient was that of AD (unpublished data). In contrast to classical AD, large amounts of vascular βA4 amyloid were seen. Also, senile plaques contained larger amounts of βA4 amyloid.

APP Mutations

Hitherto six different single base mutations have been identified in the APP gene (*for review* see Van Broeckhoven 1994): four mutations in classical AD patients – three different mutations involving codon 717 and a double mutation involving codons 670/671; one mutation in HCHWA-D patients at codon 693; and one mutation in AD and CAA patients at codon 692 (Table 1). Each of these mutations causes single amino acid substitutions localized, respectively, at the N- and C-terminus of the βA4 amyloid sequence in AD patients or within the βA4 amyloid sequence in HCHWA-D and AD/CAA patients. Only the APP717(Val to Ile) mutation has been observed in distinct presenile AD families of different ethnic backgrounds. APP mutations are the genetic cause of AD or related disorders in only a small fraction of cases, i.e., approximately 5% of all familial presenile AD cases. The importance of these mutations lies, however, in the fact that they can be used as genetic tools to study the underlying mechanisms of these diseases. One approach is to transfect cells with an APP cDNA bearing the respective mutations and to study the effect of the mutation on APP processing and βA4 amyloid production, aggregation and deposition.

Table 1. Pathogenic mutations in AD and CAA patients

Codon[a]	Substitution	Phenotype	Reference
APP670/671	Lys to Asn and Met to Leu	AD	Mullan et al. 1992
APP692	Ala to Gly	AD + CAA	Hendriks et al. 1992
APP693	Glu to Gln	HCHWA-D	Levy et al. 1990
APP717	Val to Ile	AD	Goate et al. 1991
	Val to Phe	AD	Murrell et al. 1991
	Val to Gly	AD	Chartier-Harlin et al. 1991

[a] Codon numbering according to the APP770 isoform.

Processing of APP

There are two major ways to process APP (*for review* see Haass and Selkoe 1993): an amyloidogenic and a non-amyloidogenic pathway (Fig. 1A). In the non-amyloidogenic pathway APP is cut within the βA4 amyloid sequence, producing a membrane-bound 10 kDa (P10) fragment containing only part of βA4 amyloid and a soluble APP (Esch et al. 1990). Several potential amyloidogenic pathways have been described. An endosomal/lysosomal pathway produces βA4 amyloid-containing, 11 kDa (P11) C-terminal fragments of APP (Estus et al. 1992; Golde et al. 1992). In this pathway full length APP may be reinternalized from the cell surface and is then degraded into potentially amyloidogenic fragments accumulating in the lysosomes (Haass et al. 1992a). In addition it has been shown that intact βA4 amyloid is secreted in the medium of different cell cultures and in cerebrospinal fluid of nondemented control subjects (e.g., Haass et al. 1992b; Seubert et al. 1992, 1993; Shoji et al. 1992; Busciglio et al. 1993). In this secretory pathway the cleavage of APP to generate βA4 amyloid takes place between Met671 and Asp672 at the N-terminus of βA4 amyloid (Seubert et al. 1993).

The proteolytic enzymes involved in the different APP metabolic pathways have not yet been identified. The enzymes have been termed α-secretase, cutting APP within βA4 amyloid, and β-secretase and γ-secretase, cutting APP at, respectively, the N- and C-terminal site of βA4 amyloid (Fig. 1A). The APP mutations that were identified in AD and AD-related CAA patients interfere with the proteolytic activity of each one of these APP secretases.

Influence of APP Mutations on βA4 Amyloid Deposition

APP670/671 Mutation

The effect of the pathogenic APP mutations on the production of βA4 amyloid is depicted schematically in Figure 1B. A five- to eight-fold overproduction of βA4 amyloid has been reported for the double mutation

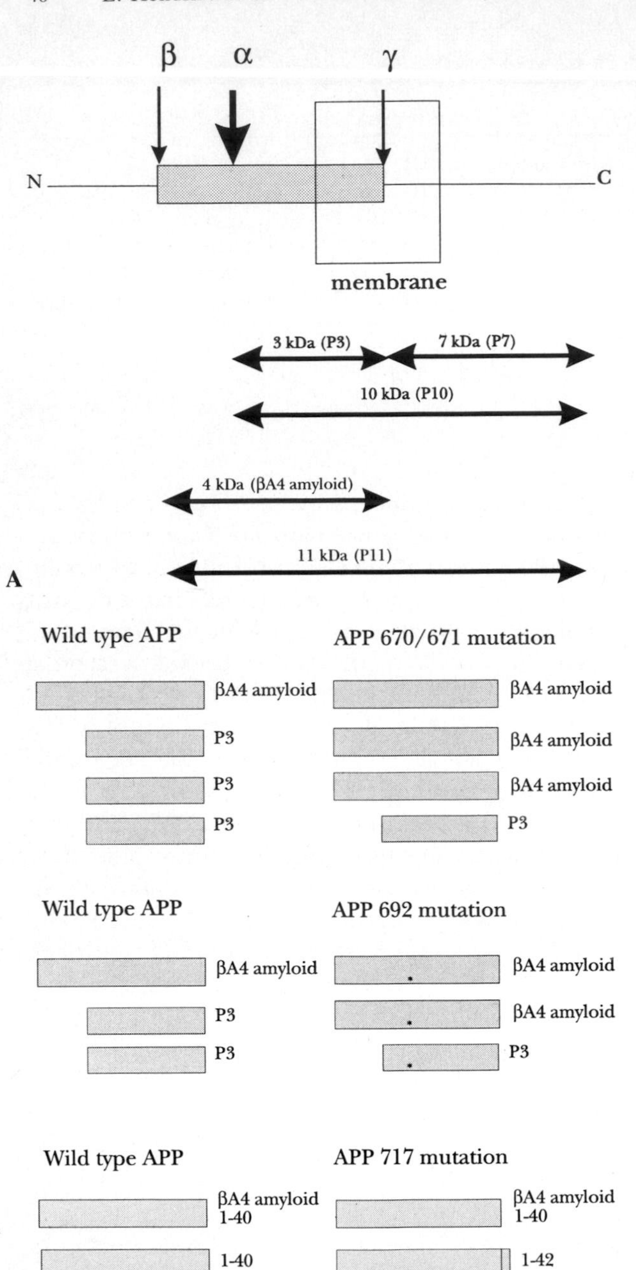
β
α
γ
N
C
membrane
3 kDa (P3)
7 kDa (P7)
10 kDa (P10)
4 kDa (βA4 amyloid)
11 kDa (P11)
A
Wild type APP
βA4 amyloid
P3
P3
P3
APP 670/671 mutation
βA4 amyloid
βA4 amyloid
βA4 amyloid
P3
Wild type APP
βA4 amyloid
P3
P3
APP 692 mutation
βA4 amyloid
βA4 amyloid
P3
Wild type APP
βA4 amyloid 1-40
1-40
1-42
APP 717 mutation
βA4 amyloid 1-40
1-42
1-42
B

APP670/671 (Cai et al. 1993; Citron et al. 1992). Citron and coworkers (1992) found that the APP671(Met to Leu) substitution is responsible for this increase in βA4 amyloid secretion. These authors also showed that the 3 kDa (P3) fragment, the C-terminal subpeptide of βA4 amyloid resulting from both α- and γ-secretase activity (Fig. 1A), was decreased in the medium of cells transfected with a APP cDNA containing the APP670/671 mutation. Both the increase in βA4 amyloid and decrease in P3 was seen in the media of transiently and stably transfected human kidney 293 cells and Chinese hamster ovarian (CHO) cells. Cai and coworkers (1993) transfected human neuroblastoma M17 cells with APP cDNA constructs containing the APP670/671 and APP717(Val to Ile) mutation, respectively. In the case of the APP670/671 mutant, they observed a six-fold increase in the release into the cell culture medium of the P11 fragment and βA4 amyloid, whereas no such effect could be detected for the APP717 mutation. Similar results were obtained in H4 neuroglioma cells by Felsenstein et al. (1994). With the use of specific antibodies these authors demonstrated a qualitative change in the APP secreted products, most likely resulting from a shift in the APP secretase activities favoring β-secretase activity in the APP670/671 mutant. Cheung et al. (1994) and Dovey et al. (1993) sequenced the P11 fragments and βA4 amyloid from both wild type APP and the APP670/671 mutant and showed that the latter is cleaved at the same location as wild type APP. Thus the increased production of P11 and βA4 amyloid indicated an accelerated β-secretase activity. It is likely that P11 is further processed into βA4 amyloid by γ-secretase. Mass spectrometric analysis and sequencing of the C-terminus of βA4 amyloid produced by APP cDNA transfected human kidney 293 cells showed that the predominant form of βA4 amyloid is 1–40 amino acids long. However, a significant amount of βA4 amyloid of 1–42 amino acids was also present (Dovey et al. 1993).

APP692 and APP693 Mutations

In vitro aggregation studies with synthetic βA4 amyloid comprising the APP693 mutation showed accelerated amyloid fibril formation (Wisniewski et al. 1991; Clements et al. 1993) as well as enhanced amyloid fibril stability

Fig 1. APP processing in wild type and mutated APP. **A** Schematic representation of APP with the extracellular N-terminus and intracellular C-terminus. The βA4 amyloid part is shown by the horizontal shaded box. The transmembrane domain is indicated by the vertical open box. The proteolytic cleavage sites of α-secretase in the non-amyloidogenic pathway and of β- and γ-secretases in the amyloidogenic pathway are indicated by *arrows*. The proteolytic products are shown together with their molecular mass and names as *solid horizontal lines*. **B** Influence of APP mutations on βA4 amyloid production. The main differences between the processing of wild type APP (*left*) and mutated APP (*right*) are shown. The intra βA4 amyloid mutation APP692 is shown as an *asterisk*

(Fraser et al. 1992). For the APP692 mutation, however, no significant change in rate of in vitro amyloid aggregation was detected. In fact, the aggregation of βA4 amyloid fibrils bearing the APP692 mutation was slightly slower than that of wild type βA4 amyloid (Clements et al. 1993).

Recently, transfection experiments in human kidney 293 cells with an APP cDNA containing the APP692 mutation demonstrated a relative increase in the secretion of βA4 amyloid and a decrease in P3 production (Haass et al. 1994). These data suggested a lowered activity for the α-secretase due to the presence of the APP692 mutation. In APP693 cDNA transfection experiments in COS-1 cells no altered processing of the mutated APP cDNA was observed (Maruyama et al. 1991). Also, in APP693 cDNA transfected H4 neuroglioma cells no altered production of P11 compared to the wild type APP cDNA was detected (Felsenstein and Lewis-Higgins 1993).

APP717 Mutation

As discussed above, there was no increase in the βA4 amyloid production when neuroblastoma M17 or H4 neuroglioma cells were transfected with a APP cDNA bearing the APP717(Val to Ile) mutation (Cai et al. 1993; Felsenstein et al. 1994). Previous studies showed that no difference could be observed in the processing of APP cDNAs bearing the APP717 mutation compared to wild type APP cDNA when transfected in COS-1 cells (Maruyama et al. 1991). Recently, it has been shown that, compared to wild type APP, the APP717(Val to Ile) and APP717(Val to Phe) mutations produce at least 1.5 times more of the longer, i.e., 1–42 βA4 amyloid, peptides (Suzuki et al. 1994). Longer βA4 amyloid peptides tend to aggregate faster (Jarrett et al., 1993), and could therefore be responsible for AD in patients carrying these APP mutations.

Results

To investigate whether pathogenic APP mutations influence APP or βA4 amyloid production, transfection experiments with mutated APP cDNA constructs were performed. Hereto we used the site-directed mutagenesis protocol described by Deng and Nickoloff (1992) to introduce the nucleotide changes in the APP770 and APP695 cDNAs corresponding to the following amino acid substitutions: Ala692Gly, Glu693Gln, Val717Ile and LysMet670/671AsnLeu. The APP cDNAs were cloned in the HindIII site of a sligthly modified pSG5 plasmid provided by F. Van Leuven, University of Leuven (KUL), Belgium. Wild type and mutated APP cDNAs were transfected in COS-1 cells using the DEAE dextran method, followed by chloroquine treatment (Luthman and Magnusson 1983). The production of βA4 amyloid was studied.

βA4 amyloid was precipitated from the medium with the polyclonal antisera B7/6 (1:50; gift of F. Van Leuven, Leuven, Belgium) or SGY2134 (1:50; gift of S. Younkin, Cleveland, Ohio, USA) and protein A Sepharose beads. After fractionation on a 4–20% Tris/tricine gel and blotting on a Immobilon PVDF membrane, the membrane was first incubated with either B7/6 (1:500) or the monoclonal antibody 4G8 (1:500; Kim et al. 1988) followed by treatment with goat-anti-rabbit horsh radish peroxidase (GARHRP; 1:3000) or with goat-anti-mouse (GAMHRP; 1:3000). B7/6 precipitates βA4 amyloid (Figs. 2 and 3) while SGY2134 also precipitates the P3 fragment (Fig. 3). Detection was done by chemoluminescence.

COS-1 cells transfected with the APP770 cDNA bearing the APP670/671 double mutation produced two- to three-fold more βA4 amyloid when com-

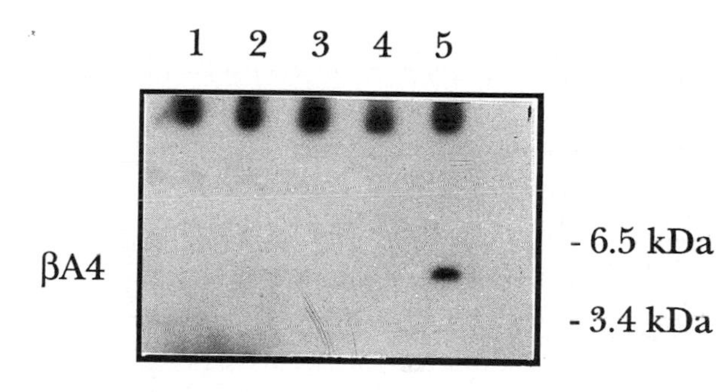

Fig 2. Double immunoprecipitation of βA4 amyloid from medium of COS-1 cells transfected with wild type or mutated APP770 cDNAs using polyclonal antiserum B7/6 (1:50) and detection with monoclonal antibody 4G8 (1:500). Lane 1, mock; lane 2, wild type APP; lane 3, APP692; lane 4, APP693; lane 5, APP670/671

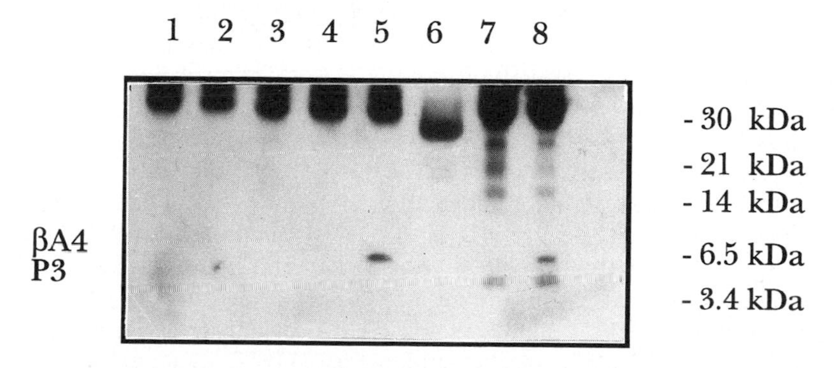

Fig 3. Double immunoprecipitation of βA4 amyloid from medium of COS-1 cells transfected with wild type or mutated APP770 cDNAs using B7/6 (1:50) (lanes 1–5) and SGY2134 (1:50) (lanes 7–8) and detection with B7/6 (1:500) and GARHRP (1:3000). Lane 1, mock; lane 2, wild type APP; lane 3, APP692; lane 4, APP693; lane 5, APP670/671; lane 6, molecular weight marker; lane 7, wild type APP; lane 8, APP670/671

pared with COS-1 cells transfected with the wild type APP cDNA (Figs. 2 and 3). Similar results were obtained when mutated APP695 cDNAs were used in the transfection experiments (data not shown). In different transfection experiments the APP670/671 cDNA always gave a strong signal whereas the APP692 and APP693 cDNAs and the APP717 (Val to Ile) cDNA (data not shown) always gave a very weak or no signal, comparable to that of wild type APP cDNA. When there is no overproduction of βA4 amyloid the latter is difficult to detect, since COS-1 cells produce very small amounts of βA4 amyloid. We estimated that COS-1 cells transfected with the wild type APP cDNA produce less than 0.3 ng βA4 amyloid per ml of medium.

The overproduction of βA4 amyloid in COS-1 cells transfected with the APP670/671 cDNA confirms previous findings obtained in transfection experiments using several other cell lines (Cai et al. 1993; Citron et al. 1992; Felsenstein et al. 1994). Also the observation that the other APP cDNA mutants do not overproduce βA4 amyloid conforms with previously published data (Cai et al. 1993; Felsenstein and Lewis-Higgins 1992; Maruyama et al. 1991). In one transfection experiment we could not observe a decrease in P3 fragment for the APP670/671 cDNA compared to wild type APP cDNA as suggested by the study of Citron et al. 1993.

Discussion

Recent studies have suggested a biological role for apolipoprotein E (apoE) in AD pathogenesis, based on the observation that apoE is present in AD brain lesions and that ApoE binds with high avidity to βA4 amyloid (Strittmatter et al. 1993a,b). Possibly the binding of apoE to βA4 amyloid makes the latter less soluble and thus more prone to deposition (Wisniewski and Frangione 1992). ApoE, a polymorphic protein mainly functioning in the metabolism of cholesterol and triglycerides, has three major isoforms, E2, E3 and E4, whose biosynthesis is controlled by a single apoE gene (APOE) at chromosome 19q13.2. The E2 and E4 isoforms differ from the most common E3 isoform by single amino acid substitutions at positions 112 and 158: E4 (Cys112, Arg158), E2 (Arg158Cys) and E4 (Cys112Arg). The E4 allele of the apoE gene (APOE*4) significantly increases the risk for both early-onset (van Duijn et al. 1994) and late-onset AD (Saunders et al. 1993; Strittmatter et al. 1993b) and the risk is highest for APOE*4 homozygotes. Also, in familial AD the onset age of AD decreased as the number of APOE*4 alleles increased (Corder et al. 1993; van Duijn et al. 1994). The E2 allele of the apoE gene (APOE*2) decreased the risk for AD, suggesting that APOE*2 carriers are protected against AD (Chartier-Harlin et al. 1994; Corder et al. 1994). The opposite actions of the APOE*2 and APOE*4 alleles support the conclusion that apoE is directly involved in AD pathogenesis. It was already been shown that apoE4 has a greater affinity for βA4 amyloid than apoE3 (Strittmatter et al. 1993b), and in patients homozygous for the APOE*4 allele there is an increase in β A4 amyloid deposition in senile plaques and blood vessels (Schmechel et al. 1993).

A similar effect on onset age of the APOE*4 allele was observed in the AD family segregating the APP670/671 mutation (Hardy et al. 1993). Although the APP670/671 mutation is a sufficient cause for AD in this family, patients with the APOE*4 allele had the earliest onset age and patients with the APOE*2 allele had the latest onset age. In families with APP717 mutations the APOE allele frequencies did not differ significantly from population frequencies (Saunders et al. 1993; Hardy et al. 1993). In the HCHWA-D families and the family with AD/CAA patients segregating, respectively, the APP693 and APP692 mutations, a high APOE*4 allele frequency was observed (Haan et al. 1994). However, no correlation was found between the number of APOE*4 alleles and age at onset, age at death, occurrence of dementia or hemorrhage and number of strokes. In the case of the APP692 and APP693 families, the mutation in APP is localized within that part of the βA4 amyloid sequence that showed the strongest affinity for apoE, i.e., amino acids 12–28 (Strittmater et al. 1993a,b). Abolition of the binding of apoE to βA4 amyloid may explain why the APOE genotype does not modulate the clinical expression of the disease in the APP692 and APP693 families. The APP717 ad APP670/671 mutations are located outside the βA4 amyloid sequence leaving the βA4 amyloid sequence intact, and thus no interference with the apoE binding is expected. In future transfection experiments with APP cDNAs bearing AD and/or CAA mutations we will use brain-related cell types such as neuroblastomas, astrocytes or gliomas. Also, we will take into account the possible effect of the APOE genotype on the expression of the AD and/or CAA phenotypes.

Acknowledgments. The authors wish to acknowledge that parts of this work were done in collaboration with investigators at the Erasme University Rotterdam and the University of Leiden, The Netherlands, and at the University of Leuven, Belgium. This work was funded by the Flemish Biotechnology Programme of the Ministry of Economy, Belgium. CVB is a research associate of the National Fund for Scientific Research (NFSR), Belgium.

References

Busciglio J, Gabuzda D, Matsudeira P, Yanker B (1993) Generatin of β-amyloid in the secretory pathway in neuronal and nonneuronal cells. Proc Natl Acad Sci USA 90: 2092–2096

Cai X-D, Golde TE, Younkin SG (1993) Release of excess amyloid β protein from a mutant amyloid β protein precursor. Science 259: 514–516

Chartier-Harlin M-C, Crawford F, Houlden H, Warren A, Hughes D, Fidani L, Goate A, Rossor M, Roques P, Hardy J, Mullan M (1991) Early-onset Alzheimer's disease caused by mutations at codon 717 of the β-amyloid precursor protein gene. Nature 353: 844–846
Chartier-Harlin M-C, Parfitt M, Legrain S, Pérez-Tur J, Brousseau T, Evans A, Berr C, Vidal O, Roques P, Gourlet V, Fruchart J-C, Delacourte A, Rossor M, Amouyel P (1994) Apolipoprotein E, ε4 allele as a major risk factor for sporadic early and late-onset forms of Alzheimer's disease: analysis of the 19q13.2 chromosomal region. Human Mol Genet 3: 569–574
Cheung TT, Ghiso J, Shoji M, Cai X-D, Golde T, Gandy S, Frangione B, Younkin S (1994) Characterization by radiosequencing of the carboxyl-terminal derivatives produced from normal and mutant amyloid β protein precursors amyloid. Int J Exp Clin Invest 1: 30–38
Citron M, Oltersdorf T, Haass C, McConlogue L, Hing AY, Seubert P, Vigo-Pelfrey C, Lieberburg I, Selkoe DJ (1992) Mutation of the β-amyloid precursor protein in familial Alzheimer's disease increases β-protein production. Nature 360: 672–674
Clements A, Walsh DM, Williams CH, Allsop D (1993) Effects of the mutations Glu22 to Gln and Ala21 to Gly on the aggregation of a synthetic fragment of the Alzheimer's amyloid β/A4 peptide. Neurosci Lett 161: 17–20
Corder EH, Saunders AM, Strittmatter WJ, Schmechel DE, Gaskell PC, Small GW, Roses AD, Haines JL, Pericak-Vance MA (1993) Gene dose of Apolipoprotein E type 4 allele and the risk of Alzheimer's disease in late onset families. Science 261: 921–923
Corder EH, Saunders AM, Risch NJ, Strittmatter WJ, Schmechel DE, Gaskell PC, Rimmler JB, Locke PA, Conneally PM, Schmader KE, Small GW, Roses AD, Haines JL, Pericak-Vance MA (1994) Protective effect of apolipoprotein E type 2 allele for late onset Alzheimer disease. Nature Genet 7: 180–184
Deng WP, Nickoloff JA (1992) Site directed mutagenesis of virtually any plasmid by eliminating a unique side. Anal Biochem 200: 81–88
Dovey HF, Suomensaari-Chrysler S, Lieberburg I, Sinha S, Keim PS (1993) Cells with a familial Alzheimer's disease mutation produce authentic β-peptide. Neurochemistry 4: 1039–1042
Esch FS, Keim PS, Beattie EC, Blacher RW, Culwell AR, Oltersdorf T, McClure D, Ward PJ (1990) Cleavage of amyloid β peptide during constitutive processing of its precursor. Science 248: 1122–1124
Estus S, Golde TE, Kunishita T, Blades D, Lowery D, Eisen M, Udiak M, Qu X, Tabira T, Greenberg BD, Younkin SG (1992) Potentially amyloidogenic, carboxyl-terminal derivatives of the amyloid protein precursor. Science 255: 726–728
Felsenstein KM, Lewis-Higgins L (1993) Processing of the β-amyloid precursor protein carrying the familial, Dutch-type, and a novel recombinant C-terminal mutation. Neurosci Lett 152: 185–189
Felsenstein KM, Hunihan LW, Roberts SB (1994) Altered cleavage and secretion of a recombinant β-APP bearing the Swedish familial Alzheimer's disease mutation. Nature Genet 6: 251–256
Fraser PE, Nguyen JT, Inouye H, Surewicz WK, Selkoe DJ, Podlisny MB, Kirschner DA (1992) Fibril formation by primate, rodent and Dutch-hemorrhagic analogues of Alzheimer amyloid β-protein. Biochem 31: 10716–10723
Goate A, Chartier-Harlin MC, Mullan M, Brown J, Crawford F, Fidani L, Giuffra L, Haynes A, Irving N, James L, Mant R, Newton P, Rooke K, Roques P, Talbot C, Pericak-Vance M, Roses A, Williamson R, Rossor M, Owen M, Hardy J (1991) Segregation of a missence mutation in the amyloid precursor protein gene with familial Alzheimer's disease. Nature 349: 704–706
Golde T, Estus S, Younkin LH, Selkoe DJ, Younkin SG (1992) Processing of the amyloid protein precursor to potentially amyloidogenic derivatives. Science 255: 728–730
Haan J, Hardy JA, Roos RA (1991) Hereditary cerebral hemorrhage with amyloidosis-Dutch type: its importance for Alzheimer research. Trends Neurosci 14: 231–234

Haan J, van Broeckhoven C, van Duijn CM, Voorhoeve E, van Harscamp F, van Swieten JC, Maat-Schieman MLC, Roos RAC, Bakker E (1994) The apolipoprotein E ε4 allele does not influence the clinical expression of the amyloid precursor protein-gene codon 693 or 692 mutations. Ann Neurol 36: 434–437

Haass C, Selkoe DJ (1993) Cellular processing of β-amyloid precursor protein and the genesis of amyloid β-peptide. Cell 75: 1039–1042

Haass C, Koo EH, Mellon A, Hung AY, Selkoe DJ (1992a) Targeting of cell-surface β-amyloid precursor protein to lysosomes: alternative processing into amyloid-bearing fragments. Nature 357: 500–503

Haass C, Schlossmacher MG, Hung AY, Vigo-Pelfrey C, Mellon A, Ostaszewski BL, Lieberburg I, Koo EH, Schenk D, Teplow DB, Selkoe DJ (1992b) Amyloid β-peptide is produced by cultured cells during normal metabolism. Nature 359: 322–327

Haass C, Hung AY, Selkoe DJ, Teplow DB (1994) Mutations assosiated with a locus for familial Alzheimer's disease result in alternative processing of amyloid β-protein precursor. J Biol Chem 269: 17741–17748

Hardy J, Houlden H, Collinge J, Kennedy A, Newman S, Rossor M, Lannfelt L, Lilius L, Winblad B, Crook R, Duff K (1993) Apolipoprotein E genotype and Alzheimer's disease. Lancet 342: 737–738

Hendriks L, van Duijn C, Cras P, Cruts M, Van Hul W, van Harskamp F, Warren A, McInnis M, Antonarakis SE, Martin JJ, Hofman A, Van Broeckhoven C (1992) Presenile dementia and cerebral haemorrhage linked to a mutation at codon 692 of the β-amyloid precursor protein gene. Nature Genet 1: 218–221

Jarrett JT, Berger EP, Lansbury PT Jr (1993) The carboxy terminus of the β amyloid protein is critical for the seeding of amyloid formation: implications for the pathogenesis of Alzheimer's disease. Biochemistry 32: 4693–4697

Kang J, Lemaire HG, Unterbeck A, Salbaum JM, Masters CL, Grzeschik KH, Multhaup F, Beyreuther K, Muller-Hill B (1987) The precursor of Alzheimer's disease amyloid A4 protein resembles a cell-surface receptor. Nature 325: 733–736

Kitaguchi N, Takahashi Y, Tokushima Y, Shiojiri S, Ito H (1988) Novel precursor of Alzheimer's disease amyloid protein shows protease inhibitory activity. Nature 331: 530–532

Levy E, Carman M, Fernandez-Madrid IJ, Power M, Lieberburg I, van Duinen S, Bots G, Luyendijk W, Frangione B (1990) Mutation of the Alzheimer's disease amyloid gene in hereditary cerebral haemorrhage, Dutch type. Science 248: 1124–1126

Luthman H, Magnussson G (1983) High efficiency polyoma DNA transfection of chloroquine treated cells. Nucl Acids Res 11: 1295–1308

Maruyama K, Usami M, Yamao-Harigaya W, Tagawa K, Ishiura S (1991) Mutation of Glu693 to Gln or Val717 to Ile has no effect on the processing of Alzheimer amyloid precursor protein expressed in COS-1 cells by cDNA transfection. Neurosci Lett 132: 97–100

Mullan M, Crawford F, Axelman K, Houlden H, Lilius L, Winblad B, Lannfelt L (1992) A pathogenic mutation for probable Alzheimer's disease in the APP gene at the N-terminus of β-amyloid. Nature Genet 1: 345–347

Murrell J, Farlow M, Ghetti B, Benson M (1991) A mutation in the amyloid precursor protein associated with hereditary Alzheimer's disease. Science 254: 97–99

Ponte P, DeWhitt PG, Schilling J, Miller J, Hsu D, Greenberg B, Davis K, Wallace W, Lieberburg I, Fuller F, Cordell B (1988) A new A4 amyloid mRNA contains a domain homologous to serine proteinase inhibitors. Nature 331: 525–527

Saunders AM, Strittmatter WJ, Schmechel D, St. George-Hysolp PH, Pericak-Vance MA, Joo SH, Rosi BL, Gusella JF, Crapper-McLachlan DR, Alberts MJ, Hulette C, Crain B, Goldgaber D, Roses AD (1993) Association of apolipoprotein E allele E4 with late-onset familial and sporadic Alzheimer's disease. Neurology 43: 1467–1472

Seubert P, Vigo-Pelfrey C, Esch F, Lee M, Dovey H, Davis D, Sinha S, Schlossmacher M, Whaley J, Swindlehurst C, McCormack R, Wolfert R, Selkoe D, Lieberburg I, Schenk D

(1992) Isolation and quantification of soluble Alzheimer's β-peptide from biological fluids. Nature 359: 325–327
Seubert P, Oltersdorf T, Lee MG, Barbur R, Blomquist C, Davis DL, Bryant K, Fritz LC, Galasko D, Thal LJ, Lieberburg I, Schenk DB (1993) Secretion of β-amyloid precursor protein cleaved at the amino terminus of the β-amyloid peptide. Nature 361: 260–263
Shoji M, Golde TE, Ghiso J, Cheung TT, Estus S, Shaffer LM, Cai X-D, McKay DM, Tintner R, Frangione B, Younkin SG (1992) Production of the Alzheimer's amyloid β protein by normal proteolytic processing. Science 258: 126–129
Schmechel DE, Saunders AM, Strittmatter WJ, Crain B, Hulette C, Joo SH, Pericak-Vance MA, Goldgaber D, Roses AD (1993) Increased amyloid β-peptide deposition in cerebral cortex as a consequence of apolipoprotein E genotype in late-onset Alzheimer disease. Proc Natl Acad Sci USA 90: 9646–9653
Strittmatter WJ, Weisgraber KH, Huang D, Dong L, Salvesen GS, Pericak-Vance M, Schmechel D, Saunders AM, Goldgaber D, Roses AD (1993a) Binding of human apolipoprotein E to synthetic amyloid β peptide: isoform-specific effects and implications for late-onset Alzheimer disease. Proc Natl Acad Sci USA 90: 8098–8102
Strittmatter WJ, Saunders AL, Schmechel D, Pericak-Vance M, Enghild J, Salvesen GS, Roses AD (1993b) Apolipoprotein E: high avidity binding to β-amyloid and increased frequency of type 4 allele in late-onset familial Alzheimer disease. Proc Natl Acad Sci USA 90: 1977–1981
Suzuki N, Cheung TT, Cai X-D, Odaka A, Otvos Jr L, Eckman C, Golde TE, Younkin SG (1994) An increased percentage of long amyloid β protein secreted by famililal amyloid β protein precuresor (βAPP_{717}) mutants. Science 264: 1336–1340
Tanzi RE, Gusella JF, Watkins PC, Bruns GAP, St George-Hyslop P, Van Keuren ML, Patterson D, Pagan S, Kurnit DM, Neve RL (1987) Amyloid ß protein gene: cDNA, mRNA distribution, and genetic linkage near the Alzheimer locus. Science 235: 880–884
Tanzi RE, McClatchey AI, Lamperti ED, Villa-Komaroff L, Gusella JF, Neve RL (1988) Protease inhibitor domain encoded by an amyloid protein precursor mRNA associated with Alzheimer's disease. Nature 331: 528–530
van Duijn C, de Knijf P, Cruts M, Wehnert A, Havekes LM, Hofman A, Van Broeckhoven C (1994) Apolipoprotein E4 allele in a population-based study of early-onset Alzheimer's disease. Nature Genet 7: 74–78
Van Broeckhoven CL (1994) Molecular genetics of Alzheimer disease: identification of genes and gene mutations. Eur Neurol (in press)
Wisniewski T, Ghiso J, Frangione B (1991) Peptides homologous to the amyloid protein of Alzheimer's disease containing a glutamine for glutamic acid substitution have accelerated amyloid fibril formation. Biochem Biophys Res Commun 179: 1247–1254
Wisniewski T, Frangione B (1992) Apolipoprotein E: a pathological chaperone protein in patients with cerebral and systemic amyloid. Neurosci Lett 135: 235–238
Yoshikai S-I, Sakaki H, Doh-ura K, Furuya H, Sakaki Y (1990) Genomic organization of the human amyloid beta-protein precursor gene. Gene 87: 257–263

Combining In Vitro Cell Biology and In Vivo Mouse Modelling to Study the Mechanisms Underlying Alzheimer's Disease

*B. De Strooper**, *D. Moechars, K. Lorent, I. Dewachter*, and *F. Van Leuven*

The pathology of Alzheimer's disease (AD) as a major neurodegenerative disorder is characterized by the presence of senile plaques, neurofibrillary tangles and cerebrovascular deposits. Neither the molecular mechanisms of the formation of these structures nor their direct bearing on the neurodegeneration per se are understood. The major component of the senile plaques and vascular deposits is the β-amyloid peptide (βA4), a 39–43 amino acid fragment derived from a larger precursor, the amyloid precursor protein (APP; Goldgaber et al. 1987; Kang et al. 1987). APP is proteolytically processed by at least three as yet unidentified proteinases, named secretases, into several proteins and peptides, including the proteinase Nexin II or ectodomain of APP and the βA4-peptide, recovered from the amyloid plaques in AD (Haass and Selkoe 1993). The nature of the mechanisms involved requires an experimental approach that combines in vitro and in vivo techniques, that is a combination of cell biology with an experimental animal model system.

Amyloid Precursor Protein Processing in Polarized and Unpolarized Cells In Vitro

The characterization of the α-, β- and γ-secretases is central to our understanding of the mechanism of formation of the secreted form of APP and of the βA4-peptide. This in turn would yield clues to the origin and pathogenesis of the amyloid lesions in AD. The α-secretase is the proteinase that cleaves APP in the β-amyloid sequence and thereby releases the soluble APP ectodomain. It displays a very relaxed sequence specificity when tested with mutated APP substrates in transfected cells (Maruyama et al. 1991; Sisodia 1992; De Strooper et al. 1993). We therefore postulated other molecular signals in the APP sequence to be determinative in the specific interaction of APP with α-secretase. These molecular determinants could be envisaged to be either responsible for the direct binding of APP to α-

* Experimental genetics group, Center for Human Genetics, K.U. Leuven, Campus Gasthuisberg O&N 6,3000 Leuven, Belgium

K.S. Kosik et al. (Eds.)
Alzheimer's Disease: Lessons from Cell Biology

secretase or, alternatively, could target newly synthesized APP specifically towards an as yet not further defined "α-secretase compartment" (De Strooper et al. 1993). To delineate the minimal structural regions in the APP sequence needed for either type of α-secretase processing, we have introduced specific deletions and mutations in the APP cDNA sequence and have evaluated the resulting effect on cellular APP processing and routing in unpolarized COS1 cells and in polarized MDCK cells. In addition, potential α-secretase candidates of the furin family of prohormone processing proteinases were tested for their ability to process APP in a vaccinia virus-based protein expression system in PK(15) cells.

APP Mutants Used

Several mutants of APP have been constructed, first by introducing stop codons at different positions in the cytoplasmic domain to delete the potential Y . . . EVD, GY, and NPTY sorting signals (see Fig. 1). Secondly, by introducing stop codons at the border of the extracellular and the integral membrane domain, at the α- and at the β-secretase cleavage sites, we follow the intracellular transport of soluble APP. Thirdly, by mutating amino acids around the α-secretase cleavage site, we attempt to prevent the cleavage in the βA4 sequence. Finally, by introducing mutations as observed in the familial forms of AD, we explore their effect on APP processing.

To investigate the cellular pathways followed by APP and to overcome the high background expression of endogenous APP in neuronal cells, we

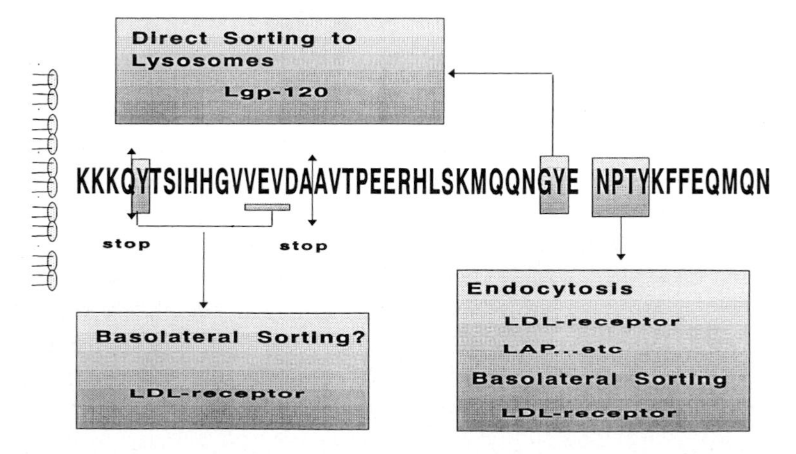

Fig. 1. Putative sorting signals in the cytoplasmic domain of APP. The cytoplasmic domain of APP is represented by the one letter code for amino acid residues. Several candidate sorting signals are boxed and proteins in which these signals were originally identified are indicated

prepared constructs in which the cytoplasmic, the transmembrane and the βA4 domains of APP were fused to a peroxidase reporter gene. These different mutants were transfected and expressed in COS cells as well as in MDCK cells, and the effect of these mutations on the α-proteolytic processing of APP was investigated.

APP Processing in COS Cells

The complete or partial deletion of the cytoplasmic domain of APP (constructs APP/695M/Y653* and APP/695M/A666*, numbering as in APP695) does not inhibit secretion of the soluble form of APP (APPs). On the contrary, the truncated version of APP (stop codon at position Ala666) displays an increased secretion of 180% compared to wild type APP (De Strooper et al. 1993; Haass et al. 1992). Interestingly, the mutant APP/695M/A666*, truncated immediately in front of the integral membrane domain, is secreted without being proteolytically processed. This was demonstrated by the fact that the antibody 4G8, the epitope of which is located carboxy-terminal from the α-secretase cleavage site on the β-amyloid sequence (Kim et al. 1988), reacted with both the cell and secreted forms of this mutant APP. This finding confirmed that the membrane anchor is needed for processing by the α-secretase pathway (Sisodia et al. 1992).

The peroxidase reporter fused to the βA4 region, the transmembrane domain and the cytoplasmic domain of APP is secreted into the culture medium at a rate which is similar to that of wild type APP. The introduction of a stop codon at position Ala666, which deletes the cytoplasmatic GY/NPTY sorting signals in the fusion protein, resulted in increased secretion. Treatment of the cells with methylamine inhibited the secretion of the peroxidase reporter protein, similar to what was observed with wild type APP (De Strooper et al. 1993). Preliminary results from immunoblotting experiments indicated that the hybrid fusion protein is cleaved at the same position as wild type APP, which supports the conclusion that the extracellular domain contains no molecular determinants that are of importance for the interaction with α-secretase or for the sorting of APP towards an α-secretase compartment. Together with the results obtained with the truncated cytoplasmic domain mutants, this demonstrates conclusively the importance of the β-amyloid sequence and of the integral membrane domain of APP for successful α-secretase processing in COS cells.

The exact role of the βA4-sequence in this context is not yet understood. Although the sequence specificity of α-secretase is relaxed (Maruyama et al. 1991; Sisodia 1992), changing arginine at position P4 into aspartic acid and lysine at P1 towards glutamic acid decreases the secretion of mouse APP to less than 50% relative to wild type APP (De Strooper et al. 1993). In terms of protein-chemical structure, such mutations are very drastic and the result can hardly be explained by our current knowledge of proteinase-substrate binding characteristics. A biologically interesting alternative interpretation of our observations is that multiple α-secretases are responsible and can release APPs (Zhong et al. 1994). This hypothesis does imply, however, that the specificity of proteolytic cleavage is completely defined by molecular sorting determinants in APP situated in the integral membrane domain, which remains to be explored.

Polarized Sorting of APP in MDCK Cells

The MDCK cell line is by far the most widely used and the best studied model for polarized protein sorting (Simons and Zerial 1993; Matter et al. 1992; Rodriguez-Boulan and Powell 1992). Interestingly, aspects of the sorting machinery in these epithelial cells might be similar or comparable to those of hippocampal neurons (Dotti and Simons 1990; Dotti et al. 1991; Huber et al. 1993). Several known proteins, including those with a glycolipid anchor, are targeted to the apical surface in MDCK cells and to the axonal side in neurons. Similarly, the basolateral side of epithelia and the somatodendritic side of neurons are reached by similar proteins (Table 1).

APP turns out to be an interesting exception, since we observed endogenous as well as transfected APP to be secreted almost exclusively in the basolateral compartment of MDCK cell cultures, whereas it was demonstrated to be transported via fast axonal transport in neurons (Koo et al. 1990). Treatment of the cells with phorbol esters increases the basolateral secretion rate as in transfected COS cells. Treatment with methylamine or with other primary amines results in a dose-dependent reversal of secretion from the basolateral to the apical side of MDCK cells, which contrasts with the inhibition of cleavage and of secretion observed in unpolarized cells.

Table 1. Sorting mechanisms in hippocampal neurons and MDCK cells. Proteins that are targeted to the apical side in MDCK cells are targeted to the axons in neurons. Similarly, basolateral and somatodendritic compartments are equivalent sorting destinations (Dotti and Simons 1990; Dotti et al. 1991; Huber et al. 1993)

MDCK	APICAL	BASOLATERAL
NEURONS	AXONAL	SOMATODENDRITIC
	Influenza Virus HA	Vesicular Stomatitis Virus Glycoprotein
	Thy-1 glycoprotein (GPI-Linked)	Semliki Forest Virus Glycoprotein
		Transferrin receptor Rab8p

This effect is apparently caused by the alkalinization of an intracellular compartment, since a similar shift in secretion towards the apical compartment is also observed with drugs that are known inhibitors of the vacuolar type H^+-ATPase, such as bafilomycin A1, bafilomycin B1, concanamycin A and concanamycin C in the nanomolar concentration range. Re-targeting towards the apical surface from their normal basolateral secretion route by addition of NH_4Cl_2 was previously documented for laminin as well as for proteoglycans (Caplan et al. 1987).

We suggest, therefore, that APP is sorted as a soluble protein in MDCK cells and propose that the acidic compartment through which APP is routed is the basolateral endosome. In support of this interpretation, we observed that deletion of the putative basolateral sorting signals in the cytoplasmic domain of APP (Fig. 1) did not change the destination of APP, and that these signals are therefore not essential for the basolateral secretion of APP. Moreover, deletion of the transmembrane domain (APP/695M/A626*) resulted in only a minor effect on the polarized secretion of APP (80% basolateral, 20% apical), which means that a basolateral sorting signal must be present in the extracellular domain of APP. We therefore speculate that APP is cleaved by α-secretase in the trans-Golgi network and that APPs is then sorted by virtue of a pH-sensitive receptor to the basolateral side of MDCK cells. This model was first proposed for the basolateral sorting of laminin and proteoglycan in MDCK cells (Caplan et al. 1987). Our current research is aimed at identifying this basolateral sorting signal in the extracellular domain of APP.

Screening Candidate α Secretases

These and other observations demonstrate that α-secretase operates intracellularly in a compartment located between the Golgi complex and the cell surface (De Strooper et al. 1992; Sambamurti et al. 1992; Kuentzel et al. 1993). Although APP can be cleaved at the cell surface as well (Haass et al. 1992; Sisodia 1992), this fraction of APP processing seems not to be preponderant, because a wide range of proteinase inhibitors, either synthetic or natural such as alpha-2-macroglobulin, added extracellularly, fail to inhibit significantly the processing and secretion of APP (De Strooper et al. 1992). In contrast, methylamine, a primary amine that accumulates in acidic compartments such as lysosomes but also the late compartments of the constitutive secretion pathway, inhibits in a dose-dependent manner the cleavage and secretion of APP_s (Soluble APP) in unpolarized cells, similar to what has been observed for the conversion and secretion of proalbumin, procomplement factor C3 and murine α2-macroglobulin (Oda and Ikehara 1985; Oda et al. 1986; De Strooper et al. 1993). Deletion of endocytosis and lysosomal sorting determinants in the cytoplasmic domain of APP does not interfere with cleavage and secretion, indicating that the α-cleavage does not occur in the endosomal/lysosomal pathways (Caparoso et al. 1992; De Strooper et al. 1993; Da Cruz e Silva 1994).

Proteinases of the furin family of proprotein processing enzymes are potential candidates for a function as α-secretase (Van De Ven et al. 1992; Steiner et al. 1992; Halban and Irminger 1994). Furin, the prototype of this family, is an ubiquitously expressed, subtilisin-like proteinase that is membrane anchored in the trans-Golgi network, but which can also appear at the cell surface (Molloy et al. 1994). Substrates for this enzyme are membrane-bound glycoproteins like the insulin receptor precursor, and soluble precursors of secreted proteins like Von Willebrands Factor, the pro-βA-chain of activin and complement factor pro-C3 (Roebroek et al. 1993; Van de Ven et al. 1990; Wise et al. 1990; Misumi et al. 1991). Interestingly, cleavage and secretion of the latter were demonstrated to be inhibited by methylamine, similar to what was observed for APP and indicating that an acidic compartment is involved in these activation or converting pathways (Oda and Ikehara 1985; Oda et al. 1986; De Strooper et al. 1993).

Although the prohormone processing proteinase furin cleaves substrates on the carboxyl side of the consensus sequence Arg-X-Lys/Arg-Arg, which is not present in APP, the basic residue at the P2 position has been shown to be of lesser importance (Misumi et al. 1991; Molloy et al. 1992; Creemers et al. 1993). The sequence specificity of the other members of the furin family, namely PC1/PC3, PC2, PC4, PC5/PC6 (with two isoforms PC5/PC6A and PC5/PC6B) and PACE 4 has not been studied as completely (Van De Ven et al. 1992; Steiner et al. 1992; Halban and Irminger 1994). The expression of these prohormone processing enzymes is more restricted and some of them are not membrane anchored.

We have tested the possibility that, of the known members of the furin family, one or more could display α-secretase activity in vaccinia virus-infected PK(15) cells. These cells do not display appreciable furin-like proteolytic activity, while secretion of proteins is essentially normal. Co-transfection of constructs coding for one of the processing enzymes with vectors coding for one of different types of substrate made it possible to study the proteolytic processing in a cellular context. Transfection of the APP cDNA, contained in the pGEM 13zf(+) vector, resulted in the cellular expression of APP, but secretion of APPs was much lower relative to transfected COS cells, which are able to proteolytically process APP. Co-transfection with a construct coding for any of seven different members of the furin family of prohormone convertases did not lead to an increased release of APPs, which suggests that none of the known members of the furin family of proteinases is involved in the processing of APP. This does not exclude the possibility that other, as yet to be discovered members of this family display α-secretase activity. Moreover, the currently developed assay system will be put to use to test these and other candidate α-secretases for their proteolytic activity towards APP in a cellular context.

APP Metabolism in Transgenic Mice

In the APP the βA4-amyloid peptide sequence is located partially in the extracellular and partially in the transmembrane domain. The processing of APP by α-secretase involves cleavage in the amyloid sequence and precludes the subsequent formation of the βA4 peptide (Weidemann et al. 1989; Esch et al. 1990; Wang et al. 1991; Sisodia et al. 1992). Since this cleavage and amyloid formation are mutually exclusive processes, we are testing whether APP that escapes α-secretase activity can be channelled and processed in an amyloidogenic pathway. To this end the α-secretase cleavage site in APP was mutated as described above (APP/RK mutant). The substitution of Arg609 for Asp and Lys612 for Glu resulted in an APP mutant that was secreted at a rate of only about 50% relative to wild type APP (De Strooper et al. 1993).

This slow secretion mutant APP695/RK was incorporated in a construct and placed under the control of either the human APP promoter region with or without a viral enhancer element or under the control of the mouse Thy1 gene promoter. The elements of the mouse Thy1 gene promoter used were employed to direct expression to the brain, without expression in the T-lymphocyte. Expression levels are especially high in brain cortex and in the hippocampus.

The human APP promoter was cloned from a human placenta genomic library in lambda and the promoter region was subcloned. In the final construct, a 3.8 kb BamHI fragment of the APP gene 5' flanking region was used. This fragment ends at position 55 in the APP exon 1 and includes the transcription initiation site. A similar fragment was used and shown by others to mediate neuron-specific expression in the CNS of transgenic mice with an expression pattern resembling that of the endogenous APP mRNA (Wirak et al. 1991). The enhancer element used is derived from the polyoma virus as present in the pMC1 vector (Stratagene).

The APP695/RK mutant was placed under the control of the three different promoter cassettes and was used to generate transgenic mice by microinjection of prenuclear embryos isolated from superovulated mice from the FVB/N strain or from F1 (CBA/J × C57Bl/6). Thus, a total of 23 transgenic founder mice were identified by Southern blot analysis on tail biopt DNA using a probe specific for the mouse APP cDNA. Seventeen founders were genotyped containing the APP695/RK mutant under the control of the human APP promotor with (10) or without (7) the enhancer. Six founders containing the same mutant APP cDNA under the control of the Thy1 promoter were created. As judged by Southern blotting, all founders transmitted the transgene to their offspring, albeit largely different copies of the transgene were integrated: between one and 20 copies per haploid genome segregated in subsequent generations, indicating tandem integration in one site.

Transcription of the transgene in the brain was detected by Northern blotting and RT-PCR. Brains were removed and total RNA or polyA mRNA was isolated. In four of the six lines with the Thy1-APP695/RK transgene, transcription of the transgene was detected by Northern blotting: the 3.4 kb endogenous APP695 transcript was present alongside a 3.9 kb APP695/RK mutant transcript. In the transgenic lines resulting from the APP695/RK mutant under the control of the human APP promoter, three and five lines, respectively without and with the enhancer, transcription was demonstrated by RT-PCR. In this instance both transcripts are of about the same size (3.4 kb), which necessitated the development of an RT-PCR procedure that discriminated between endogenous APP and the APP695/RK mutant mRNA.

Further biochemical and immunochemical analysis will show whether these animals indeed produce more βA4-peptide or other degradation products relative to non-transgenic controls and whether, as a result, amyloid deposits develop.

To construct mouse models for other aspects of the etiology of AD, a number of other mini-genes are injected in prenuclear embryos to obtain founder mice with expression in the mouse brain, making use of the above-mentioned promoters. A direct gene dose effect of the APP gene is proposed for the appearance of senile plaques in the brain of Down's syndrome patients. The mechanism is unknown and might be either saturation of the α-secretase pathway or escape from this processing step, resulting in lesions typical for AD. To test this hypothesis, constructs with the human APP cDNA driven by either of the three promoters were injected in prenuclear embryos.

Similarly, a direct increase in the amount of βA4-peptide in the brain is attempted by expressing the βA4 coding sequence or the C-terminal part of APP, including the βA4 coding sequence in mouse brain. Finally, three naturally occurring mutants, identified in familial cases of AD, are being expressed in transgenic mice: APP/Ala692Gly (Hendriks et al. 1992), APP/Glu693Gln (Wisniewski et al. 1991) and APP/Lys670Asn-Met671Leu (Mullan et al. 1992). These mutants are directly linked to the cerebral amyloid angiopathy and/or early onset AD.

Founder mice are available for all of these constructs and are being used to produce offspring homozygous for the transgene; they are screened for expression of the transgene in their brain at the RNA and protein level.

Expression of APP, APLP 1, APLP 2, A2MR/LRP, HBP-44, AOE, LPL and A2M During Mouse Embryogenesis and in Adult Tissues

Recent work has indicated that APP appears to be a member of a multigene family or at least to have close relatives, as judged from sequence homologies.

In Drosophila, an APP-like protein (APPL) has been described (Rosen et al. 1989), while in human and mouse a nervous system-specific amyloid precursor-like protein 1 (APLP 1) was identified (Wasco et al. 1992, 1993). In human, rat and mouse, another member of this family, APLP 2, was identified as a nuclear protein which was proposed to function in the physical processes of chromosome stabilization and segregation (Blangy et al. 1991; Vidal et al. 1992; von der Kammer et al. 1994).

Immunochemical analysis of the senile plaques in AD brain has shown not only the presence of the β-amyloid peptide but also the association of APP (Cras et al. 1991; Joachim et al. 1991), Apolipoprotein E (ApoE; Namba et al. 1991), the wide-spectrum proteinase inhibitor Alpha-2-Macroglobulin (A2M; Van Gool et al. 1991) and its receptor, the low density lipoprotein receptor related protein or A2MR/LRP (Tooyama et al. 1993; Rebeck et al. 1993), as well as several other proteinases and proteinase inhibitors (Cataldo et al. 1991). It is useful to indicate here that A2MR/LRP not only binds A2M-proteinase complexes (Ashkom et al. 1991; Moestrup and Gliemann 1989), but also ApoE-enriched lipoproteins (Kowal et al. 1989) and lipoprotein lipase (LPL; Chappell et al. 1992), which thereby enhances the binding of chylomicrons and very low density lipoproteins (VLDL) by A2MR/LRP (Beisiegel et al. 1991). The 44 kDa heparin binding protein HBP-44 (Furukawa et al. 1988) is the mouse equivalent of the A2M receptor-associated protein, which is found associated with affinity-isolated or immunoprecipitated A2MR/LRP. Recombinant or naturally isolated HBP-44 is able to inhibit the binding of all ligands to the receptor (Herz et al. 1991; Williams et al. 1992). However, its precise function in vivo is unknown.

ApoE immunoreactivity in AD brain is co-localized with A2MR/LRP staining in senile plaques (Rebeck et al. 1993) and closely imitates the distribution of βA4 amyloid (Strittmatter et al. 1993a). The epidemiological finding of the increased frequency of the ApoE4 allele in AD (Strittmatter et al. 1993a; Saunders et al. 1993a,b) presents another challenge for fundamental research to explain and explore. Possible indications are to be found in the increased ApoE mRNA content in AD brains (Diedrich et al. 1991) and in the avid binding of ApoE to synthetic βA4 peptide (Strittmatter et al. 1993a), whereas different ApoE isoforms might vary in their ability to interact with βA4 (Strittmatter et al. 1993b). Fundamentally, the role of ApoE metabolism and function in brain needs clarification.

Evidence thus indicates that the A2MR/LRP-ApoE system might be of importance in the pathogenesis of AD. As one aspect of the unravelling of the mechanism of this complex system in an animal model, we have studied the expression of these components and of the two APP-like proteins, APLP 1 and APLP 2, in normal mice during embryonic development and in adult tissues by Northern blotting and in situ hybridization. Transgenic mice with overexpression and mutants of APP, those with expression of the human ApoE4 isoform, and siblings of APP transgenics mated with the ApoE4-expressing animals will be analyzed similarly.

Expression in Embryonic Tissues

During embryogenesis APP, APLP 1, APLP 2, ApoE and LPL total embryo mRNA levels were compared by Northern blotting. All components examined became increasingly expressed from embryonal day 10 (E10) onwards, the earliest stage examined. Although APP mRNA levels increased steadily, APLP 1 expression was most pronounced between E13 and E15. This period corresponds to the stage of development in which the nervous system is formed and expands dramatically. At E10, expression of APLP 2 was more prominent than that of APP and APLP 1, but it did not increase as much during further development.

A2M mRNA first appeared on E13, with a sharp increase in levels thereafter and beyond birth. Whereas embryonal A2MR/LRP mRNA levels decreased during development, the HBP-44 message was nearly constant except for the period between E10 and E13, during which time the 1.8 kb species was increased (Lorent et al. 1994). The highest levels of expression of ApoE and LPL were observed between E17 and E19, immediately prior to birth.

The results of in situ hybridization on paraffin sections of E12, E13 and E15 embryos confirmed and extended the data obtained by Northern blotting of entire embryo extracts. APP expression was demonstrated to be widespread in the central and peripheral nervous system, but was also observed in thyroid, thymus, submandibular gland, lung, kidney, intestine, blood vessels, cartilage primordia of vertebrae and bone structures in the head and the limbs, and in the E12 placenta.

APLP 1 mRNA was prominent in the developing central and peripheral nervous system and its distribution in the brain was more even than that of APP. APLP 2 was expressed even more ubiquitously than APP.

A2MR/LRP also exhibited expression throughout the embryo, whereas synthesis of HBP-44 seemed more restricted and occurred mainly in kidney, brain cortex, lung, liver and thymus. Remarkably, as was stressed previously, the cellular localization of HBP-44 mRNA in kidney and in most other tissues differed completely from that of A2MR/LRP (Lorent et al. 1994). This fact clearly points to a role for HBP-44 which might be unrelated to its receptor binding activity.

Very strong hybridization signals were observed with the ApoE probe in liver, choroid plexus and in the visceral yolk sac of E12 embryos. Other sites of synthesis were Rathke's pouch, the epithelial lining of the ventricles, ganglia, spinal cord, skeletal muscle, lung, kidney, and pancreas.

In contrast to adult mice, in embryos the liver is one of the main sites of LPL mRNA synthesis. Heart, skeletal muscle, thyroid, and submandibular gland also displayed clearly positive signals.

Expression of A2M was never observed in any other tissue than the embryonal liver, and the message was not detected before E13.

Expression in Adult Tissues

In adult tissues the APP mRNA synthesis was highest in brain, intermediate in kidney and lung, and lowest in heart and liver. Although APLP 1 mRNA was abundant in brain (with very weak signals in extracts from heart, lung and kidney confirming that this protein is brain-specific), APL 2 was expressed in all tissues examined at more or less comparable levels, indicating a fundamental and generalized role in the nucleus.

A remarkable difference between the A2MR/LRP-and HBP-44 expression patterns was also obvious in these experiments. The A2MR/LRP message was most abundant in liver, while brain was the second major site of synthesis. Lesser amounts of mRNA were detected in heart, lung and kidney. The expression of HBP-44 was highest in kidney and lowest in liver, again underlining the conclusion proposed above.

The ApoE mRNA synthesis pattern more or less parallelled that of A2MR/LRP, which is not entirely obvious from the known functions of both proteins. Large amounts of LPL mRNA were present in heart, but only very small amounts in brain and liver and, finally, the A2M message was again detected exclusively in the liver.

Discussion

Many of the fundamental, physiologically important functions of most of the molecules studied here are largely unknown, especially when relating to the brain in vivo. Concerning APP, multiple functions in cellular contacts and growth control have been proposed, but none has been extended to the in vivo situation. It is conceivable that the various domains and isoforms of APP have different functions.

On the basis of its expression pattern, the role for APLP 1 is more than likely restricted to functioning in the nervous system. The ubiquitous presence of APLP 2 mRNA is congruent with its apparent essential role in chromosomal organization or re-organization and extends this role to most if not all cell types in vivo. It would be of some interest to examine whether APLP 1 and/or 2 are also associated with the neuropathological lesions in AD, especially since they harbour neither the βA4 nor an equivalent domain.

Because of its multifunctional role, A2MR/LRP was believed to be important in many different physiological systems, a thesis that was experimentally proven by the inactivation of the mouse gene coding for this receptor (Herz et al. 1993). However, the resulting embryonic lethal phenotype does not allow one to explore these functions further in later stages of life. It has been proposed that, in the brain, A2MR/LRP might be responsible for clearance of ApoE-βA4 complexes, but this has not yet been proven experimentally (Rebeck et al. 1993). The presence of A2MR/LRP in all brain areas is much more pronounced than that of the LDLR, indicating that lipoprotein metabolism in brain is mainly mediated by A2MR/LRP, hence by lipoproteins carrying ApoE.

The results of ApoE expression in normal mice confirm published data on rat and marmoset (Elshourbagy et al. 1985). Besides demonstrating that the brain is active in the synthesis of ApoE, these results do not allow us to speculate any further about the function of ApoE in the brain, let alone its contribution to the development of AD. Mice expressing the human ApoE4 isoform are being created and these, alone or in crosses with mice over-expressing APP, will probably shed more light on this intriguing and important phenomenon.

The changing expression pattern of LPL from embryonic to adult life confirms what was described before in other animals (Auwerx et al. 1992; Enerbäck and Gimble 1993). Since A2M is present in senile plaques (Van Gool et al. 1993), the absence of A2M mRNA in mouse brain was unexpected. Several explanations are possible: the expression in human and mouse may be different or secondary factors might trigger onset of A2M expression in AD brain. Under certain conditions, A2M might even pass the blood-brain barrier to become lodged in the areas described (Van Gool et al. 1993). In all these instances, an experimental approach as well as analysis of normal and transgenic mice will be implemented to understand the exact mechanisms involved.

References

Ashkom JD, Tiller SE, Dickerson K, Cravens JL, Argraves WS, Strickland DK (1990) The human Alpha-2-macroglobulin receptor: identification of a 420 kDa cell surface glycoprotein specific for the activated conformation of Alpha-2-Macroglobulin. J Cell Biol 110: 1041–1048

Auwerx J, Leroy P, Schoonjans K (1992) Lipoprotein lipase: recent contributons from molecular biology. Crit Rev Lab Sci 29: 243–268

Beisiegel U, Weber W, Bengtsson-Olivecrona G. (1991) Lipoprotein lipase enhances the binding of chylomicrons to low density lipoprotein receptor-related protein. Proc Natl Acad Sci USA 88: 8342–8346

Blangy A, Leopold P, Vidal F, Rassoulzadegan M, Cuzin F (1991) Recognition of the CDEI otif GTCACATG by mouse nuclear proteins and interference with the early development of the mouse embryo. Nucl Acids Res 19: 7243–7250

Caparoso G, Gandy S, Buxbaum J, Greengard P (1992) Choloroquine inhibits intracellular degradation but not secretion of Alzheimer β/A4 amyloid precursor protein. Proc Natl Acad Sci USA 89: 2252–2256

Caplan M, Stow J, Newman A, Madri J, Andrson HC, Farquhar M, Palade G, Jamieson JD (1987) Dependence on pH of polarized sorting of secreted proteins. Nature 329: 632–635

Cataldo AM, Nixon RA (1990) Enzymatically active lysosomal proteases are associated with amyloid deposits in Alzheimer brain. Proc Natl Acad Sci USA 87: 3861–3865

Chappell DA, Fry GL, Waknitz MA, Iverius P-H, Williams SE, Strickland DK (1992) The Low Density Receptor related protein/α2Macroglobulin receptor binds and mediates catabolism of bovine milk lipase. J Biol Chem 267: 25764–25767

Chen W, Goldstein J, Brown M (1990) NPXY, a sequence often found in cytoplasmic tails, is required for coated pit-mediated internalization of the low density lipoprotein receptor. J Biol Chem 265: 3116–3123

Corder EH, Saunders AM, Strittmatter WJ, Schmechel DE, Gaskell PC, Small GW, Roses AD, Haines JL, Pericak-Vance MA (1993) Gene doses of apolipoprotein E type 4 allele and the risk of Alzheimer's disease in late onset families. Science 261: 921–923

Cras P, Kawai M, Lowery D, Gonzalez-DeWhitt P, Greenberg B, Perry G (1991) Senile plaque neurites in Alzheimer disease accumulate amyloid precursor protein. Proc Natl Acad Sci USA 88: 7552–7556

Creemers JWM, Siezen RJ, Roebroek AJM, Ayoubi TAY, Huylebroeck D, Van de Ven, WJM (1993) Modulation of furin-mediated proprotein processing by site-directed mutagenesis. J Biol Chem 268: 21826–21834

Da Cruz Silva OAB, Lverfeldt K, Oltersdorf T, Sinha S, Lieberburg I, Ramabhadran, T, Suzuki T, Sisodia S, Gandy S, Greengard P (1993) Regulated cleavage of Alzheimer β-amyloid precursor protein in the absence of the cytoplasmic tail. Neuroscience 57: 873–877

De Strooper B, Umans L, Van Ieuven F, Van den Berghe H (1993) Study of the synthesis and secretion of normal and artificial mutants of murine amyloid precursor protein (APP): Cleavage of APP occurs in a late compartment of the default secretion pathway. J Cell Biol 121: 295–304

De Strooper B, Van Leuven F, Van den Berghe H (1992) α-2-Macroglobulin and other proteinase inhibitors do not interfere with the secretion of Amyloid Precursor Protein in mouse neuroblastoma cells. FEBS Lett 308: 50–53

Diedrich JF, Minnigan J, Carp RI, Whitaker JN, Race R, Frey W, II, Haase AT (1991) Neuropathological changes in Scrapie and Alzheimer's disease are associated with increased expression of apolipoprotein E and cathepsin D in astrocytes. J Virol 65: 4759–4768

Dotti C, Simons K (1990) Polarized sorting of viral glycoproteins to the axon and dendrites of hippocampal neurons in culture. Cell 62: 63–72

Dotti C, Parton R, Simons K (1991) Polarized sorting of glypiated proteins in hippocampal neurons. Nature 349: 158–161

Elshourbagy NA, Liao WS, Mahley RW, Taylor JM (1985) Apolipoprotein E mRNA is abundant in the brain and adrenals, as well as in the liver, and is present in other peripheral tissues of rats and marmosets. Proc Natl Acad Sci USA 82: 203–207

Emi M, Wu LL, Robertson MA, Myers RL, Hegele RA, Williams RR, White R, Lalouel JM (1988) Genotyping and sequence analysis of apolipoprotein E isoforms. Genomics 3: 373–379

Enerbäck S, Gimble JM (1993) Lipoprotein lipase gene expression: physiological regulators at the transcriptional and post-transcriptional level Biochim. Biophys. Acta 1169: 107–125

Esch FS, PS, Keim EC, Beattle RW, Blacher AR, Culwell T, Oltersdorf D, McClure, Ward PJ (1990) Cleavage of Amyloid β peptide during constitutive processing of its precursor. Science 248: 1122–1124

Feurst TO, Niles EG, Studier FW, Moss B, Loh YP (1986) Eukaryotic transient expression system based on recombinant vaccinia virus that synthesizes bacteriophage T7 RNA polymerase. Proc. Natl. Acad Sci USA 83: 8122–8126

Furukawa T, Masayuki O, Huang R-P, Muramatsu T (1990) A heparin binding protein whose expression increases during differentiation of embryonal carcinoma cells to parietal endoderm cells: cDNA cloning and sequence analysis. J Biochem. 108: 297–302

Giguére V, Isobe K, Grosveld F (1985) Structure of the murine Thy-1 gene. EMBO 4: 2017–2024

Glenner GG, Wong CW (1984) Alzheimer's disease: initial report of the purification and characterization of a novel cerebrovascular amyloid protein. Biochem. Biophys Res Commun 120: 885–890

Goldgaber D, Lerman MI, McBridge OW, Saffiot U, Gajdusek DC (1987) Characterization and chromosomal localization of a cDNA encoding brain amyloid of Alzheimer's disease. Science 235: 877–880

Haass C, Selkoe D (1993) Cellular processing of β-amyloid precursor protein and the genesis of amyloid β-peptide. Cell 75: 1039–1042

Haass C, E, Koo A, Mellon A, Hung A, Selkoe D (1992) Targeting of cell-surface β-amyloid precursor protein to lysosomes: alternative processing into amyloid-bearing fragments. Nature 357: 500–502

Halban P, Irminger JC (1994) Sorting and processing of secretory proteins. Biochem. J. 299: 1–18

Harter C, Mellman I (1992) Transport of the lysosomal membrane glycoprotein lgp 120 (lgp-A) to lysosomes does not require appearance on the plasma membrane. J Cell Biol 117: 311–325

Hendriks L, CM, van Duijn P, Cras M, Cruts W, Van Hul F, van Harskamp A, Warren MG, Mclnnis SE, Antonorakis JJ, Martin A, Hofman Van Broekhoven C (1992). Presenile dementia and cerebral haemorrhage linked to a mutation at codon 692 of the β-amyloid precursor protein gene. Nature Gen 1: 218–222

Herz J, Goldstein JL, Strickland DK, Ho YK, Brown MS (1991) 39-kDa Protein modulates binding of ligands to low density lipoprotein receptor-related protein/α2-macroglobulin receptor. J Biol Chem 266: 21232–21238

Herz J, Clouthier DE, Hammer RE (1992) LDL receptor-related protein internalizes and degrades uPA-PAl-1 complexes and is essential for embryo implantation. Cell 71: 411–421

Huber L, De Hoop M, Dupree P, Zerial M, Simons K, Dotti C (1993) Protein transport to the dendritic plasma membrane of cultured neurons is regulated by rab8p. J Cell Biol 123: 47–55

Ingraham H, Lawless G, Evans G (1986) The mouse Thy-1.2 glycoprotein gene: complete sequence and identification of an unusual promoter. J Immunol 136: 1482–1489

Joachim CL, Games D, Morris J, Ward P, Frenkel D, Selkoe D (1991) Antibodies to non-β regions of the β-amyloid precursor protein detect a subset of senile plaques. Am J Pathol 138: 373–384

Kang J, Lemaire H, Unterbeck A, Salbaum JM, Masters CL, Grzeschik KH, Multhaup G, Beyreuther K, Muller-Hill B (1987) The precursor of Alzheimer's disease amyloid A4 protein resembles a cell-surface receptor. Nature 325: 733–736

Kim SK, Miller D, Sapienza V, Chen CM, Bai C, Grundke-lqbal I, Currie, J, Wisniewski H. (1988) Production and characterization of monoclonal antibodies reactive to synthetic cerebrovascular amyloid peptide. Neurosci Res Commun 2: 121–130

Koo EH, Sisodia S, Archer D, Martin L, Weidemann A, Beyreuther K, Fischer P, Masters C, Price D (1990). Precursor of amyloid protein in Alzheimer disease undergoes fast anterograde axonal transport. Proc. Natl Acad Sci USA 87: 1561–1565

Kowal RC, Herz J, Goldstein JL, Esser V, Brown MS (1989) Low density lipoprotein receptor-related protein mediates uptake of cholesteryl esters derived from apoprotein E-enriched lipoproteins. Proc Natl Acad Sci USA 86: 5810–5814

Kuentzel S, Ali S, Altman R, Greenberg B, Raub T (1993) The Alzheimer β-amyloid protein precursor/protease nexin II is cleaved by secretase in a trans-Golgi secretory compartment in human neuroglioma cells. Biochem J 295: 367–378

Lorent K, Overbergh L, Delabie J, Van Leuven F, Van Den Berghe H (1994) Study of the distribution of mRNA coding for Alpha-2-Macroglobulin, the Murinoglobulins, the Alpha-2-Macroglobulin Receptor and the Alpha-2-Macroglobulin receptor associated protein during mouse embryogenesis and in adult tissues. Differentiation 55: 213–223

Maruyama K, Kametani F. Usami M, Yamao-Harigaya W, Tanaka K (1991) "Secretase," Alzheimer Amyloid protein precursor secreting enzyme is not sequence-specific. Biochem Biophys Res Commun 179: 1670–1676

Matter K, Hunziker W, Mellman I (1992) Basolateral sorting of LDL receptor in MDCK cells: the cytoplasmic domain contains two tyrosine-dependent targeting determinants. Cell 71: 741–753

Misumi Y, Oda K, Fujiwara T, Takami N, Tashiro K, and Ikehara Y (1991) Functional expression of furin demonstrating its intracellular localization and endoprotease activity

for processing of proalbumin and complement pro-C3. J Biol Chem 266: 16954–16959
Moestrup SK, Gliemann J (1989) Analysis of ligand recognition by the purified α2-macroglobulin receptor/low density lipoprotein receptor-related protein. J Biol Chem 264: 15574–15577
Molloy SS, Brenahan PA, Leppla L. Klimpel K, Thomas G (1992) Human furin is a calcium-dependent serine endoprotease that recognizes the sequence Arg-X-X-Arg and efficiently cleaves anthrax toxin protective antigen. J. Biol. Chem. 267: 16396–16402
Molloy SS, Thomas L, Van Slyke JK, Stenberg PE, Thomas G (1994) Intracellular trafficking and activation of the furin proprotein convertase: localization to the TGN and recycling from the cell surface. EMBO J 13: 18–33
Mullan M, Houlden H, Windelspecht M, et al (1992) A locus for familial early-onset Alzheimer's disease on the long arm of chromosome 14, proximal to the α1-antichymotrypsin gene. Nature Gen 2: 340–342
Namba Y, Tomonaga M, Kawasaki H, Otomo E, Ikeda K (1991) Apolipoprotein E immunoreactivity in cerebral amyloid deposits and neurofibrillary tangles in Alzheimer's disease and Kuru plaque amyloid in Creutzfeld-Jacob disease. Brain Res 541: 163–166
Oda K, Ikehara Y (1985) Weakly basic amines inhibit the proteolytic conversion of proalbumin to serum albumin in cultured rat hepatocytes. Eur J Biochem 152: 605–609
Oda K, Koriyama Y, Yamada Y, Ikehara Y (1986) Effects of weakly basic amines on proteolytic processing and terminal glycosylation of secretory proteins in cultured rat hepatocytes. Biochem J 240: 739–745
Pericak-Vance MA, Bebout JL, Gaskell PC, Yamaoka LH, Hung WY, Alberts MJ, Walker AP, Bartlett RJ, Haynes CA, Welsh KA, Earl NL, Heyman A, Clark CM, Roses AD (1991) Linkage studies in familial Alzheimer disease: Evidence for chromosome 19 linkage. Am J Human Genet 48: 1034–1050
Rebeck GW, Reiter JS, Strickland DK, Hyman BT (1993) Apolipoprotein E in sporadic Alzheimer's disease: allelic variation and receptor interactions. Neuron 11: 575–580
Rodriguez-Boulan E, Powell S (1992) Polarity of epithelial and neuronal cells. Ann Rev Cell Biol 8: 395–427
Roebroek AJM, Creemers JWM, Pauli IGL, Bogaert T, Van De Ven WJM (1993) Generation of structural and functional diversity in furin-like proteins in Drosophila melanogaster by alternative splicing of the DFUR1 gene. EMBO J 12: 1853–1870
Rosen DR, Martin-Morris L, Luo L, White K (1989) A Drosophila gene encoding a protein resembling the human β-amyloid protein precursor. Proc Natl Acad. Sci USA 86: 2478–2482
Saunders AM, Schmader K, Breitner JCS, Benson MD, Brown WT, Goldfarb L, Goldgaber D, Manwaring MG, Szymanski MH, McCown N, Dole KC, Schmechel DE, Strittmatter WJ, Pericak-Vance M.A. Roses AD (1993) Apolipoprotein E ε4 allele distributions in late-onset Alzheimer's disease and in other amyloid-forming diseases. Lancet 342: 710–711
Saunders AM, Strittmatter WJ, Schmechel D, St. George-Hyslop MD, Perivak-Vance MA, Joo SH, Rosi BL, Gusella JF, Crapper-McLachlan DR, Alberts M. J, Hulette C, Crain B, Goldgaber D, Roses AD (1993) Association of apolipoprotein E allele ε4 with late onset familial and sporadic Alzheimer's disease. Neurology 43: 1467–1472
Schmechel DE, Saunders AM, Strittmatter WJ, Crain BJ, Hulette CM, Joo SH, Pericak-Vance MA, Goldgaber D, Roses AD (1993) Increased amyloid β-peptide deposition in late cerebral cortex as a consequence of apolipoprotein E genotype in late-onset Alzheimer disease. Proc Natl Acad Sci USA 90: 9649– 9653
Simons K, Zerial M (1993) Rab proteins and the road maps for intracellular transport. Neuron 11: 789–799
Sisodia S (1992) β-Amyloid precursor protein cleavage by a membrane-bound protease. Proc Natl Acad Sci USA 89: 6075–6079
Steiner DF, Smeekens SP. Ohagi S, Chan SJ (1992) The new enzymology of precursor processing endoproteases. J Biol Chem 267: 23435–23438

Strittmatter WJ, Saunders AM, Schmechel D, Pericak-Vance M, Enghild J, Salvesen GS, Roses AD (1993a) Apolipoprotein E: High-avidity binding to β-amyloid and increased frequency of type 4 allele in late-onset familial Alzheimer disease Proc Natl Acad Sci USA 90: 1977–1981

Strittmatter WJ, Weisgraber KH, Huang DY, Dong L-M, Salvesen GS, Pericak-Vance M, Schmechel D, Saunders AM, Goldgaber D, Roses AD (1993b) Binding of apolipoprotein E to βA4 peptide: isoform-specific effects and implications for late-onset Alzheimer disease. Proc Natl Acad Sci USA 90: 8098–8102

Strittmatter WJ, Weisgraber KH, Goedert M, Saunders AM, Huang D, Corder EH, Dong L-M, Jakes R, Alberts MJ, Gilbert JR, Han S-H, Hulette C, Einstein G, Schmechel DE, Pericak-Vance MA, Roses AD (1994) Hypothesis: microtubule instability and paired helical filament formation in the Alzheimer disease brain are related to apolipoprotein E genotype. Exp Neurol 125: 163–171

Tooyama I, Kawamata T, Akiyama H, Moestrup SK, Gliemann J, McGeer PL (1993) Immunohistochemical study of α2 macroglobulin receptor in Alzheimer and control postmortem human brain. Mol Chem. Neuropathol. 18: 153–160

Van Broeckhoven C,H, Backhovens M, Cruts G, De Winter M, Bruyland P, Cras J-J. Martin (1992) Mapping of a gene predisposing to early-onset Alzheimer's disease to chromosome 14q24.3. Nature Gen 2: 335–339

Van De Ven WJM, Voorberg J, Fontijn R, Pannekoek H, Van den Ouweland AMW, Siezen RJ. (1990) Furin is a subtilisin-like proprotein-processing enzyme in higher eukaryotes. Mol Biol Rep 14: 265–275

Van De Ven WJM, Van Duynhoven JLP, Roebroek AJM (1992) Structure and function of mammalian proprotein processing enzymes of the subtilisin family of serine proteases. Crit Rev Oncogen 4: 115–136

Van Gool D, De Strooper B, Van Leuven F, Triau E, Dom R (1993) α2-macroglobulin expression in neuritic-type plaques in patients with Alzheimer's disease. Neurobiol. Aging 14: 233–237

Vidal F, Blangy A, Rassoulzadegan M, Cuzin F (1992) A murine sequence-specific DNA binding protein shows extensive local similarities to the amyloid precursor protein. Biochem Biophys Res Comm 189: 1336–1341

von der Kammer H, Hanes J, Klaudiny J, Scheit KH (1994) A human amyloid precursor-like protein is highly homologous to a mouse sequence-specific DNA binding protein L. DNA Cell Biol

Wang R, JF, Meschia RJ, Cotter, Sisodia SS (1991) Secretion of the β/A4 amyloid precursor protein. Identification of a cleavage site in cultured mammalian cells J Biol Chem 266: 16960–16964

Wasco W, Brook JD, Tanzi RE (1993) The amyloid precursor-like protein (APLP) gene maps to the long arm of human chromosome 19. Genomics 15: 237–239

Wasco W, Bupp K, Magendantz M, Gusella JF, Tanzi RE, Solomon F (1992) Identification of a mouse brain cDNA that encodes a protein related to the Alzheimer disease-associated amyloid β protein precursor. Proc Natl Acad Sci USA 89: 10758–10762

Weidemann A, G, König D, Bunke P, Fischer JM, Salbaum CL, Masters Beyreuther K (1989). Identification, biogenesis, and localization of precursors of Alzheimer's disease A4 amyloid protein. Cell 57: 115–126

Williams SE, Ashcom JD, Argraves WS, Strickland DK (1992) A novel mechanism for controlling the activity of α2-macroglobulin receptor/low density lipoprotein receptor-related protein. Multiple regulatory sites for 39-kDa receptor-associated protein. J Biol Chem 267: 9035–9040

Wirak DO, Bayney R, Kundel CA, Lee A, Scangos GA, Trapp BD, Unterbeck AJ (1991) Regulatory region of human amyloid precursor protein (APP) gene promotes neuron-specific gene expression in the CNS of transgenic mice. EMBO J 10: 289–296

Wise RJ, Barr PJ, Wong PA, Kiefer MC, Brake AJ, Kaufman RJ (1990) Expression of a human proprotein processing enzyme: correct cleavage of the von Willebrand factor precursor at a paired basic amino acid site. Proc Natl Acad Sci USA 87: 9378–9382

Wisniewski T, Ghiso J, Frangione B (1991) Peptides homologous to the amyloid protein of Alzheimer's disease containing a glutamine for glutamic acid substitution have accelerated amyloid fibril formation. Biochem Biophys Res Comm 179: 1247–1254

Zhong Z, Higaki J, Murakami K, Wang Y, Catalano R, Quon D, Cordell B (1994) Secretion of β-amyloid Precursor Protein involves multiple cleavage sites. J. Biol. Chem. 269: 627–632

Implication of the Amyloid Precursor Protein in Neurite Outgrowth

B. Allinquant, P. Hantraye, C. Bouillot, K.L. Moya,*
and *A. Prochiantz*

Introduction

The amyloidogenic fragment βA4 is derived from a larger precursor, the amyloid precursor protein (APP), which is normally expressed in the brain as a larger transmembrane glycoprotein which can be subject to several cleavage events and secreted. The predominant processing pathway cleaves APP within the βA4 peptide sequence, while other events can result in minor quantities of βA4. Thus it is believed that the accumulation of βA4 results from one or several errors in the normal metabolism of its precursor (for review, see Selkoe et al. 1994). However, the normal physiological function of APP and its trafficking remains to be elucidated.

In previous studies, we reported that in-vivo, APP is sent down the axon to the terminals in fast axonal transport and various isoforms have a characteristic developmental time course (Moya et al. 1994). In other experiments, we showed that APP associates with the cytoskeleton and is distributed in two pools, one highly enriched in the soma and the axons of differentiating neurons in vitro (Allinquant et al. 1994). These data suggest that APP could be involved in neurite elongation, an idea consistent with the results observed when APP is used as a substrate (Milward et al. 1992; Kibbey et al. 1993; Small et al. 1994). To investigate this possibility more directly, we have used an original strategy to internalize APP antisense oligonucleotides to examine the effects on neuronal morphology.

Linkage of Oligonucleotides to pAntp Vector Peptide

The pAntp peptide corresponds to the homeodomain of Antennapedia and we have demonstrated that it translocates through biological membranes (Joliot et al. 1991; Bloch-Gallego et al. 1993; Le Roux et al. 1993). pAntp accumulates directly in the cytoplasm and nucleus, thus decreasing considerably its targetting into endosomal and lysosomal compartments and its subsequent degradation. Two antisense oligonucleotides (15 and 25 mers) as well as the corresponding sense oligonucleotides were coupled to pAntp or

* CNRS URA 1414, Ecole Normale Supérieure, 46 rue d'Ulm, 75230 Paris Cédex 05, France

K.S. Kosik et al. (Eds.)
Alzheimer's Disease: Lessons from Cell Biology

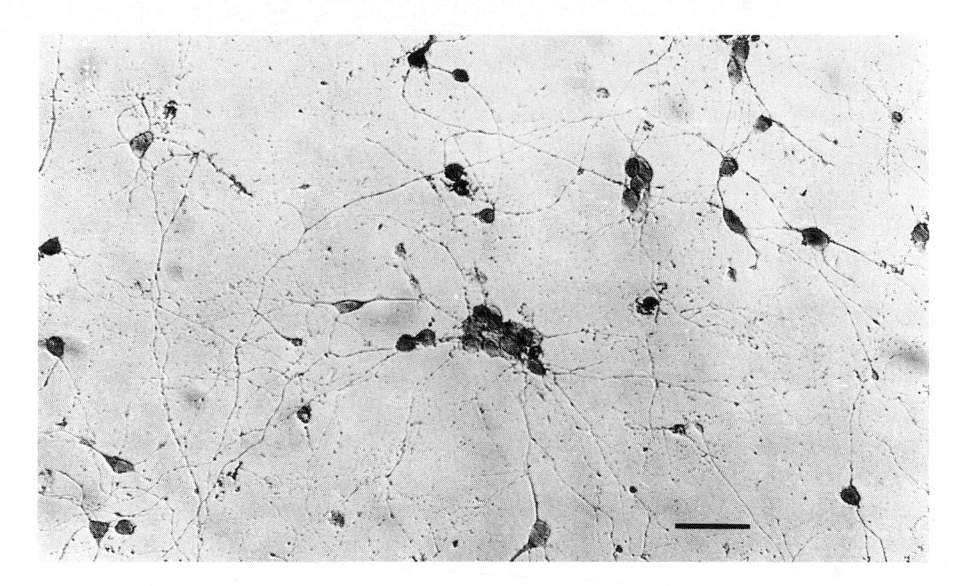

Fig. 1. pAntp coupled to APP antisense or sense oligonucleotides is internalized by neurons in vitro. The internalization of biotinylated APP oligonucleotides coupled to pAntp was detected using streptavidin-alkaline phosphatase. Note that all neurons have internalized the oligonucleotides after two hours of incubation. Scale bar, 10 μm

to pAntp50A, a mutant version of pAntp that can translocate through biological membranes but is devoid of intrinsic biological properties. The linkages between the peptides and oligonucleotides were quantitative and were followed by internalization of 100% of the cells (Fig. 1).

Inhibition of APP Synthesis by the Antisense Oligonucleotides

Table 1 summarizes the inhibition of APP synthesis after antisense internalization. To quantitate the degree of inhibition, cells from E15 rat cortex grown in conditions permitting neuronal survival and differentiation (Lafont et al. 1993) were incubatd with the oligonucleotides for one hour prior to addition of ^{35}S methionine for an additional hour. The methionine was added to the cells at different times after the addition of oligonucleotides. As summarized in Table 1, we found that the inhibition of APP synthesis

Table 1. Percent inhibition of APP neosynthesis after the addition of pAntp coupled to APP antisense (AS) oligonucleotides (ONS)

Time after addition of pAntp- AS ONS (hours)	2.5	4.5	7.5	19.5
% inhibition of APP synthesis	46	30	9.4	7.8

was transient and lasted for no more that seven hours. Inhibition was never complete, even when the concentration of antisense sequence was raised from 40 nM (our standard concentration) to 200 nM. We observed no inhibition by the tested sense oligonucleotides at any concentration between 40 and 200 nM.

Inhibition of Neuritogenesis by the Antisense Oligonucleotides

The addition of the antisense or sense oligonucleotides had no effect on neuronal survival. Indeed, compared to the number of cells plated, the survival rate, estimated by trypan blue exclusion, was always superior to 85%, even after 24 hours in vitro. The number of neurite-bearing cells was reduced by 50% six hours after the addition of the antisense oligonucleotides compared to the cells treated with the sense oligonucleotides or the vector alone. The effect of the antisense was also examined after 24 and 42 hours in culture. We found no effect at 42 hours, although a 50% inhibition of neurite length was clearly visible at 24 hours (not shown).

Conclusion

The results reported here demonstrate that an antisense oligonucleotide directed against APP mRNA transiently decreases APP synthesis and affects neurite elongation. Although this finding was predicted by experiments done with neuronal cell lines (Kibbey et al. 1993), this report is the first that addresses this question using normal neuronal cells in primary cultures.

References

Allinquant B, Moya KL, Bouillot C, Prochiantz A (1994) Amyloid precursor protein in cortical neurons: Co-existence of two pools differentially distributed in axons and dendrites and association with cytoskeleton. J Neurosci 14: 6842–6854

Bloch-Gallego E, Le Roux I, Joliot AH, Volovitch M, Henderson CE, Prochiantz A (1993) Antennapedia homeobox peptide enhances growth and branching of embryonic chicken motoneurons in vitro. J Cell Biol 120: 485–492

Kibbey MC, Jucker M, Weeks BS, Neve RL, Van Nostrand WE, Kleinman HK (1993) β-amyloid precursor protein binds to the neurite-promoting IKVAV site of laminin. Proc Natl Acad Sci USA 90: 10150–10153

Lafont F, Rouget M, Rousselet A, Valenza C, Prochiantz A (1993) Specific responses of axons and dendrites to cytoskeleton pertubations: an in vitro study. J Cell Sci 104: 433–443

Le Roux I, Joliot AH, Bloch-Gallego E, Prochiantz A, Volovitch M (1993) Neurotrophic activity of the antennapedia homeodomain depends on its specific DNA-binding properties. Proc Natl Acad Sci USA 90: 9120–9124

Milward EA, Papadoulos R, Fuller SJ, Moir RD, Small D, Beyreuther K, Masters CL (1992) The amyloid protein precursor of Alzheimer's disease is a mediator of the effects of nerve growth factor on neurite outgrowth. Neuron 9: 129–137

Moya KL, Benowitz LI, Schneider GE, Allinquant B (1994) The amyloid precursor protein is developmentally regulated and correlated with synaptogenesis. Dev Biol 161: 597–603

Selkoe DJ (1994) Normal and abnormal biology of the β-amyloid precursor protein. Annu Rev Neurosci 17: 489–517

Small DH, Nurcombe V, Reed G, Clarris H, Moir R, Beyreuther K, Masters CL (1994) A heparin-binding domain in the amyloid protein precursor of Alzheimer's disease is involved in the regulation of neurite outgrowth. J Neurosci 14: 2117–2127

Physiological Production and Polarized Secretion of the Amyloid β-Peptide in Epithelial Cells: A Route to the Mechanism of Alzheimer's Disease

*D.J. Selkoe**

Summary

Converging evidence from many laboratories has implicated altered metabolism of the amyloid β protein precursor (βAPP) and progressive deposition of its amyloid β-peptide (Aβ) fragment as an early and constant feature of the pathogenesis of Alzheimer's disease. βAPP is a widely expressed type I integral membrane glycoprotein that undergoes processing by two principal routes: exocytic (secretory) and endocytic (reinternalization and lysosomal targeting). Either or both of these routes is capable of generating the intact Aβ fragment following cleavages by as yet unknown proteases. This leads to constitutive secretion of Aβ as a soluble peptide from a variety of neural and non-neural cells that express βAPP. One compartment in which Aβ generation appears to occur is the early endosome. Studies of βAPP trafficking in polarized epithelial cells grown in monolayer cultures suggests that Aβ is generated at least in part from βAPP molecules that traffic asymmetrically to the cell surfaces, with the major portion of Aβ being released from the basolateral surface. These and other studies of the detailed cellular mechanism for generation of Aβ should provide insights into specifically inhibiting cellular production and secretion of Aβ as a therapeutic approach toward slowing the progression of Alzheimer's disease.

Introduction

Progressive cerebral dysfunction in Alzheimer's disease and Down's syndrome is accompanied by the formation of innumerable extracellular amyloid deposits in the form of senile plaques and microvascular amyloid. The amyloid fibrils are composed of the 40–43 residue amyloid β-protein (Aβ), a fragment of the integral membrane polypeptide, β-amyloid precursor protein (βAPP). Evidence from several laboratories has shown that amorphous, largely nonfilamentous deposits of Aβ ("diffuse or "preamyloid" plaques) precede the development of fibrillary amyloid, dystrophic neurites, neuro-

* Harvard Medical School, Brigham and Women's Hospital, Boston, MA 02115, USA

K.S. Kosik et al. (Eds.)
Alzheimer's Disease: Lessons from Cell Biology

fibrillary tangles, and other cytopathological changes in Down's syndrome and, by inference, in Alzheimer's disease. This finding suggests that β-amyloidosis, like numerous systemic amyloidoses, does not occur secondary to local cellular pathology (e.g., dystrophic neurites) but rather precedes it. The clearest evidence that the processing of βAPP into Aβ can actually cause Alzheimer's disease has come from the identification by several laboratories of missense mutations in the βAPP gene within and flanking the Aβ region in affected members of certain families having Alzheimer's disease or hereditary cerebral hemorrhage with amyloidosis of the Dutch type.

Normal Proteolytic Processing of βAPP Leads to Secretion of the Aβ Peptide

The mechanism of proteolytic release of the Aβ fragment from βAPP is incompletely understood. Because the normal secretion of the large extramembranous portion of βAPP (APP_s) from cells involves a proteolytic cleavage within Aβ, thus precluding amyloid formation, we searched for evidence of an alternate pathway of βAPP processing that leaves Aβ intact. In view of the presence of a consensus sequence (NPXY) in the cytoplasmic tail of βAPP that could mediate internalization of the protein from the cell surface and its targeting to endosomes/lysosomes, we looked specifically for evidence of endocytotic trafficking of βAPP. Incubation of an antibody to the extracellular region of βAPP with living human cells led to binding of the antibody to cell-surface βAPP and trafficking of the antigen antibody complex to endosomes/lysosomes (Haass et al. 1992a). The resultant βAPP-immunoreactive pattern closely resembled that seen after incubating the same cells with rhodamine-tagged albumin, a marker for fluid-phase pinocytosis. Late endosomes/lysosomes purified from the cells contained abundant full-length βAPP plus an array of low molecular weight fragments ranging from ~10 to ~22 kDa, most of which are of a size and immunoreactivity suggesting they contain the intact Aβ peptide (Haass et al. 1992a). These results provided direct evidence that some βAPP molecules are normally reinternalized from the cell surface and targeted to lysosomes. This second normal pathway for βAPP processing is capable of producing potentially amyloidogenic fragments. However, it is not yet clear whether this pathway, an alternative proteolytic cleavage occurring within the secretory pathway (Seubert et al. 1993), or another, yet undescribed trafficking pathway is actually responsible for Aβ formation.

During the aforementioned studies, we searched intensively for evidence of the production and release of the Aβ peptide itself during normal cellular metabolism, based in part on the hypothesis that some Aβ deposits (e.g., those in capillary walls and the subpial cortex) might arise from a circulating (plasma or cerebral spinal fluid) form of the peptide. To this end, a series of antibodies to Aβ were used to screen the conditioned media of several cell types for the presence of soluble Aβ. These experiments demonstrated that Aβ is continuously produced as a soluble 4 kDa peptide and is released into the media of normal cells (Haass et al. 1992b). Moreover, Aβ-immunoreactivity has also been detected in human cerebrospinal fluid (CSF; Seubert et al. 1992; Shoji et al. 1992) and plasma (Seubert et al. 1992). The form in CSF has been purified and sequenced, confirming that it is authentic Aβ (Seubert et al. 1992; Shoji et al. 1992; Haass et al. 1993), whereas the plasma form is not yet characterized. Aβ peptides of varying length are released by all βAPP-expressing cells studied to date under normal culture conditions (Haass et al. 1992b, 1993; Seubert et al. 1992; Shoji et al. 1992; Busciglio et al. 1993), Aβ in culture supernatants is entirely soluble and generally present in high picomolar to low nanomolar concentrations (Seubert et al. 1992). Pulse-chase and biological toxin experiments suggest that Aβ is produced following full maturation of βAPP and involves an acidic compartment other than lysosomes, e.g., early endosomes or the late Golgi (Haass et al. 1993). The two proteolytic cleavages generating Aβ may therefore occur in an acidic vesicle near the cell surface, after which Aβ is rapidly released into the medium, with very little or no Aβ detected intracellularly (Haass et al. 1992b, 1993).

Excessive Aβ Production Is Implicated in Some Forms of Familial Alzheimer's Disease

The relevance of such *in vitro* Aβ production to the pathogenesis of Alzheimer's disease is demonstrated by the finding that a βAPP missense mutation causing a Swedish form of familial Alzheimer's disease (FAD), when expressed in cultured cells, leads to a marked increase in Aβ production (Citron et al. 1992). The amyloidogenic mechanisms of other FAD-linked βAPP mutations are now being elucidated in both transfected and primary (donor) cells. Effects of the recently identified FAD defect on chromosome 14 (Schellenberg et al. 1992) on βAPP processing and Aβ production can be searched for in cultured primary cells from these patients, even before the responsible gene is identified and characterized. Importantly, transfected or primary cells expressing the Swedish mutant gene can readily be used to screen a variety of compounds and identify those capable of decreasing Aβ secretion to normal levels in the absence of significant cytotoxicity. Such agents can then be tested in laboratory rodents or in aged animals (e.g., dogs, monkeys) that spontaneously develop Aβ plaques to determine the agents' effects on brain and CSF levels of Aβ, as well as their safety.

Mechanism of Generation of Aβ

Information about the characteristics of the proteolytic cleavage that creates the N-terminus of Aβ (caused by an unknown enzyme designated "β-secretase") has begun to emerge. A truncated APP_s species which ends at

met_{670}, i.e., just prior to the start of Aβ, has been detected immunochemically in the medium of βAPP-transfected cells (Seubert et al. 1993). This finding, coupled with the detection in cell lysates of a ~12 kD carboxyl terminal fragment (CTF) beginning at or near asp_{671} under some experimental circumstances (Cai et al. 1993; Golde et al. 1992) suggests that alternative secretory processing of the holoprotein could serve as an initiating step in Aβ formation. The amount of the truncated APP_s species appears to be much lower than that of conventional APP_s in the cell types examined to date. As in the case of the α-secretase processing that generates conventional APP_s, cell-surface βAPP can serve as a substrate for the β-secretase cleavage (P. Seubert, personal communication). This finding suggests that surface-inserted βAPP could be one source of Aβ. Given the evidence reviewed above that agents that alter intravesicular pH block Aβ formation, we have postulated that one site for Aβ generation may be within an early endosome following reinternalization of βAPP (or a ~12 kD CTF) from the cell surface (Haass and Selkoe 1993). Such a mechanism has now been demonstrated directly by selective labeling of surface βAPP in living cells with radioiodine followed by incubation at 37° to allow internalization and processing (Koo and Squazzo 1994). Radioiodinated Aβ was released into the medium. The use of potassium depletion to inhibit endocytosis led to a marked decrease in the production of iodinated Aβ in pulse-chase experiments; restoration of normal potassium levels restored Aβ secretion (Koo and Squazzo 1994). If the YENPTY sequence in the cytoplasmic tail was deleted, Aβ release was markedly decreased (Koo and Squazzo 1994), consistent with a similar decrease observed previously when the entire cytoplasmic domain was deleted (Haass et al. 1993). These experiments have provided the first demonstration of a specific pathway that generates Aβ: internalization of surface βAPP into early endosomes followed by excision and release of the Aβ peptide and its rapid release from cells.

That endosomes containing βAPP (and fragments thereof) can rapidly recycle to the cell surface has been demonstrated directly by labeling of the surface precursor with monoclonal antibodies at 4°C, warming briefly to initiate internalization, then acid-stripping residual surface label and incubating at 37°C for varying times in the presence of secondary antibody. In this paradigm, any internalized βAPP tagged with primary antibody that recycles to the cell surface will pick up secondary antibody and reinternalize, and this can then be demonstrated by fixation and fluorescent immunocytochemistry. These experiments showed that βAPP is internalized and rapidly recycled within 5–10 min to the cell surface (Yamazaki et al. 1993). A portion of the internalized βAPP moved progressively from smaller vesicles near the cell surface to larger vesicles localized immediately perinuclear, suggesting that some internalized molecules traffic to lysosomes, as expected. Repetition of these studies with cells expressing βAPP having almost the entire cytoplasmic domain deleted showed far less βAPP internalization and recycling (Yamazaki et al. 1993).

As regards the substrate requirements for β-secretase cleavage, mutagenesis around the N-terminus of Aβ has suggested a high sequence specificity of the enzyme(s) (Citron et al. 1993). Besides the methionine at βAPP_{670} that occurs in wild-type molecules, only the leu substitution at this position which occurs in the Swedish FAD kindred allowed proper cleavage and Aβ generation. All other substitutions at this position examined to date led to little or no detectable Aβ. Similarly, a variety of substitutions at asp_{671} resulted in a loss of the 4 kD Aβ peptide and variable production of an intermediate 3.5 kDa species which begins at glu_{11} (Citron et al. 1993). Mutations at βAPP_{669} or $_{672}$ also interfered markedly with Aβ generation. Deletion of 4–5 residues between asp_{671} and the membrane did not significantly alter β-secretase cleavage, which still occurred at the met-asp bond. Therefore, in contrast to α-secretase, β-secretase does not appear to cleave at a specified distance from the membrane, at least in the cell types studied (Citron et al. 1993). Importantly, cleavage by β-secretase and generation of Aβ required βAPP to be membrane inserted. When stop codons were introduced either 40 residues or 51 residues after the Aβ start site, only the latter truncated molecule underwent membrane insertion and could generate Aβ; the former showed no production of Aβ or the smaller related peptide (p3) (Citron et al. 1993). These experiments suggest that β-secretase is a highly sequence-specific enzyme(s) with properties distinct from α-secretase. The N-terminal heterogeneity of Aβ species seen both *in vitro* (Haass et al. 1992b, 1994a) and *in vivo* (Masters et al. 1985; Miller et al. 1990) probably represents cleavages by distinct but related proteases, since the mutagenesis experiments just reviewed make it unlikely that a single β-secretase enzyme could effect these varied cleavages.

The generation of the C-terminus of Aβ is an unusual event in that it requires access by the protease to the hydrophobic transmembrane region. Aβ may be the first demonstrated example of a peptide released by such intramembranous proteolysis in normal cells. It will be of interest to determine whether peptides containing the transmembrane domains of other proteins are secreted in a similar fashion. Exactly how and when the protease(s) accesses the Aβ C-terminal region remains a mystery. What is known is that protease(s) capable of generating this cleavage in cultured cells show considerable nonspecificity, as mutagenesis of residue 42 of the Aβ sequence to various other hydrophobic residues, deletion of residues 39–42 or insertion of 4 leucines between positions 39 and 40 all do not substantially alter the generation of a ~4 kDa Aβ peptide (Selkoe et al. 1993).

APP Trafficking in Polarized Epithelial Cells

Virtually all of the information about βAPP processing described above has been generated in nonpolarized cell types. Such studies have recently been extended to analyses conducted in Madin Darby canine kidney (MDCK) cells, which display a polarized phenotype with distinct apical and basolateral plasma membranes when grown as monolayers on polycarbonate or nitrocellulose filters (Matlin and Simons 1984; Matter et al. 1992). In addition to the benefit of extensive comparative data on the polarized trafficking of other surface proteins in MDCK cells, studies of βAPP processing in these epithelial cells could be relevant to the preferential deposition of the Aβ fragment in the abluminal basement membrane of endothelial cells in Alzheimer's disease brain (Yamaguchi et al. 1992).

The endogenous expression of βAPP in MDCK cells is sufficiently high that initial observations about its polarized secretion could be made without the need for transfection. The conventional APP_s generated by α-secretase cleavage was released ~80–90% from the basolateral surface, as were the Aβ and p3 peptides (Haass et al. 1994). Stable transfection of wild-type βAPP into these cells resulted in the same strong basolateral secretion of APP_s, Aβ and p3 as occurs for the endogenous molecule. The enhanced βAPP signal produced by transfection enabled surface antibody labeling studies which demonstrated ~80–90% insertion of the holoprotein on the basolateral membrane surface. Moreover, surface biotinylation of the transfectants revealed release of biotinylated APP_s overwhelmingly from the basolateral membrane (Haass et al. 1994).

Treatment of the transfected MDCK cells with 10mM ammonium chloride abolishes the polarized secretion of APP_s so that it is released almost equally from apical and basolateral surfaces (Haass et al. submitted for publication). Nonetheless, surface βAPP distribution still remains ~90% basolateral. These findings suggest that APP_s may exist in part in intracellular acidic vesicles whose trafficking can be regulated independently of the routing of holoβAPP to the cell surface. Further support for this hypothesis derives from the detection of intraluminal APP_s in membrane vesicles purified from MDCK cells and extracted in sodium carbonate buffer (Haass et al., submitted). The signals which could direct the polarized trafficking of the non-membrane-inserted APP_s molecule are not known. In this regard, the expression of the Swedish mutant βAPP molecule in MDCK cells has led to the finding that the truncated APP_s species (ending at leu_{670}) released abundantly from this precursor by β-secretase is principally secreted apically rather than basolaterally (S. Sisodia et al., unpublished data). Similarly, the small amount of this truncated APP_s released from wild-type transfected MDCK cells is also predominantly released apically (S. Sisodia et al., unpublished data). These results indicate that β-secretase-generated APP_s traffics differently than α-secretase-generated APP_s and raise the possibility that there is a signal in the last 16 residues of the latter derivative that help in part to direct it basolaterally.

Information about the signals which regulate the polarized trafficking of holoβAPP has begun to emerge. A series of increasingly larger truncations of the cytoplasmic domain which remove the last 12, 22 or 32 residues caused little detectable change in the polarized basolateral sorting of βAPP

(Haass et al., submitted). However, removal of the last 42 residues, i.e., almost the entire cytoplasmic tail, led to a shift in βAPP polarity: 40–50% of the molecules were now inserted on the apical membrane (Haass et al., submitted). This result suggests that a signal in or near the region 729–738 of βAPP_{770} participates in directing the precursor basolaterally. Additional mutagenesis experiments will now be needed to elucidate fully the signals which mediate polarized trafficking of βAPP. Importantly, in all of the MDCK studies to date, the pattern of Aβ secretion has followed the trafficking of the holoprotein, e.g., Aβ is ~50% apically released from molecules lacking the last 42 residues of βAPP (Haass et al., submitted).

The various findings just reviewed lead to a model for βAPP trafficking in polarized epithelial cells that includes cleavage of a small population of precursor molecules by β-secretase at a relatively early point in βAPP processing to generate a truncated APP_s form that is principally secreted apically. In contrast, conventional APP_s generated by α-secretase may arise somewhat later during βAPP sorting and is always directed basolaterally. The holoprotein is normally trafficked preferentially to the basolateral membrane, but alteration of signals in the N-terminal portion of the cytoplasmic tail can partially redirect it to the apical surface. Aβ secretion appears to depend on the pattern of trafficking of holoβAPP to the surface and thus may arise following endocytosis of the cell-surface molecule, consistent with studies in non-polarized cells discussed previously.

Conclusion

Studies of *in vitro* Aβ production, such as those summarized above, should advance the fundamental understanding and ultimately the pharmacological treatment of β-amyloidosis in Alzheimer's disease. Therapeutic approaches could ultimately include: 1) inhibiting the amyloid-generating proteases; 2) diverting βAPP from amyloidogenic (β-secretase) to non-amyloidogenic (α-secretase) processing pathways; 3) preventing the gradual aggregation of soluble extracellular Aβ into fibrils; 4) interfering with the toxic response of neurons to aggregated Aβ and its tightly associated proteins; and 5) inhibiting the chronic inflammatory process (including microgliosis and astrocytosis) that develops around amyloid plaques.

References

Busciglio J, Gabuzda DH, Matsudaira P, Yankner BA (1993) Generation of β-amyloid in the secretory pathway in neuronal and nonneuronal cells. Proc Natl Acad Sci USA 90: 2092–2096

Cai X-D, Golde TE, Younkin GS (1993) Release of excess amyloid β protein from a mutant amyloid β protein precursor. Science 259: 514–516

Citron M, Oltersdorf T, Haass C, McConlogue L, Hung AY, Seubert P, Vigo-Pelfrey C, Lieberburg I, Selkoe DJ (1992) Mutation of the β-amyloid precursor protein in familial Alzheimer's disease causes increased β-protein production. Nature 360: 672–674

Citron M, Teplow DB, Schmitt FO, Selkoe DJ (1993) The N-terminus of amyloid β-peptide is cleaved by a sequence specific proteins. Soc Neurosci Abstr 19(1): 18

Golde TE, Estus S, Younkin LH, Selkoe DJ, Younkin SG (1992) Processing of the amyloid protein precursor to potentially amyloidogenic carboxyl-terminal derivatives. Science 255: 728–730

Haass C, Koo EH, Mellon A, Hung AY, Selkoe DJ (1992a) Targeting of cell surface β-amyloid precursor protein to lysosomes: Alternative processing into amyloid-bearing fragments. Nature 357: 500–503

Haass C, Schlossmacher MG, Hung AY, Vigo-Pelfrey C, Mellon A, Ostaszewski BL, Lieberburg I, Koo EH, Schenk D, Teplow DB, Selkoe DJ (1992b) Amyloid β-peptide is produced by cultured cells during normal metabolism. Nature 359: 322–325

Haass C, Selkoe DJ (1993) Cellular processing of β-amyloid precursor protein and the genesis of amyloid β-peptide. Cell 75: 1039–1042

Haass C, Hung AY, Schlossmacher MG, Teplow DB, Selkoe DJ (1993) β-Amyloid peptide and a 3-kDa fragment are derived by distinct cellular mechanisms. J Biol Chem 268: 3021–3024

Haass C, Koo EH, Teplow DB, Selkoe DJ (1994) Polarized secretion of β-amyloid precursor protein and amyloid β-peptide in MDCK cells. Proc Natl Acad Sci USA 91: 1564–1568

Koo EH, Squazzo S (1994) Evidence that production and release of amyloid β-protein involves the endocytic pathway. J Biol Chem 269: 17386–17389

Masters CL, Simms G, Weinman NA, Multhaup G, McDonald BL, Beyreuther K (1985) Amyloid plaque core protein in Alzheimer disease and Down syndrome. Proc Natl Acad Sci USA 82: 4245–4249

Matlin KS, Simons K (1984) Sorting of an apical plasma membrane glycoprotein occurs before it reaches the cell surface in cultured epithelial cells. J Cell Biol 99: 2131–2139

Matter K, Hunziker W, Mellman I (1992) Basolateral sorting of LDL receptor in MDCK cells: the cytoplasmic domain contains two tyrosine-dependent targeting determinants. Cell 71: 741–753

Miller D, Potempska A, Papayannopoulos IA, Iqbal K, Styles J, Kim KS, Mehta PR, Currie JR (1990) Characterization of the cerebral amyloid peptides and their precursors. J Neuropathol Exp Neurol 49: 267

Schellenberg GD, Bird TD, Wijsman EM, Orr HT, Anderson L, Nemens E, White JA, Bonnycastle L, Weber JL, Alonso ME, Potter H, Heston LH, Martin GM (1992) Genetic linkage evidence for a familial Alzheimer's disease locus on chromosome 14. Science 258: 668–671

Selkoe DJ, Watson D, Hung AY, Teplow DB, Haass C (1993) Influence of the Aβ C-terminal sequence and of a FAD mutation at βAPP_{692} on the formation of amyloid β-peptide. J Neurosci Abstr 19: 431

Seubert P, Vigo-Pelfrey C, Esch F, Lee M, Dovey H, Davis D, Sinha S, Schlossmacher M, Whaley J, Swindlehurst C, McCormack R, Wolfert R, Selkoe D, Lieberburg I, Schenk D (1992) Isolation and quantification of soluble Alzheimer's β-peptide from biological fluids. Nature 359: 325–327

Seubert P, Oltersdorf T, Lee MG, Barbour R, Blomquist C, Davis DL, Bryant K, Fritz LC, Galasko D, Thal LJ, Lieberburg I, Schenk DB (1993) Secretion of β-amyloid precursor protein cleaved at the amino-terminus of the β-amyloid peptide. Nature 361: 260–263

Shoji M, Golde TE, Ghiso J, Cheung TT, Estus S, Shaffer LM, Cai X, McKay DM, Tintner R, Frangione B, Younkin SG (1992) Production of the Alzheimer amyloid β protein by normal proteolytic processing. Science 258: 126–129

Yamaguchi H, Yamazaki T, Lemere CA, Frosch MP, Selkoe DJ (1992) Beta amyloid is focally deposited within the outer basement membrane in the amyloid nagiopathy of Alzheimer's disease. Am J Pathol 141: 249–259

Yamazaki H, Koo EH, Hedley-White ET, Selkoe DJ (1993) Intracellular trafficking of cell surface βAPP in living cells. Soc Neurosci Abstr 19(1): 396

Regulation and Structure of the MAP Kinases ERK1 and ERK2

*M.H. Cobb**, *J.E. Hepler*, *E. Zhen*, *D. Ebert*, *M. Cheng*, *A. Dang*, and *D. Robbins*

Summary

The MAP kinases ERK1 and ERK2 represent a subfamily of the protein kinases with a significant role in hormonal signal transduction. They are ubiquitous, growth factor-stimulated protein kinases that phosphorylate and thereby modulate the properties of many proteins that have key regulatory functions. These include other protein kinases, transcription factors, membrane enzymes, and cytoskeletal proteins (e.g., tau, MAP2). This wide array of phosphorylations results in their pleiotropic properties. Aberrant activation of the MAP kinases may contribute to pathological changes in the phosphorylation of tau protein associated with Alzheimer's disease. X-ray crystallographic studies of ERK2 have revealed important features of its phosphoregulatory mechanism and should be valuable in developing specific MAP kinase inhibitors.

Introduction

The MAP kinase pathway represents one branch of the complex cellular regulatory machinery that leads to pleiotropic changes in cell function. This regulatory machinery may be impaired or activated inappropriately in disease states such as Alzheimer's disease. Thus, determining the molecular mechanisms controlling pathways such as the MAP kinase pathway has implications for understanding disease states. In this article aspects of the mechanisms of regulation and the function of the MAP kinase pathway are discussed.

The MAP kinases ERK1 and ERK2 (extracellular signal-regulated protein kinases 1 and 2) comprise a subfamily of the protein kinases (Ray and Sturgill 1987, 1988; Boulton et al. 1990, 1991b) and play a major role in signal transduction. ERK1 and ERK2 are acutely activated by many extracellular stimuli, leading to proliferation or differentiation depending on the cell state (reviewed in Cobb et al. 1991; Blenis 1993). They act in the nucleus to alter gene expression and also in the cytoplasm, at the membrane,

* The University of Texas Southwestern Medical Center at Dallas, Department of Pharmacology, 5323 Harry Hines Blvd., Dallas, TX 75235-9041, USA

K.S. Kosik et al. (Eds.)
Alzheimer's Disease: Lessons from Cell Biology

and on the cytoskeleton. Substrates of these enzymes include microtubule-associated protein 2 (MAP2; Ray and Sturgill 1987), tau (Drechsel et al. 1992), and caldesmon, proteins intimately involved in organization of the cytoskeleton. Interestingly, two groups have demonstrated that MAP kinase influences interactions among microtubule-associated proteins. In lysates of *Xenopus* oocytes, phosphorylation by MAP kinase altered microtubule dynamics by inducing a pattern characteristic of metaphase, whereas phosphorylation of tau on three to four sites in vitro was associated with a decrease in the affinity of tau for the microtubule lattice (Drechsel et al. 1992). While MAP kinases clearly participate in regulation of these structures, other protein kinases may be at least as important in changing cell shape and structure through effects on the cytoskeleton. At least one of these has been implicated in alterations in tau protein that occur in Alzheimer's disease (Biernat et al. 1993).

The MAP Kinase Subfamily

The 43 and 41 kDa MAP kinases ERK1 and ERK2 were the first two members of this protein kinase subfamily to be purified and cloned (Gotoh et al. 1991; Boulton et al. 1990, 1991a,b). They are 83% identical overall. Other less closely related enzymes have been identified and include the 62 kDa ERK3 (Boulton et al. 1991b; Gonzalez et al. 1992), an immunologically related protein ERK4 (Boulton and Cobb 1991), the JNK enzymes (Dérijard et al. 1994), the p54 (Kyriakis et al. 1991) and p57 MAP kinases (Lee et al. 1993), and the HERA kinase (Williams et al. 1993). One common feature of these kinases is their specificity for substrates with phosphorylation sites followed by proline, a specificity shared with the cyclin-dependent kinases (cdk). Among both the ERK and cdk families of enzymes there are conserved amino acids (e.g., Arg or Lys four residues preceding the conserved APE motif of subdomain VIII) found only in these families.

The Crystal Structure of ERK2 Reveals Features of MAP Kinases That Contribute to Their Substrate Specificity

The three-dimensional structure of ERK2 has been determined at 2.3 Å resolution (Zhang et al. 1994). Many of the residues involved in determining substrate specificity are conserved in other ERKs, suggesting that all of the MAP kinases phosphorylate proteins containing the consensus Ser/Thr-Pro. These residues are also similar for cdk2, which shares the Ser/Thr-Pro specificity. The specificity of ERK2 for proline at the P + 1 position of the substrate (the residue following the phosphorylation site) comes from bulky side chains filling the pocket available for binding the P + 1 residues and an alanine (in place of glycine on cAMP-dependent protein kinase) in the phosphorylation lip that restricts backbone conformation in this region.

Comparison of Biochemical and Structural Information Concerning Phosphoregulation of the MAP Kinases

Two events are required to stimulate the MAP kinases (Anderson et al. 1990; Boulton and Cobb 1991; Ahn et al. 1991). A tyrosine and a threonine that lie one residue apart on each enzyme (Y185 and T183 of ERK2, Y204 and T202 of ERK1; Payne et al. 1991) at the mouth of the catalytic site are phosphorylated. Phosphorylation of tyrosine precedes phosphorylation of threonine (Robbins and Cobb 1992; Haystead et al. 1992), but both residues must be phosphorylated for high enzymatic activity.

Comparison of the crystal structure of ERK2 to the active structure of cAMP-dependent protein kinase (Knighton et al. 1991) reveals that phosphorylation activates ERK2 by inducing both global and local conformational changes. A rotation of the N- and C-terminal domains towards each other causes closure of the active site; this leads to a productive arrangement of residues in the active site so that catalysis can take place efficiently. This closed state is likely to be stabilized by the interaction of the phosphate group of Thr183 with arginine side chains in the N-terminal domain.

The phosphorylation lip of ERK2 contains the Thr183 and Tyr185 phosphorylation sites. In the inactive form of the enzyme, this lip blocks the access of substrates, in part because the side chain of Tyr185 lies at the bottom of the active site. A local conformational change of this tyrosine residue must precede phosphorylation of the tyrosine and threonine residues, leading to a structure in which the protein backbone of the residues in the lip has been reordered. The requirement for two conformational changes resulting from the dual phosphorylation suggests that strict control of ERK2 activity may be necessary for normal cell function (Robbins et al. 1993).

The MAP Kinases Are Most Highly Expressed in Brain

Northern analysis using probes specific for ERK1 and ERK2 indicate that they are ubiquitously distributed, with highest levels in brain and spinal cord (Boulton et al. 1991b). ERK1 mRNA appears to be enriched in astrocytes and glia. Multiple ERK2 mRNAs are detected by Northern blot with an ERK2 probe, perhaps due to alternative splicing and to cross hybridization to closely related mRNAs. The larger two mRNAs are highly expressed in hippocampus, whereas the smallest mRNA is highly expressed in retina (Boulton et al. 1991b). ERK2 mRNA abundance increases during differentiation of P19 teratocarcinoma cells into neuronal cells but remains unchanged during muscle differentiation. ERK1 mRNA abundance decreases during both differentiation programs. All tissues and cell types examined contain immunoreactive ERK1 or ERK2 (Boulton and Cobb 1991).

Upstream Participants in the ERK/MAP Kinase Cascade

MAP kinases are regulated by autocrine and paracrine growth factors, such as nerve growth factor (NGF) and ciliary neurotrophic factor (CNTF; Boulton and Yancopoulos, personal communication), and by neurotransmitter stimulation as occurs during electroconvulsive shock (Stratton et al. 1991). These factors stimulate neuronal proliferation, development, and differentiated functions. The biochemical events leading to activation of this cascade are currently being elucidated, although it is apparent that this cascade may be controlled by distinct upstream regulators. The best understood, biochemically, is the pathway transmitted by growth factor receptors that contain intrinsic tyrosine kinase activity such as the NGF receptor (see White and Kahn 1994 for a review). Ligand-enhanced tyrosine kinase activity results in receptor autophosphorylation and the subsequent association of the SH2 domains of adaptor proteins such as Grb2 (Margolis et al. 1992; Rozakis-Adcock et al. 1992) with the autophosphorylated receptors. These adaptor molecules are believed to recruit guanine nucleotide exchange protein such as Sos (Li et al. 1993) to appropriate sites on the membrane so that they may activate Ras by increasing its association with GTP. The GTP-bound form of Ras binds to the protein kinase Raf-1 (Zhang et al. 1993; Warne et al. 1993; Vojtek et al. 1993; Moodie et al. 1993), thereby targeting Raf-1 to the membrane for activation (Zhang et al. 1993; Warne et al. 1993; Vojtek et al. 1993).

Raf-1 activates the MAP kinase kinases (Ahn et al. 1991; Seger et al. 1991, 1992a; Kosako et al. 1992; L'Allemain et al. 1992; Nakielny et al. 1992) or MAP kinase/ERK kinases (MEKs), which in turn directly activate the MAP kinases ERK1 and ERK2. Two functional mammalian MEKs have been purified and cloned (Seger et al. 1992a,b; Crews et al. 1992; Zheng and Guan 1993a; Wu et al. 1993b; Otsu et al. 1993). Unlike the MAP kinases themselves, MEKs are highly specific, phosphorylating only ERK1 and ERK2 (Seger et al. 1992a). MEKs are also activated by at least two other protein kinases besides Raf, including the enzyme MEK kinase (Lange-Carter et al. 1993), which is also highly expressed in brain. Both Raf-1 and MEK kinase phosphorylate each of the MEKs in vitro and cause their phosphorylation and activation in transfected cells (Dent et al. 1992; Lange-Carter et al. 1993; Wu et al. 1993b; Xu and Cobb, unpublished data). Mechanisms regulating MEK kinase are unknown. This enzyme may not be activated in a Ras-dependent manner or may be Ras-dependent only under certain circumstances.

Activation of ERKs In Vitro

In vitro MEK phosphorylates both the tyrosine and the threonine in ERK1 and ERK2, thereby causing a ~1000-fold increase in activity of the two ERKs (Robbins et al. 1993) up to the specific activity of enzyme purified from insulin-stimulated cells (Boulton et al. 1991a). This finding strongly suggests that these two phosphorylations are sufficient to account for in vivo activation of the MAP kinases. ERK mutants lacking either one or both of the activating phosphorylation sites are not activated by MEK in vitro (L'Allemain et al. 1992; Robbins et al. 1993). If either of the phosphorylated residues in ERK2 is replaced with glutamate, the enzyme can be partially activated following a single phosphorylation (Robbins et al. 1993; L'Allemain et al. 1992; Ebert and Cobb, unpublished data). If both residues are replaced with glutamate, the enzyme is inactive (L'Allemain et al. 1992; Ebert and Cobb, unpublished data). Thus far, no mutations have been identified that greatly increase protein kinase activity of ERKs in the absence of phosphorylation; however, mutations have been found in Drosophila and yeast MAP kinases that increase function in a genetic assay (Brunner et al. 1994; Brill et al. 1994).

Other Regulatory Features of This Cascade

No substrate has yet been identified that is phosphorylated by only one of the ERKs. However, use of dominant inhibitory ERK constructs suggests that they may have distinct functions (Frost et al. 1994). Differences in cellular and subcellular distribution of ERK1 and ERK2 (see below) suggest that they may have unique properties. In addition, stimulation of only one of the two enzymes has been documented under certain circumstances, even though both ERK1 and ERK2 are present (Raines et al. 1993; Kortenjann et al., personal communication). Thus there may be mechanisms for the flux of information selectively through individual members of the cascade.

Other signal transduction systems influence this cascade. In PC12 cells cAMP enhances MAP kinase activation by NGF (Frödin et al. 1994). This finding is consistent with the synergy between cAMP and NGF in regulating neuronal differentiation of PC12 cells, but the mechanism of the synergistic response in these cells has not been determined. In contrast, in fibroblasts and smooth muscle cells cAMP inhibits activation of MAP kinases (Chen et al. 1991; Wu et al. 1993a; Cook and McCormick 1993; Graves et al. 1993) by PDGF and other factors. Raf-1 is a likely site of action of cAMP-dependent protein kinase (Wu et al. 1993a). Phosphorylation of Raf-1 on Ser43 in intact cells and by cAMP-dependent protein kinase in vitro is associated with reduced activity. Other targets, such as the small G protein known as Rap-1b or k-rev, another cAMP-dependent protein kinase substrate, may also be involved in the cAMP-mediated antagonism of ERK activation (Burgering et al. 1993). Rap-1b competes with Ras for binding to Raf (Zhang et al. 1993) and is a logical site for negative regulation.

Inactivation of ERK1 and ERK2

Because both tyrosine and threonine phosphorylations are required to activate ERK1 and ERK2, inactivation of these enzymes may occur by dephosphorylation by a phosphotyrosine phosphatase, a phosphoserine/threonine phosphatase, or a dual specificity phosphatase. Recently the latter two of these three types of phosphatase have been implicated in MAP kinase regulation in cells.

A dual specificity phosphatase of the VH1 type (Alessi et al. 1993; Sun et al. 1993; Zheng and Guan 1993b) inactivates ERK1 and ERK2 with great selectivity. A catalytically defective form of the phosphatase binds tightly to MAP kinases in intact cells (Sun et al. 1993). The phosphatase is induced many-fold by serum and growth factors. Treatment of cells with cycloheximide, to block phosphatase induction, causes prolonged activation of MAP kinase (Sun et al. 1993). Thus, the rapid inactivation of MAP kinases in serum and growth factor-treated cells may result at least in part from the action of this dual specificity enzyme.

Phosphatase 2a may also dephosphorylate MAP kinases under certain circumstances (Sontag et al. 1993). The regulatory B subunit in the phosphatase 2a heterotrimers can be replaced by small t antigen. The small t-phosphatase 2a complex displays reduced activity towards ERK1 and MEK in vitro. Transfection of cells with small t results in activation of MAP kinase and MEK. Phosphorylation site mutants of ERK2 inhibit transformation caused by small t, suggesting that the transforming capability of small t is due at least in part to the activation of the MAP kinase cascade that occurs as a consequence of interfering with the normal function of phosphatase 2a.

Intracellular Distribution of the MAP Kinases

Intracellular localization undoubtedly plays an important part in determining the functions of ERK1 and ERK2. The enzymes are found mainly in soluble fractions of lysed cells (Boulton and Cobb 1991), but ERK1 and ERK2 (Chen et al. 1992; Lenormand et al. 1993; Gonzalez et al. 1993) are translocated to the nucleus following cellular stimulation. The kinetics of activation of these enzymes may determine the efficiency of their nuclear translocation. If the kinases are rapidly inactivated, as occurs following EGF stimulation of PC12 cells, they may not enter or remain in the nucleus. A more prolonged activation, as occurs with NGF in PC12 cells (Traverse et al. 1992), causes nuclear retention. Using an overexpression strategy Gonzalez et al. (1993) found localization of ERK2 and a shorter ERK2 isoform to membrane ruffles. These findings suggest that the unique N-terminal sequences on ERK1 and ERK2 isoforms may contribute to their subcellular distributions and may endow them with distinct cellular functions.

Association of ERK1 and ERK2 with cytoskeletal structures has also been suggested. Angelastro and Greene find immunoreactive yet distinct ERK-related proteins tightly associated with microtubules (Volonté et al. 1993). ERK1 but not ERK2 has been found to associate with Trk, the high affinity NGF receptor (Loeb et al. 1992), whereas both ERK1 and ERK2 associate with the low affinity p75 receptor in response to NGF (Volonté et al. 1993). Results of these localization studies support the idea that the MAP kinases play a significant role in the control of the cytoskeleton. Recently, ERK2 has been found in neuronal processes (Trojanowski et al. 1993) by immunofluorescence. In addition, the enzyme is found in neurofibrillary tangles and senile plaques from Alzheimer's disease tissue. This finding further suggests that the MAP kinases may contribute to this pathological state.

Acknowledgments. The authors thank other members of the Cobb lab for helpful discussions and Jo Hicks for preparation of the manuscript. Grant support was provided by grant DK34128 from the National Institutes of Health and a grant from The Welch Foundation.

References

Ahn NG, Seger R, Bratlien RL, Diltz CD, Tonks NK, Krebs EG (1991) Multiple components in an epidermal growth factor-stimulated protein kinase cascade. In vitro activation of a myelin basic protein/microtubule-associated protein 2 kinase. J Biol Chem 266: 4220–4227

Alessi DR, Smythe C, Keyse SM (1993) The human CL100 gene encodes a Tyr/Thr-protein phosphatase which potently and specifically inactivates MAP kinase and suppresses its activation by oncogenic Ras in Xenopus oocyte extracts. Oncogene 8: 2015–2020

Anderson NG, Maller JL, Tonks NK, Sturgill TW (1990) Requirement for integration of signals from two distinct phosphorylation pathways for activation of MAP kinase. Nature 343: 651–653

Biernat J, Gustke N, Drewes G, Mandelkow E-M, Mandelkow E (1993) Phosphorylation of Ser^{262} strongly reduces binding of tau to microtubules: distinction between PHF-like immunoreactivity and microtubule binding. Neuron 11: 153–163

Blenis J (1993) Signal transduction via the MAP kinases: proceed at your own RSK. Proc Natl Acad Sci USA 90: 5889–5892

Boulton TG, Cobb MH (1991) Identification of multiple extracellular signal-regulated kinases (ERKs) with antipeptide antibodies. Cell Reg 2: 357–371

Boulton TG, Yancopoulos GD, Gregory JS, Slaughter C, Moomaw C, Hsu J, Cobb MH (1990) An insulin-stimulated protein kinase similar to yeast kinases involved in cell cycle control. Science 249: 64–67

Boulton TG, Gregory JS, Cobb MH (1991a) Purification and properties of extracellular signal-regulated kinase 1, an insulin-stimulated microtubule-associated protein 2 kinase. Biochemistry 30: 278–286

Boulton TG, Nye SH, Robbins DJ, Ip NY, Radziejewska E, Morgenbesser SD, DePinho RA, Panayotatos N, Cobb MH, Yancopoulos GD (1991b) ERKs: a family of protein-serine/threonine kinases that are activated and tyrosine phosphorylated in response to insulin and NGF. Cell 65: 663–675

Brill JA, Elion EA, Fink GR (1994) A role for autophosphorylation revealed by activated alleles of *FUS3*, the yeast MAP kinase homolog. Mol Cell Biol 5: 297–312

Brunner D, Oellers N, Szabad J, Biggs III WH, Zipursky SL, Hafen E (1994) A gain-of-function mutation in Drosophila MAP kinase activates multiple receptor tyrosine kinase signaling pathways. Cell 76: 875–888

Burgering BMTh, Pronk GJ, van Weeren PC, Chardin P, Bos JL (1993) cAMP antagonizes $p21^{ras}$-directed activation of extracellular signal-regulated kinase 2 and phosphorylation of mSos nucleotide exchange factor. EMBO J 12: 4211–4220

Chen R-H, Chung J, Blenis J (1991) Regulation of pp90^{rsk} phosphorylation and S6 phosphotransferase activity in Swiss 3T3 cells by growth factor-, phorbol ester-, and cyclic AMP-mediated signal transduction. Mol Cell Biol 11: 1861–1867

Chen R-H, Sarnecki C, Blenis J (1992) Nuclear localization and regulation of *erk*- and *rsk*-encoded protein kinases. Mol Cell Biol 12: 915–927

Cobb MH, Boulton TG, Robbins DJ (1991) Extracellular signal-regulated kinases: ERKs in progress. Cell Regulation 2: 965–978

Cook SJ, McCormick F (1993) Inhibition by cAMP of Ras-dependent activation of Raf. Science 262: 1069–1072

Crews C, Alessandrini A, Erikson R (1992) The primary structure of MEK, a protein kinase that phosphorylates the ERK gene product. Science 258: 478–480

Dent P, Haser W, Haystead TAJ, Vincent LA, Roberts TM, Sturgill TW (1992) Activation of mitogen-activated protein kinase kinase by v-Raf in NIH 3T3 cells and in vitro. Science 257: 1404–1407

Dérijard B, Hibi M, Wu I-H, Barrett T, Su B, Deng T, Karin M, Davis RJ (1994) JNK1: A protein kinase stimulated by UV light and Ha-Ras that binds and phosphorylates the c-Jun activation domain. Cell 76: 1025–1037

Drechsel DN, Hyman AA, Cobb MH, Kirschner MW (1992) Modulation of the dynamic instability of tubulin assembly by the microtubule-associated protein tau. Mol Biol Cell 3: 1141–1154

Frost JA, Geppert TD, Cobb MH, Feramisco JR (1994) A requirement for extracellular signal-regulated king ERK function in the activation of AP-1 by H-ras, phorbol 12-myristate 13-acetate and serum. Proc Natl Acad Sci USA 91: 3844–3848

Frödin M, Peraldi P, vanObberghen E (1994) Cyclic AMP activates the mitogen-activated protein kinase cascade in PC12 cells. J Biol Chem 269: 6207–6214

Gonzalez FA, Raden DL, Rigby MR, Davis RJ (1992) Heterogeneous expression of four MAP kinase isoforms in human tissues. FEBS Lett 304: 170–178

Gonzalez FA, Seth A, Raden DL, Bowman DS, Fay FS, Davis RJ (1993) Serum-induced translocation of mitogen-activated protein kinase to the cell surface ruffline membrane and the nucleus. J Cell Biol 122: 1089–1101

Gotoh Y, Nishida E, Matsuda S, Shiina N, Kosako H, Shiokawa K, Akiyama T, Ohta K, Sakai H (1991) In vitro effects on Microtubule dynamics of purified *Xenopus* M phase-activated MAP kinase. Nature 349: 251–254

Graves LM, Bornfeldt KE, Raines EW, Potts BC, MacDonald SG, Ross R, Krebs EG (1993) Protein kinase A antagonizes platelet-derived growth factor-induced signaling by mitogen-activated protein kinase in human arterial smooth muscle cells. Proc Natl Acad Sci USA 90: 10300–10304

Haystead TAJ, Dent P, Wu J, Haystead CMM, Sturgill TW (1992) Ordered phosphorylation of p42^{mapk} by MAP kinase Kinase. FEBS Lett 306: 17–22

Knighton DR, Zheng J, Ten Eyck LF, Ashford VA, Xuong N-H, Taylor SS, Sowadski JM (1991) Crystal structure of the catalytic subunit of cyclic adenosine monophosphate-dependent protein kinase. Science 253: 407–413

Kosako H, Gotoh Y, Matsuda S, Ishikawa M, Nishida E (1992) *Xenopus* MAP kinase activator is a serine/threonine/tyrosine kinase activated by threonine phosphorylation. EMBO J 11: 2903–2908

Kyriakis JM, Brautigan DL, Ingebritsen TS, Avruch J (1991) pp54 Microtubule-associated protein-2 kinase requires both tyrosine and serine/threonine phosphorylation for activity. J Biol Chem 266: 10043–10046

L'Allemain GL, Her J-H, Wu J, Sturgill TW, Weber MJ (1992) Growth factor-induced activation of a kinase activity which causes regulatory phosphorylation of p42/microtubule-associated protein kinase. Mol Cell Biol 12: 2222–2229

Lange-Carter CA, Pleiman CM, Gardner AM, Blumer KJ, Johnson GL (1993) A divergence in the MAP kinase regulatory network defined by MEK kinase and Raf. Science 260: 315–319

Lee H, Ghose-Dastidar J, Winawer S, Friedman E (1993) Signal transduction through extracellular signal-regulated kinase-like pp57 blocked in differentiated cells having low protein kinase Cβ activity. J Biol Chem 268: 5255–5263

Lenormand P, Sardet C, Pages G, L'Allemain G, Brunet A, Pouysségur J (1993) Growth factors induce nuclear translocation of MAP kinases ($p42^{mapk}$) but not of their activator MAP kinase kinase ($p45^{mapkk}$) in fibroblasts. J Cell Biol 122: 1079–1088

Li N, Batzer A, Daly R, Yajnik V, Skolnik E, Chardin P, Bar-Sagi D, Margolis B, Schlessinger J (1993) Guanine-nucleotide-releasing factor hSos1 binds to Grb2 and links receptor tyrosine kinases to Ras signalling. Nature 363: 85–88

Loeb DM, Tsao H, Cobb MH, Greene LA (1992) Nerve growth factor induces association between extracellular signal-regulated kinase 1 (ERK1), but not ERK2, and the NGF receptor, $gp140^{prototrk}$. Neuron 9: 1053–1065

Margolis B, Silvennoinen O, Comoglio F, Roonprapunt C, Skolnik E, Ulrich A, Schlessinger J (1992) High-efficiency expression/cloning of epidermal growth factor-receptor-binding proteins with Src homology 2 domains. Proc Natl Acad Sci USA 89: 8894–8898

Moodie SA, Willumsen BM, Weber MJ, Wolfman A (1993) Complexes of Ras-GTP with Raf-1 and mitogen-activated protein kinase kinase. Science 260: 1658–1660

Nakielny S, Cohen P, Wu J, Sturgill T (1992) MAP kinase activator from insulin-stimulated skeletal muscle is a protein threonine/tyrosine kinase. EMBO J 11: 2123–2130

Otsu M, Terada Y, Okayama H (1993) Isolation of two members of the Rat MAP kinase kinase gene family. FEBS Lett 320: 246–250

Payne DM, Rossomando AJ, Martino P, Erickson AK, Her J-H, Shananowitz J, Hunt DF, Weber MJ, Sturgill TW (1991) Identification of the regulatory phosphorylation sites in pp42/mitogen-activated protein kinase (MAP kinase). EMBO J 10: 885–892

Raines MA, Kolesnik RN, Golde DW (1993) Sphingomyelinase and ceramide activate mitogen-activated protein kinase in Myeloid HL-60 cells. J Biol Chem 268: 14572–14575

Ray LB, Sturgill TW (1987) Rapid stimulation by insulin of a serine/threonine kinase in 3T3-L1 adipocytes that phosphorylates microtubule-associated protein 2 in vitro. Proc Natl Acad Sci USA 84: 1502–1506

Ray LB, Sturgill TW (1988) Characterization of insulin-stimulated microtubule-associated protein kinase. Rapid isolation and stabilization of a novel serine/threonine kinase from 3T3-L1 cells. J Biol Chem 263: 12721–12727

Robbins DJ, Cobb MH (1992) ERK2 autophosphorylates on a subset of peptides phosphorylated in intact cells in response to insulin and nerve growth factor: analysis by peptide mapping. Mol Biol Cell 3: 299–308

Robbins DJ, Zhen E, Okami H, Vanderbilt C, Ebert D, Geppert TD, Cobb MH (1993) Regulation and properties of extracellular signal-regulated protein kinases 1 and 2 in vitro. J Biol chem 268: 5097–5106

Rozakis-Adcock M, McGlade J, Mbamalu G, Pelicci G, Daly R, Li W, Batzer A, Thomas S, Brugge J, Pelicci PG, Schlessinger J, Pawson T (1992) Association of the Sch and Grb2/Sem5 SH2-containing proteins is implicated in activation of the Ras pathway by tyrosine kinases. Nature 360: 689–692

Seger R, Ahn NG, Boulton TG, Yancopoulos GD, Panayotatos N, Radziejewska E, Ericsson L, Bratlien RL, Cobb MH, Krebs EG (1991) Microtubule-associated protein 2 kinases, ERK1 and ERK2, undergo autophosphorylation on both tyrosine and threonine residues: implications for their mechanism of activation. Proc Natl Acad Sci USA 88: 6142–6146

Seger R, Ahn NG, Posada J, Munar ES, Jensen AM, Cooper JA, Cobb MH, Krebs EG (1992a) Purification and characterization of mitogen-activated protein kinase activator(s) from epidermal growth factor-stimulated A431 cells. J Biol Chem 267: 14373–14381

Seger R, Seger D, Lozeman FJ, Ahn NG, Graves LM, Campbell JS, Ericsson L, Harrylock M, Jensen AM, Krebs EG (1992b) Human T-cell mitogen-activated protein kinase kinases are related to yeast signal transduction kinases. J Biol Chem 267: 25628–25631

Sontag E, Federov S, Robbins D, Cobb M, Mumby M (1993) The interaction of SV40 Small tumor antigen with protein phosphatase 2A stimulates the MAP kinase pathway and induces cell proliferation. Cell 75: 887–897

Stratton KR, Worley PF, Litz JS, Parsons SJ, Huganir RL, Baraban JM (1991) Electroconvulsive treatment induces a rapid and transient increase in tyrosine phosphorylation of a 40-kilodalton protein associated with microtubule-associated protein 2 kinase activity. J Neurochem 56: 147–152

Sun H, Charles CH, Lau LF, Tonks NK (1993) MKP-1 (3CH134), an immediate early gene product, is a dual specificity phosphatase that dephosphorylates MAP kinase in vivo. Cell 75: 487–493

Traverse S, Gomez N, Paterson H, Marshall C, Cohen P (1992) Sustained activation of the mitogen-activated protein (MAP) kinase cascade may be required for differentiation of PC12 cells. Comparison of the effects of nerve growth factor and epidermal growth factor. Biochem J 288: 351–355

Trojanowski JQ, Mawal-Dewan M, Scmidt ML, Martin J, Lee VM-Y (1993) Localization of the mitogen activated protein kinase ERK2 in Alzheimer's Disease neurofibrillary tangles and senile plaque neurites. Brain Res 618: 333–337

Vojtek AB, Hollenberg SM, Cooper JA (1993) Mammalian Ras interacts directly with the serine/threonine kinase Raf. Cell 74: 205–214

Volonté C, Angelastro JM, Greene LA (1993) Association of protein kinases ERK1 and ERK2 with p75 nerve growth factor receptors. J Biol Chem 268: 21410–21415

Warne PH, Viciana PR, Downward J (1993) Direct interaction of Ras and the amino-terminal region of Raf-1 in vitro. Nature 364: 353–355

White MF, Kahn CR (1994) The insulin signaling system. J Biol Chem 269: 1–4

Williams R, Sanghera J, Wu F, Carbonaro-Hall D, Campbell DL, Warburton D, Pelech S, Hall F (1993) Identification of a human epidermal growth factor receptor-associated protein kinase as a new member of the mitogen-activated protein kinase/extracellular signal-regulated protein kinase family. J Biol Chem 268: 18213–18217

Wu J, Dent P, Jelinek T, Wolfman A, Weber MJ, Sturgill TW (1993a) Inhibition of the EGF-activated MAP kinase signaling pathway by adenosine 3′,5′-monophosphate. Science 262: 1065–1068

Wu J, Harrison JK, Dent P, Lynch KR, Weber MJ, Sturgill TW (1993b) Identification and characterization of a new mammalian mitogen-activated protein kinase kinase, MKK2. Mol Cell Biol 13: 4539–4548

Zhang F, Strand A, Robbins D, Cobb MH, Goldsmith EJ (1994) Atomic structure of the MAP kinase ERK2 at 2.3 Å Resolution. Nature 367: 704–710

Zhang X, Settleman J, Kyriakis JM, Takeuchi-Suzuki E, Elledge SJ, Marshall MS, Bruder JT, Rapp UR, Avruch J (1993) Normal and oncogenic $p21^{ras}$ proteins bind to the amino-terminal regulatory domain of c-Raf-1. Nature 364: 308–313

Zheng C-F, Guan K-L (1993a) Cloning and characterization of two distinct human extracellular signal-regulated kinase activator kinases, MEK1 and MEK2. J Biol Chem 268: 11435–11439

Zheng C-F, Guan K-L (1993b) Dephosphorylation and inactivation of the mitogen-activated protein kinase by a mitogen-induced thr/tyr protein phosphatase. J Biol Chem 268: 16116–16119

Calcineurin as a Pivotal Ca^{2+}-Sensitive Switching Element in Biological Responses: Implications for the Regulation of Tau Phosphorylation in Alzheimer's Disease

*R.L. Kincaid**

Summary

Reversible protein phosphorylation provides a key strategy for biological control, as it allows the informational status of a cell to be modified dynamically *en route* to changes in cellular response. A hallmark feature of Alzheimer's disease is the presence of "hyperphosphorylated" forms of the microtubule-associated protein, tau, in paired helical filaments. Although the relationship of this biochemical abnormality to disease etiology and pathology is unknown, it may reflect aberrant cytoskeletal regulation. One hypothesis to explain tau hyperphosphorylation is a dysregulation of phosphoprotein phosphatases. Consistent with this idea, incubation of human brain tissue slices with the phosphatase inhibitor okadaic acids leads to production of the characteristic tau epitope (Harris et al. 1993). Studies in vitro have shown that each of the three major serine-threonine phosphatases inhibited by okadaic acid (PP-1, PP-2A, PP-2B) can dephosphorylate tau, with PP-2A having the highest activity. Recently, studies carried out with cerebellar macroneuron model cultures showed that the protein phosphatase, calcineurin (PP-2B), is intimately associated with developing microtubule/microfilament structures; thus, calcineurin may have "privileged access" to substrates that modulate the cytoskeleton. Provocatively in this study, specific calcineurin inhibitors blocked elements of neuronal "decision-making" (axonal determination) while, at the same time, causing tau hyperphosphorylation (Ferreira et al. 1993). These data suggest that this Ca^{2+}-sensitive phosphatase may play a role in regulating tau function in vivo.

Calcineurin is uniquely suited to influence signal transduction events because it is under the direct control of the second messenger, Ca^{2+}. However, because it shows rather narrow substrate specificity in vitro, it seems likely that it acts on relatively few target proteins which may affect broader signaling cascades. This appears to be true in skeletal muscle, where calcineurin mediates a reversal of epinephrine-induced glycogen breakdown by dephosphorylating an inhibitor of the broad specificity protein phos-

---

* Department of Cell Biology, Human Genome Sciences, Inc., 9620 Medical Center Drive, Rockville, MD 20850, USA

K.S. Kosik et al. (Eds.)
Alzheimer's Disease: Lessons from Cell Biology

phatase (PP-1). A related scenario may occur in T cell lymphokine responses, where multiple *trans*-acting factors are activated through events controlled by calcineurin, although the details of these events are not clear. Similarly, the regulation of tau dephosphorylation described above may also be indirect, operating via pathways that are tightly linked to calcineurin.

Introduction

Reversible protein phosphorylation provides a useful biochemical paradigm for controlling cellular responses because it takes into account changes in the activities of protein kinases and protein phosphatases that are modulated by external stimuli. Such information can be integrated in a hierarchical fashion, controlling events that are important for different levels of cellular function, e.g., energy utilization, cytoskeletal dynamics and gene transcription (Fig. 1). Key to the coordination of such signaling events is the interplay with second messenger systems, such as those mediated by cyclic nucleotides and Ca^{2+}. Given the reliance of protein phosphorylation on second messengers, it seems important to characterize the obvious positions of intersection that might help direct the flow of information via protein phosphate. For this reason, we have focused our attention on the role of the

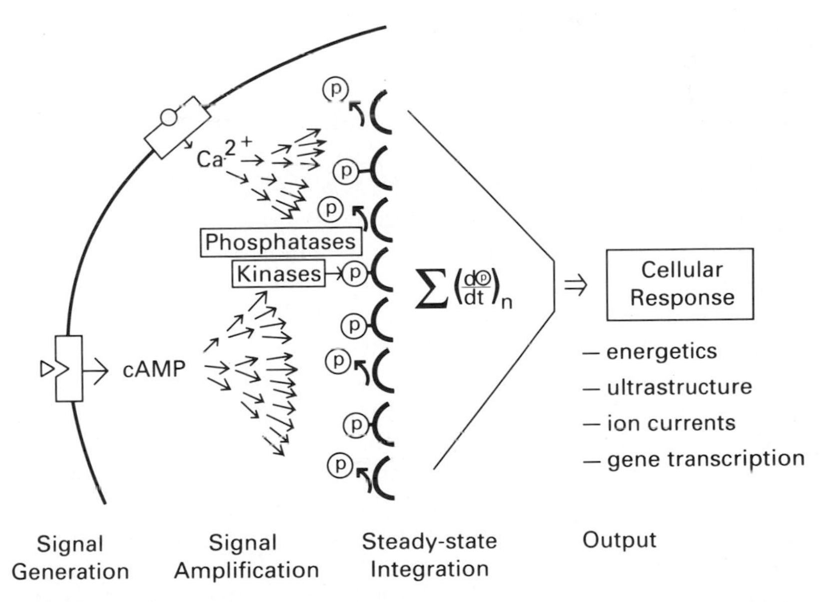

Fig. 1. Second messengers and phosphoproteins in signal transduction. A schematic portrayal of the various stages of cellular signaling events, mediated through cAMP- and Ca^{2+}-regulated phosphorylation/dephosphorlyation, that culminate in a biological response. Adapted with permission from Kincaid (1993)

Ca^{2+}- and calmodulin-regulated serine/threonine phosphatase ("calcineurin," "protein phosphatase-2B").

In skeletal muscle, calcineurin appears to mediate the reversal of epinephrine-induced glycogen utilization by inactivating a cAMP-regulated inhibitory protein that controls the broad-specificity protein phosphatase-1. In the immune system, this enzyme has emerged as the pivotal regulator of the interleukin-2 gene transcription, and thus of T-lymphocyte proliferation (Kincaid and O'Keefe 1993). Although superficially this may suggest disparate signaling roles for this Ca^{2+}-dependent protein phosphatase, there may be a common philosophy in its actions. It may be that, rather than directly control the phosphoprotein substrates responsible for the biological end point, glycogen regulation or lymphokine production, this enzyme may activate key components of signaling cascades needed to initiate and support such actions.

This chapter will attempt to relate this putative role as a "switching element" to the control of microtubule assembly via the microtubule-associated protein (MAP) tau. This seems relevent to the area of Alzheimer's research, as an alteration in tau protein phosphorylation is a major hallmark of the diseased tissue. A brief background will be provided, both for the phosphatase and for tau phosphorylation, to provide a context for the discussion of this signaling hypothesis. We will then discuss how this phosphatase might regulate the dynamics of cytoskeletal assembly and, by inference, how it may mediate Ca^{2+}-dependent effects on neuronal plasticity and development that are important to maintenance of healthy tissue.

The Calmodulin-dependent Protein Phosphatase, Calcineurin

Calcineurin was originally identified as a heat-labile protein that could attentuate the calmodulin-mediated activation of cyclic nucleotide phosphodiesterase and adenylate cyclase (Wallace et al. 1978; Klee and Krinks 1978). It was shown to be the predominant calmodulin-binding protein in the brain ($\approx$0.5% of the soluble protein) and to have high affinity Ca^{2+}-binding activity, hence the name calcineurin (Klee et al. 1979). Because of its abundance and its ability to bind both Ca^{2+} and calmodulin, it was suggested that calcineurin might play a role in sequestering Ca^{2+} (Klee and Haiech 1980), in addition to inhibiting other calmodulin-dependent enzyme activities. Shortly thereafter, the protein was shown to be a Ca^{2+}-sensitive protein phosphatase (Stewart et al. 1982), the first example of a phosphatase that is directly controlled by a second messenger. Numerous studies followed that defined its specificities toward physiological and non-physiological phosphoprotein substrates (Ingebritsen and Cohen 1983; King et al. 1984). Convincing evidence for a biological role first came from work carried out in Philip Cohen's laboratory, where a primary role for calcineurin was suggested in terminating the breakdown of glycogen in skeletal muscle after epinephrine activation of adenylate cyclase (Ingebritsen et al. 1983). Significantly, this finding also provided the first suggestion that this enzyme might function to counteract events that are initiated by cyclic nucleotide kinases.

Calcineurin is a heterodimeric protein, consisting of a 60 kDa catalytic ("A") subunit and a 19 kDa regulatory ("B") subunit that binds Ca^{2+} (Fig. 2). The catalytic subunit is composed of several distinct domains that account for its phosphatase activity and confer allosteric control. The "catalytic core" region of ≈250 amino acids near the amino terminus is related to protein phosphatases 1 and 2A, and the ability of the A subunit to interact with the B subunit resides in a region of ≈40 residues on the carboxyl side of the catalytic core. Studies have shown that the mammalian enzyme has negligible catalytic activity unless reconstituted with both the regulatory subunit and calmodulin (Merat et al. 1985). Since the heterodimer's activity cannot be stimulated greatly by the addition of Ca^{2+} alone, the B subunit is unlikely to directly mediate activation in vivo.

Two allosteric control regions are found in the carboxyl one-third of the A subunit. One of these, an amphiphilic helix of ≈20 amino acids, is the site of interaction with calmodulin (Kincaid et al. 1988). A second modulatory domain near the carboxyl terminus is termed the "autoinhibitory domain." This region of 20–24 amino acids is capable of inhibiting the activity of the enzyme in a noncompetitive fashion (Hashimoto et al. 1990). Therefore, in contrast to a regulatory mechanism that mimics substrate binding (i.e., "pseudosubstrate inhibition"), this small inhibitory domain may interact with another region of the catalytic subunit to block phosphatase activity. The site(s) of substrate binding and catalysis has not yet been defined, although the protein contains intrinsically bound iron and zinc (King and Huang 1984), which may participate in the hydrolysis of the phosphate ester. In vitro, the enzyme can be stimulated markedly by divalent cations

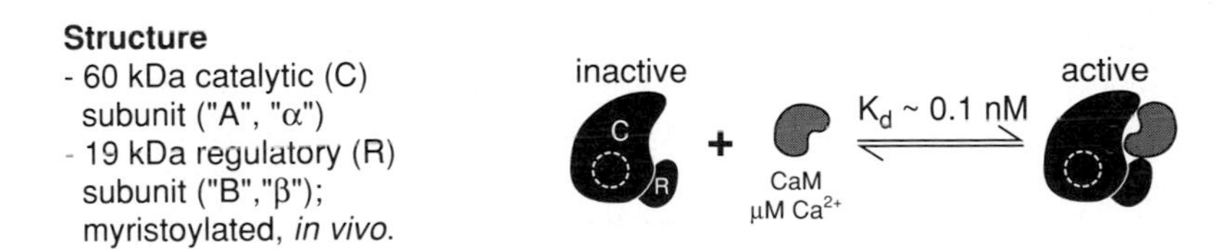

Properties

Dependent on the second messenger, Ca^{2+} unless proteolytically activated. (removal of COOH inhibitory region)
- Narrow substrate specificity; prefers substrates of cAMP kinases. (may counteract cAMP cascades)
- Fe^{2+}, Zn^{2+} metalloprotein; binuclear metal center involved in catalysis.

Fig. 2. Structural and enzymological properties of the calmodulin-dependent protein phosphatase, calcineurin (PP-2B)

such as Ni^{2+}, Mn^{2+} and Co^{2+}, which do not act synergistically with calmodulin, suggesting two independent modes by which the phosphatase can be stimulated. Treatment of calcineurin with proteases (e.g., trypsin, chymotrypsin, clostripain) generates a constitutively activated catalyst that no longer requires Ca^{2+} and calmodulin (Tallant and Cheung 1984). Cleavage of even 4 kDa from the carboxyl terminus is sufficient, presumably acting by destroying the autoinhibitory domain (Hubbard and Klee 1989).

A property of likely physiological significance is the narrow substrate specificity of calcineurin, when compared to protein phosphatases 1 and 2A, that probably results from primary sequence differences in the catalytic domain. The consequences of such selectivity for specific phosphoprotein substrates is important, as it should limit dephosphorylation in response to Ca^{2+} to a minor subset of proteins. In this manner, Ca^{2+} may act as a true "switch," effectively altering the functions of a small group of phosphoproteins, which themselves may control or influence other signal transduction cascades. Significantly, several preferred substrates of calcineurin are components of cAMP-dependent protein kinase cascades, such as Inhibitor-1, DARPP-32 and the type 2 regulatory subunit of cAMP-dependent protein kinase.

Several forms of calcineurin are present in brain and other tissues and, thus far, three mammalian genes for the catalytic subunit have been characterized, each of which can undergo alternative splicing to yield additional variants (Kincaid et al. 1990; Guerini and Klee 1989; Muramatsu et al. 1992). Broadly speaking, these can be divided into the "neural" forms, A_α and A_β, and a "non-neural" form, A_γ that is highly enriched in testis (Muramatsu et al. 1992). Of these, A_α is the most abundant form, its concentration in brain being roughly three times that of A_β (Rathna Giri et al. 1992). Such distinct isoforms may impart increased specialization of function or, conversely, may reflect a necessary redundancy. In the brain, this enzyme is ubiquitously distributed in neurons, suggestive of a fundamental role in maintaining nerve cell viability and function. Early ultrastructural studies suggested that this enzyme is localized in post-synaptic densities (Wood et al. 1980), where it may play a role in regulating various post-receptor phosphorylation events. Its expression during development is correlated with periods of synaptogenesis (Polli et al. 1991) and, in adult brain, it is concentrated in the hippocampus and striatum (Matsui et al. 1987), areas of considerable neuronal plasticity. Thus it seems plausible that the enzyme may modulate neural development.

Tau Phosphorylation and Alzheimer's Disease

Microtubules are self-assembling structures composed of tubulin, in association with several MAPs that affect the stability and properties of these biopolymers. One of the better-studied MAPs is the heat-stable 60 kDa protein, tau, the expression of which is primarily neuronal (Goedert et al. 1991), implying an important role for this protein in microtubule regulation in nerve cells. Multiple forms of tau are produced by alternative splicing and these may be important for cell-specific regulation of microtubules (Lee et al. 1988). Studies of neuronal development have indicated that microtubule assembly is a highly dynamic and regulated process (Tanaka and Kirschner 1991; Bamburg et al. 1986), providing a mechanism to control the extent and polarity of neurite growth. It seems probable that different pools of microtubules exist in all cells whose stability and association with other organellar structures may be determined in part by their binding to MAPs such as tau. Importantly, studies of neuronal differentiation have implicated tau in the events necessary for the elongation of neurites into axons. In cerebellar macroneuron studies, anti-sense oligonucleotides to tau suppressed the selective extension of a minor neurite that is necessary to form axons (Caceres and Kosik 1990). Furthermore, although this MAP was detectable in all neurites prior to axon commitment, tau remained stably associated only with axonal microtubules after detergent extraction of cells (Ferreira et al. 1989). This suggests that active elongation of microtubules requires a selectively modified form of tau protein, or its association with other components.

One process that clearly influences tau binding to microtubules is its state of phosphorylation, which provides a physiological mechanism to regulate microtubule assembly (Lindwall and Cole 1984). Tau can be phosphorylated by numerous protein kinases, including CaM-kinase II and proline-directed kinases (Yamamoto et al. 1983; Baudier and Cole 1987; Biernat et al. 1992), and more than 10 potential phosphorylation sites may be involved. Although the sites of phosphorylation in vitro have been determined in many cases, the number and sites of phosphorylation that are biologically relevant are not known. Using purified tau that has been phosphorylated in vitro, protein phosphatases 2A and 2B (calcineurin) were shown to be effective (Goto et al. 1985; Goedert et al. 1992). In general, dephosphorylation of tau is expected to increase its binding affinity, although it may be that only specific sites are responsible for regulating physiological association with microtubules.

The state of tau phosphorylation is known to be dramatically altered in tissue from Alzheimer's brains, compared to the brains of normal patients (Grundke-Iqbal et al. 1986). The characteristic neurofibrillary tangles observed in senile plaques contain unusual helical structures (paired helical filaments) that are composed of immunologically distinct forms of tau (Kosik et al. 1988). One monoclonal antibody, Alz-50, that distinguishes an antigen of 68 kDa present in the brains of Alzheimer's patients (Wolozin and Davies 1987) is, in fact, specific for the highly phosphorylated species of tau present in these helical filaments. It thus appears that a pathological hallmark of the disease state is the production (or retention) of such "hyperphosphorylated" species of this MAP (Fig. 3).

Observation:

Paired helical filaments seen in neurofibrillary tangles from AD patients contain hyperphosphorylated Tau protein

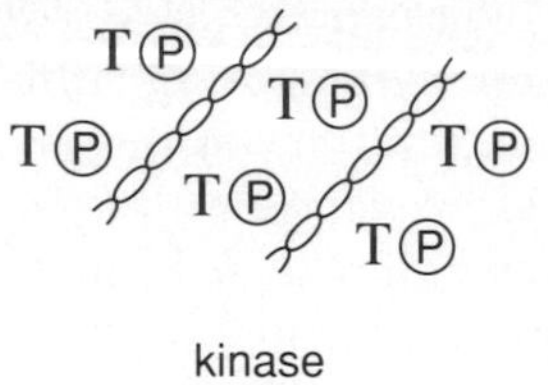

Hypothesis:

Increased Tau phosphorylation in post-mortem tissue may reflect:

- Increased and/or novel kinase activities
- Decreased protein phosphatase activities
- Aberrant regulation or expression of protein phosphatases

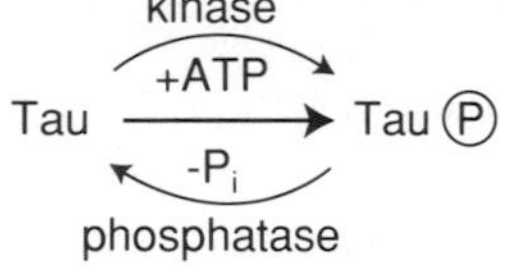

Experimental Tests:

Define phosphatases that may regulate Tau Ⓟ by using pharmacologic inhibitors in tissue extracts.

Demonstrate physiological consequence of phosphatase inhibition in model systems for cytoskeletal regulation

PP-1
PP-2A
PP-2B
inhibitors
?
Tau Ⓟ ↑

Fig. 3. Summary of hypothesis that tau hyperphosphorylation in Alzheimer's disease may reflect a defect in, or dysregulation of, serine/threonine protein phosphatases (PP). Phosphorylated protein residues are indicated by a capital P within a circle; Pi indicates inorganic phosphate

It is not known if the abnormal phosphorylation of tau plays a role in the pathogenesis of the disease or, rather, is an epiphenomenon that is associated with a later stage of cellular tissue dysfunction. Nonetheless, the detection of highly phosphorylated forms of tau in post-mortem tissues indicates that a significant change has occurred in the steady-state phosphorylation equilibrium for this protein. Such a change might suggest new or unregulated forms of protein kinase, potentially acting on a site that normally is not phosphorylated. In this regard, it is possible that unusual (polymorphic) isoforms of tau are present in Alzheimer's patients, although to date there are no data to support such an hypothesis. Alternatively, the accumulation of hyperphosphorylated tau may suggest reduced amounts of crucial protein phosphatases, or suggest that they are regulated in an aberrant manner, e.g., they are blocked by the presence of endogenous inhibitors or lack requisite regulatory subunits needed to carry out their function (Fig. 3). Because the sites of tau phosphorylation that have been described are on serine and threonine residues, it seemed plausible that inhibition of protein phosphatases-1, 2A or 2B might be involved in such a scheme.

To test the validity of such a hypothesis, two complementary approaches seem warranted. First, the state of tau phosphorylation should be assessed using pharmacologic agents that block these phosphatases; production of Alzheimer's-like forms would implicate these enzymes as likely control elements. Ideally, such inhibitor studies should be done in fresh human tissue rather than post-mortem brain to minimize concerns about altered activity or compartmentalization of key enzymes (kinases and phosphatases). Second, if hyperphosphorylated tau were to result from such in vitro treatments, it would be important to correlate phosphatase inhibition with a signficant physiological consequence. Two collaborative studies, carried out in the laboratories of Melvin Billingsley (Harris et al. 1993) and Kenneth Kosik (Ferreira et al. 1993), have sought to address these issues and, indeed, suggest that the Ca^{2+}-regulated phosphatase, calcineurin, may be important in controlling the state of tau phosphorylation and its biological function. These studies are summarized below.

Okadaic Acid Treatment of Human Brain Slices Induces Tau Hyperphosphorylation

Small amounts of brain tissue were taken from human temporal cortex during surgery for intractable epilepsy in 12 patients. This procedure made possible the preparation of fresh tissue slices, which were then incubated with varying concentrations of the phosphatase inhibitor, okadaic acid. Western blot analysis of homogenates from these samples using the 5E2 antibody against tau showed decreased electrophoretic mobility, and analysis with Alz-50 indicated clearly that Alzheimer-like epitopes were generated by inhibiting tissue phosphatases. The concentrations of okadaic acid that were needed to generate these tau epitopes were relatively high ($5-20\,\mu M$), suggesting that the phosphatase(s) responsible for dephosphorylation were relatively insensitive to this inhibitor. Of the three major serine-threonine phosphatases (1, 2A, and 2B), protein phosphatase 2B (calcineurin) has the highest inhibitory constant ($5\,\mu M$), whereas PP-1 and PP-2A have IC_{50}s in the nanomolar range (Cohen et al. 1989). This suggested that calcineurin may be responsible for dephosphorylating the sites that are associated with the Alz-50 epitope. It should be appreciated, however, that the inhibitory effects of okadaic acid in tissue slices are difficult to interpret because of the relatively high concentrations of such phosphatases ($0.2-0.5\,\mu M$) in nervous tissue and potential problems of drug permeability in tissue slices. Both of these concerns would cause the apparent inhibition to occur at higher drug concentrations.

To test the adequacy of calcineurin to remove the Alzheimer's-like epitope, tissue samples containing hyperphosphorylated forms of tau were incubated with purified calcineurin. Interestingly, calcineurin appeared to selectively dephosphorylate the sites that are characteristic of the Alzheimer's form of tau, whereas other sites were less affected. At a minimum, these data are consistent with a role for calcineurin in mediating events that regulate the Alzheimer-specific state of tau phosphorylation, although it cannot be determined whether these effects require other constituents in the tissue extract. It might be, for example, that calcineurin *enables* the dephosphorylation of tau through an intermediate pathway which alters steady-state phosphoprotein metabolism; this could occur either through activation of other protein phosphatases (such as PP-2A) and/or by inhibition of specific protein kinases.

Calcineurin Can Influence Neuronal Morphology Through Effects on Tau Dephosphorylation

The aforementioned study using human brain tissue was significant in establishing that inhibition of serine-threonine protein phosphatases, and specifically calcineurin, could account for the Alzheimer's-like state of tau. However, by itself, this does not provide any information as to the biological involvement of calcineurin in control of tau function. Such a concern is especially relevant when one considers the question of how compartmentalization might affect the accessibility of the enzyme and its potential substrates. To address the general question of enzyme localization, we initiated immunocytochemical studies to determine the cellular compartment(s) in which calcineurin is present; this might be important in predicting a role in modulating neuronal function. Cerebellar macroneuron cultures, which have proven very useful in addressing questions of development and determination of axonal polarity, were used because they exhibit progressive stages of neuronal phenotype (Ferreira and Caceres 1989). Additionally, earlier studies from Kenneth Kosik's laboratory had established functional relationships between tau protein and the ability of mammalian and insect cells to elaborate axons (Caceres and Kosik 1990; Knops et al. 1991).

The localization of calcineurin in developing macroneurons proved to be interesting for several reasons. First, this phosphatase appears to be differentially compartmentalized at various stages of development. During the initial phases of neurite outgrowth, which are characterized by the extension of the lamellopodia, calcineurin immunostaining is observed at the very edge of the veil, consistent with a possible role in regulating these domains of dynamic cytoskeletal organization. As these regions condense to form neurites, calcineurin remains at the growing tips, regardless of the absolute morphology of the growth cone structure. Throughout these early phases, immunoreactivity is also seen in the cell body, but very little is observed in the neurite shaft, suggesting a purposeful segregation of the enzyme to certain compartments. As the neurites establish clear axonal and dendritic projections, increased calcineurin staining is evident throughout these structures, suggesting that the putative factors providing for physical isolation of the enzyme are changing or being redistributed. Secondly, several observations argue for a close association between calcineurin and major cytoskeletal components. Detergent extraction of cultured cells does not markedly affect the immunolocalization of calcineurin, and pharmacologic agents that disrupt either microtubule assembly (nocodazole) or microfilaments (cytochalasin D) produce changes in the apparent cellular locale of calcineurin. Although such a co-localization might be fortuitous, it seems plausible that calcineurin is functionally important for the maintenance of interactions between these cytoskeletal systems.

Thus, this protein phosphatase is physically positioned to influence local neurite outgrowth and development. Two obvious questions that arise are how do specific inhibitors of calcineurin affect cellular morphology, and what potential cytoskeletal substrates might be involved? To test this, we decided to examine the effects of the immunosuppressant, cyclosporin A, and the autoinhibitory peptide of calcineurin, both of which selectively block the activity of this protein phosphatase (Hashimoto et al. 1990; Swanson et al. 1992). When either of these two independent types of inhibitors is incubated with the developing neuronal cultures, a rather remarkable phenotype is observed; the macroneurons fail to produce axonal projections, despite elaborating the usual array of minor neurites. These findings indicate that calcineurin is an important regulator of the decision-making machinery that commits a neurite to elongate and form an axon. Because of the important role played by tau in stablizing microtubules needed for axon development, the state of tau phosphorylation was carefully analyzed by using immunocytochemical and Western blot analyses. Cultures treated with calcineurin inhibitors display low levels of dephosphorylated tau (as evidenced by immunoreactivity toward tau-1 antibody), whereas in control macroneuron cultures, tau-1 reactive forms are present throughout. At the same time, epitopes that reflect the hyperphosphorylated species (as measured by reactivity toward AT-8 antibody) are increased greatly in the drug-treated samples.

It thus appears that calcineurin is capable of influencing the state of tau phosphorylation in vivo. More importantly, such regulation may have profound physiologic consequences to neuronal plasticity and function. Because tau hyperphosphorylation is seen in neurons that are unable to make axons, the rapid extension of microtubules seen in axons may require active (and possibly Ca^{2+}-dependent) dephosphorylation. Presumably, this regulated dephosphorylation is needed to enhance binding of tau *per se* and thereby accelerate microtubule elongation. In such a process, environmental clues may play a role in dictating the determination of the polarity by controlling local concentrations of the second messenger, Ca^{2+}.

The apparent sensitivity of tau dephosphorylation to calcineurin in these cells raises the issue of biochemical mechanism. Certainly, the most parsimonious explanation of the results would be that calcineurin acts directly on phosphorylated tau substrates. This seems reasonable especially because of the apparent cytoskeletal localization of this phosphatase, which may confer privileged access to components such as MAPs that intereact with the cytoskeleton. However, it is not necessary, *a priori*, for the regulation of tau phosphorylation to be mediated directly by calcineurin. By analogy to its mode of regulating glycogen breakdown, calcineurin may act to set in motion other broad-based amplification pathways (i.e., phosphatase activation, kinase inhibition) that rapidly convert second messenger (i.e., Ca^{2+}) information into a biological response (Fig. 4).

The hypothesis that calcineurin may affect tau dephosphorylation by an indirect means does not necessarily exclude a direct role of calcineurin on tau, but rather enables substantial amplification of response. A parallel to this possibility may exist in the biochemical mechanism underlying T cell receptor (TcR)-induced activation of lymphokine production, and in particular calcineurin's effect on the interleukin (IL)-2 promoter (O'Keefe et al. 1992; Clipstone and Crabtree 1992). The IL-2 promoter is composed of at least five sequence elements (IL-2A to IL-2E) that are required for maximal activity. Some of these sites, IL-2A, B and C, are recognized by the ubiquitously expressed transcription factors Oct-1, NF-κB and AP-1 (Singh et al. 1986; Baeuerle and Baltimore 1991; Curran and Franza 1988), whereas the IL-2E site is recognized by a factor that is T-cell specific, NF-AT (Ullman et al. 1990). Studies using artificial enhancers for these elements have clearly demonstrated that three *trans*-activation systems (OAP, NF-AT, and NF-κB) can be activated by calcineurin (Frantz et al. 1994). Because multiple enhancer elements are regulated by this phosphatase, gene activation may require a more complicated scenario than merely the dephosphorylation of primary transcription factors. Such a pleiotropic effect on transcription might involve both direct and indirect effects of the calmodulin-dependent phosphatase that serve to amplify and sustain the transcriptional signal. Given its rather narrow specificity, it is conceivable that calcineurin exerts such effects by activating (dephosphorylating) an upstream target that then ramifies to affect multiple pathways. A similar mechanism might also apply to regulation of microtubule assembly and MAP (e.g., tau) dephosphorylation.

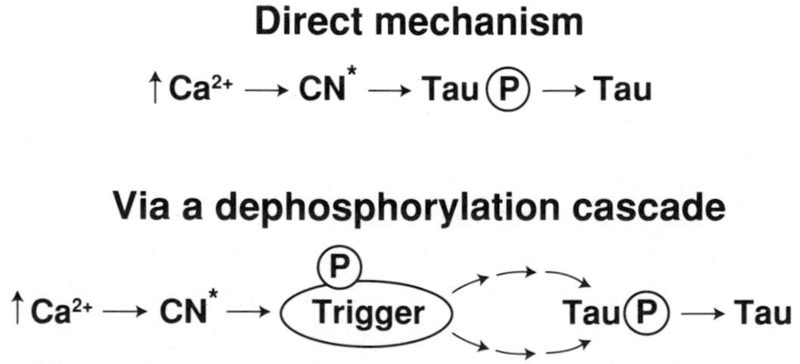

Fig. 4. A comparison of alternative biochemical mechanisms (i.e. direct or indirect) by which calcineurin may regulate the state of tau phosphorylation

Speculations on the Role of Calcineurin in Microtubule Function and Alzheimer's Disease

The observations that calcineurin is closely associated with components of the neuronal cytoskeleton, and that it may be important in controlling their states of phosphorylation, are quite provocative. Such a physical juxtaposition paradigm for regulating local assembly-disassembly equilibria would provide for tight coupling of morphological commitment to Ca^{2+} transients. By placing an enzyme at such an informational/structural interface, substantial signal amplification could be generated. If additional elements of specificity were necessary to override or modify such control, they might be exerted by inhibiting the phosphatase and/or displacing it from such sites. Although we have focused on the role such events might play in axonal determination, it is important to recognize that similar control mechanisms may be important in the "re-structuring" events that underlie cellular memory and cognition. In essence, there is likely to be a continuum of adjustments and responses to environmental information that occurs at a cytoarchitectural level, and the signaling machinery must be inherently flexible to accommodate this need.

The studies described above suggest a plausible scenario for producing the unusual phosphorylated forms of tau protein that are seen in Alzheimer's disease. At its most fundamental level, the presence of hyperphosphorylated tau forms suggests a problem in the normal mechanisms of dephosphorylation. From the perspective of those who study the biochemistry of protein phosphatases, it seems remarkable that phosphorylated sites could remain for many hours after cell death, given the large capacity of protein phosphatases to reduce the content of such modified sites. Indeed, it is usually difficult to preserve phosphorylated sites in tissues, even with the precaution of adding specific phosphatase inhibitors. One possible explanation for the resistance of such sites is that there are inhibitory components in Alzheimer's patient tissues that prevent phosphatase activation. Alternatively, the deposition of phosphorylated tau in paired helical filament structures may constitute an exclusionary compartment that is difficult for phosphatases to access.

Finally, a central question posed by these studies is whether tau hyperphosphorylation results from an underlying and causal biochemical abnormality or if it simply reflects a more generally compromised state of cellular function. It is possible that the cells accumulating these forms of tau may be damaged to such an extent that many expected biochemical processes are inoperable. However, the possibility exists that an impairment of fundamental signaling events might lead to progressive stages of pathology. Certainly, we have seen how inhibition of a key phosphatase such as calcineurin can dramatically alter cellular commitment and morphology. This is but one small lesson that cell biology has provided, but one of substantial importance to understanding the linkage of informational events to structural phenomena. Combined with future medical and genetic insights, such information may help to unravel the complex pathology of Alzheimer's disease.

Acknowledgments. I would like to thank my colleagues, Kenneth Kosik and Melvin Billingsley, in whose laboratories the calcineurin studies were conducted, for their important contributions and insights in this collaborative work. I also thank Mary Carol Gorham for her expert assistance and support in the preparaton of this manuscript.

References

Baeuerle PA, Baltimore D (1991) The physiology of the NF-κB transcription factor. In: Cohen P, Foulkes J (eds) The hormonal control of gene transcription. Amsterdam, Elsevier, pp 423–446

Bamburg JR, Bray D, Chapman K (1986) Assembly of microtubules at the tip of growing axons. Nature 321: 788–790

Baudier J, Cole RD (1987) Phosphorylation of tau proteins to a state like that in Alzheimer's brain is catalyzed by a calcium/calmodulin-dependent kinase and modulated by phospholipids. J Biol Chem 262: 17577–17583

Biernat J, Mandelkow EM, Schroter C, Lichtenberg-Kraag B, Steiner B, Berling B, Meyer H, Mercken M, Vandermeeren A, Goedert M, Mandelkow E (1992) The switch of tau protein to an Alzheimer-like state includes phosphorylation of two serine-proline motifs upstream of the microtubule binding region. EMBO J 11: 1593–1597

Caceres A, Kosik KS (1990) Inhibition of neurite polarity by tau antisense oligonucleotides in primary cerebellar neurons. Nature 343: 461–463

Clipstone NA, Crabtree GR (1992) Identification of calcineurin as a key signalling enzyme in T-lymphocyte activation. Nature 357: 695–697

Cohen P, Klumpp S, Schelling DL (1989) An improved procedure for identifying and quantitating protein phosphatases in mammalian tissues. FEBS Lett 250: 596–600

Curran T, Franza BR (1988) Fos and Jun: the AP-1 connection. Cell 55: 315–397

Ferreira A, Caceres A (1989) The expression of acetylated microtubules during axonal and dendritic growth in cerebellar macro-neurons which develop in vitro. Dev Brain Res 49: 204–213

Ferreira A, Busciglio J, Caceres A (1989) Microtubule formation and neurite growth in cerebellar macroneurons which develop in vitro: evidence for the involvement of the microtubule associated proteins, MAP-1a, HMW-MAP2, and tau. Dev Brain Res 49: 215–228

Ferreira A, Kincaid RL, Kosik K (1993) Calcineurin is associated with the cytoskeleton of cultured neurons and has a role in the acquisition of polarity. Mol Biol Cell 4: 1225–1238

Frantz B, Nordby EC, Bren G, Steffan N, Paya CV, Kincaid RL, Tocci MJ, O'Keefe SJ, O'Neill EA (1994) Calcineurin acts in synergy with PMA to inactivate IκB/MAD3, an inhibitor of NF-κB. EMBOJ 13: 861–870

Goedert M, Crowther RA, Garner CC (1991) Molecular characterization of microtubule-associated proteins tau and MAP2. Trends Neurosci 14: 193–199

Goedert M, Cohen ES, Jakes R, Cohen P (1992) p42 MAP kinase phosphorylation sites in microtubule-associated protein tau are dephosphorylated by protein phosphatase $2A_1$. FEBS Lett 312: 95–99

Goto S, Yamamoto H, Fukunaga K, Iwasa T, Matsukado Y, Miyamoto E (1985) Dephosphorylation of microtubule associated protein 2, tau factor and tubulin by calcineurin. J Neurochem 45: 276–283

Grundke-Iqbal I, Iqbal K, Tung Y-C, Quinlan M, Wisniewski H, Binder L (1986) Abnormal phosphorylation of the microtubule-associated protein tau in Alzheimer cytoskeletal pathology. Proc Natl Acad Sci USA 83: 4913–4917

Guerini D, Klee CB (1989) Cloning of human calcineurin A: Evidence for two isozymes and identification of a polyproline structural domain. Proc Natl Acad Sci USA 86: 9183–9187

Harris KA, Oyler GA, Doolittle GM, Vincent I, Lehman RAW, Kincaid RL, Billingsley ML (1993) Okadaic acid induces hyperphosphorylated forms of Tau protein in human brain slices. Ann Neurol 32: 635–645

Hashimoto Y, Perrino BA, Soderling TR (1990) Identification of an autoinhibitory domain in calcineurin. J Biol Chem 265: 1924–1927

Hubbard MJ, Klee CB (1989) Functional domain structure of calcineurin A: mapping by limited proteolysis. Biochemistry 28: 1868–1874

Ingebritsen TS, Cohen P (1983) The protein phosphatases involved in cellular regulation. I Classification and substrate specificities. Eur J Biochem 132: 255–261

Ingebritsen TS, Stewart AA, Cohen P (1983) The protein phosphatases involved in cellular regulation. 6. Measurement of type-1 and type-2 protein phosphatases in extracts of mammalian tissues: An assessment of their physiological roles. Eur J Biochem 132: 297–307

Kincaid R (1993) Calmodulin-dependent protein phosphatases from microorganisms to man: A study in structural conservatism and biological diversity. In: Shenolikar S, Nairn AC (eds) Advances in second messenger and phosphoprotein research, vol 27, pp 1–23

Kincaid RL, O'Keefe SJ (1993) Calcineurin and immunosuppression: A calmodulin-stimulated protein phosphatase acts as the "gatekeeper" to interleukin-2 gene transcription Adv. Prot. Phosphatases 7: 543–583

Kincaid RL, Nightingale MS, Martin BM (1988) Characterization of a cDNA clone encoding the calmodulin-binding domain of mouse brain calcineurin. Proc Natl Acad Sci USA 85: 8983–8987

Kincaid RL, Rathna Giri P, Higuchi S, Tamura J, Dixon SC, Marietta CA, Amorese DA Martin BM (1990) Cloning and characterization of molecular isoforms of the catalytic subunit of calcineurin using nonisotopic methods. J Biol Chem 265: 11312–11319

King MM, Huang CY (1984) The calmodulin-dependent activation and deactivation of the phosphoprotein phosphatase, calcineurin, and the effect of nucleotides, pyrophosphate and divalent metal ions. J Biol Chem 259: 8847–8856

King MM, Huang CY, Chock PB, Nairn AC, Hemmings HC, Jr, Chan K-FJ, Greengard P (1984) Mammalian brain phosphoproteins as substrates for calcineurin. J Biol Chem 259: 8080–8083

Klee CB, Krinks MH (1978) Purification of cyclic 3′,5′-nucleotide phosphodiesterase inhibitory protein by affinity chromatography on activator protein coupled to Sepharose. Biochemistry 17: 120–126

Klee CB, Haiech J (1980) Concerted role of calmodulin and calcineurin in calcium regulation. Ann NY Acad Sci 356: 43–54

Klee CB, Crouch TH, Krinks MH (1979) Calcineurin: a calcium- and calmodulin-binding protein of the nervous system. Proc Natl Acad Sci USA 79: 6270–6273

Knops J, Kosik KS, Lee G, Pardee JD, Cohen-Gould L, McConlogue L (1991) Overexpression of tau in a non-neuronal cell induces long cellular processes. J Cell Biol 114: 725–733

Kosik KS, Orecchio LD, Binder LI, Trojanowski J, Lee V, Lee G (1988) Epitopes that span the tau molecule are shared with paired helical filaments. Neuron 1: 817–825

Lee G, Cowan N, Kirschner M (1988) The primary structure and heterogeneity of tau protein from mouse brain. Science 239: 285–288

Lindwall G, Cole RD (1984) Phosphorylation affects the ability of tau protein to promote microtubule assembly. J Biol Chem 259: 5301–5305

Matsui H, Doi A, Itano T, Shimada M, Wang JH, Hatase O (1987) Immunohistochemical localization of calcineurin, the calmodulin-stimulated phosphatase, in the rat hippocampus using a monoclonal antibody. Brain Res 402: 193–196

Merat DL, Hu ZY, Carter TE, Cheung WY (1985) Bovine brain calmodulin-dependent protein phosphatase. Regulation of subunit A activity by calmodulin and subunit B. J Biol Chem 260: 11053–11059

Muramatsu T, Rathna Giri P, Higuchi S, Kincaid RL (1992) Molecular cloning of a calmodulin-dependent phosphatase from murine testis: Identification of a developmentally expressed nonneural isoenzyme. Proc Natl Acad Sci USA 89: 529–533

O'Keefe SJ, Tamura J, Kincaid RL, Tocci MJ, O'Neill EA (1992) FK-506- and CsA-sensitive activation of the interleukin-2 promoter by calcineurin. Nature 357: 692–694

Polli JW, Billingsley ML, Kincaid RL (1991) Expression of the calmodulin-dependent phosphatase, calcineurin, in rat brain: developmental patterns and the role of nigrostriatal innervation. Devel Brain Res 61: 105–119

Rathna Giri P, Marietta CA, Higuchi S, Kincaid RL (1992) Molecular and phylogenetic analysis of calmodulin-dependent protein phosphatase (calcineurin) catalytic subunit genes. DNA Cell Biol 11: 415–424

Singh H, Sen R, Baltimore D, Sharp PA (1986) A nuclear factor that binds to a conserved sequence motif in transcriptional control elements of immunoglobulin genes. Nature 319: 154–158

Stewart AA, Ingebritsen TS, Manalan A, Klee CB, Cohen P (1982) Discovery of Ca^{2+}- and calmodulin-dependent protein phosphatase. Probable identity with calcineurin ($CaM\text{-}BP_{80}$). FEBS Lett 137: 80–84

Swanson SKH, Born T, Zydowsky LD, Cho H, Chang HY, Walsh CT, Rusnak F (1992) Cyclosporin-mediated inhibition of bovine calcineurin by cyclophilins A and B. Proc Natl Acad Sci USA 89: 3741–3745

Tallant EA, Cheung WY (1984) Activation of bovine brain calmodulin-dependent protein phosphatase by limited trypsinization. Biochemistry 23: 973–979

Tanaka EM, Kirschner MW (1991) Microtubule behavior in the growth cones of living neurons during axon elongation. J Cell Biol 115: 345–363

Ullman KS, Northrop JP, Verweij CL, Crabtree GR (1990) Transmission of signals from the T lymphocyte antigen receptor to the genes responsible for cell proliferation and immune function: the missing link. Ann Rev Immunol 8: 421–452

Wallace RW, Lynch TJ, Tallant EA, Cheung WY (1978) Purification and characterization of an inhibitor protein of brain adenylate cylase and cyclic nucleotide phosphodiesterase. J Biol Chem 254: 377–382

Wolozin B, Davies P (1987) Alzheimer-related neuronal protein A68: specificity and distribution. Ann Neurol 22: 521–526

Wood JG, Wallace RW, Whitaker JN, Cheung WY (1980) Immunocytochemical localization of calmodulin and a heat-labile calmodulin-binding protein ($CaM\text{-}BP_{80}$) in basal ganglia of mouse brain. J Cell Biol 84: 66–76

Yamamoto H, Fukunaga K, Tanaka E, Miyamoto E (1983) Ca^{2+}- and calmodulin-dependent phosphorylation of microtubule-associated protein 2 and τ factor, and inhibition of microtubule assembly. J Neurochem 41: 1119–1125

Phosphorylation of Tau and Its Relationship with Alzheimer Paired Helical Filaments

E.-M. Mandelkow, J. Biernat, B. Lichtenberg-Kraag, G. Drewes, H. Wille, N. Gustke, K. Baumann,* and *E. Mandelkow*

Summary

This paper summarizes our recent studies on microtubule-associated protein tau and its pathological state resembling that of the paired helical filaments of Alzheimer's disease. The Alzheimer-like state of tau protein can be identified and analyzed in terms of certain phosphorylation sites and phosphorylation-dependent antibody epitopes. It can be induced by protein kinases which tend to phosphorylate serine or threonine residues followed by a proline; these include mitogen-activated protein kinase (MAPK), glycogen-synthase kinase-3 (GSK-3), or cyclin-dependent kinase-5 (cdk5). These kinases are tightly associated with microtubules as well as with paired helical filaments. In addition, the phosphorylation of serine 262 has a pronounced influence on the binding of tau to microtubules. All of the phosphorylation sites can be cleared by the phosphatases calcineurin and PP-2A, but not by PP-1. Structurally, tau appears as a rod-like molecule. It tends to self-associate into dimers whose monomers are antiparallel. Constructs of truncated tau made up of the microtubule binding domain can be assembled into paired helical filaments in vitro.

Introduction

Alzheimer's disease (AD) is characterized by two types of protein deposits, amyloid plaques and neurofibrillary deposits (tangles, neuropil threads). The latter are composed largely of paired helical filaments (PHF) which are in turn made up mainly of an insoluble form of the microtubule-associated protein (MAP) tau (Brion et al. 1985). The neurofibrillary deposits are particularly useful in defining the stages of AD progression because they proceed with a well-defined spatial and temporal pattern, starting with the transentorhinal region (stage 1), until eventually the frontal and temporal cortex are affected (stage 6; Braak and Braak 1991; Braak et al. 1994). Only

---

* Max-Planck-Unit for Structural Molecular Biology, c/o DESY, Notkestrasse 85, 22603 Hamburg, Germany

K.S. Kosik et al. (Eds.)
Alzheimer's Disease: Lessons from Cell Biology

the last three stages are clinically recognizable as AD. This situation explains the problem in developing a cure for AD or preventing its inception (for review, see Braak and Braak 1994).

Tau protein is thought to stabilize microtubules in axons and thus helps to maintain axonal transport, hence tau can be likened to "ties" that keep the microtubule "tracks" intact (Cleveland et al. 1977). Overexpression of tau leads to the formation of microtubule bundles and neurite extension (Kanai et al. 1992; Knops et al. 1991; Lo et al. 1993). Tau is a mixture of up to six isoforms in human brain that arise from alternative splicing and contain between 352 and 441 amino acid residues (Lee et al. 1988; Goedert et al. 1988, 1989; Himmler et al. 1989); in addition peripheral nervous tissue contains a "big tau" isoform (Couchie et al. 1992). Tau can be subdivided into several domains (Fig. 1): acidic, basic, proline-rich, repeats, pseudo-repeat, and tail. The repeats are somehow involved in microtubule binding (Ennulat et al. 1989; Butner and Kirschner 1991), although the binding depends strongly on the flanking regions as well (Kanai et al. 1992; Brandt and Lee 1993; Gustke et al. 1994).

In PHFs tau is modified in at least five ways. It is phosphorylated, aggregated, ubiquitinated, and proteolytically processed and it no longer binds to microtubules (for reviews, see Anderton 1993; Goedert 1993; Mandelkow and Mandelkow 1993). Phosphorylation, aggregation and detachment from microtubules appear to be early events in the abnormal transformation of tau and are probably linked in some fashion. Ubiquitination (Bancher et al. 1991; Morishima-Kawashima et al. 1993) and proteolysis (Kondo et al. 1988; Novak et al. 1993) are probably secondary and may reflect the cell's attempt to get rid of the precipitated protein. Here we describe recent experiments in which we have addressed the following questions: Can we identify the differences between normal and pathological

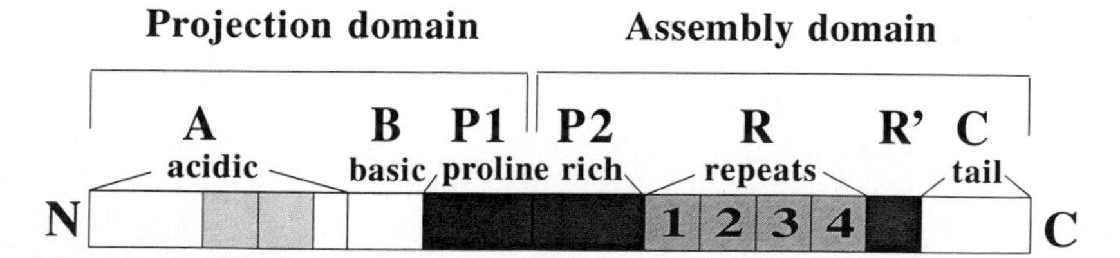

Fig. 1. Bar diagram of tau protein, isoform HTau40 (441 residues; Goedert et al. 1989). The two 29-residue inserts near the N-terminus are *lightly shaded*, the four repeats in the C-terminal half are numbered 1–4 (*medium shade*). Definition of domains (Gustke et al. 1994): projection domain (M1-Y197), does not bind to microtubules by itself; assembly domain (S198-L441), binds to microtubules, separable by chymotryptic cleavage. N-terminal domains A (acidic, M1-G120), B (basic, G120-I151), P (proline-rich and basic, I151-Q244), separated into P1 and P2 at Y197. Repeats R1-R4, Q244-N368, "5th" repeat R′ K369-S400; C, C-terminal tail G401-L441

tau? Can we generate these differences in vitro and thereby study the causes of the pathological transformation? What are the structural and biochemical consequences of the transformation? What enzymes are involved? These questions were approached using a combination of biochemical, structural, and molecular biological methods which are described in the references and will not be dealt with here.

Results and Discussion

What Is Abnormal Phosphorylation, and How Can We Recognize It?

The abnormal phosphorylation of tau in PHFs leads to a reduced electrophoretic mobility (Grundke-Iqbal et al. 1986). We searched for a protein kinase that would affect this parameter. Initially we tested several well-known kinases. For example, CaM kinase induced a clear shift by phosphorylating a single site (Ser 416 in the htau40 numbering; see Steiner et al. 1990). A small shift can also be induced by PKA, which phosphorylates Ser409 and several other sites (Scott et al. 1993; Biernat et al. 1993). However, the phosphorylated protein did not react with the PHF-specific antibodies. Other kinases such as PKC or cdc2 were even less effective with regard to electrophoretic mobility or PHF-antibodies (Steiner et al. 1990; Drewes et al. 1992).

Next we reasoned that the kinase(s) that transformed tau to an abnormal state was probably present in normal brain as well. Therefore, we prepared brain extracts and tested them for their ability to phosphorylate tau, with limited success. We then argued that the failure to obtain phosphorylation could be due either to underphosphorylation or overdephosphorylation. After all, normal brain tau is not pathologically phosphorylated, so that if there are "normal" kinases then these must be balanced by "normal" phosphatases. Thus abnormal phosphorylation could be expected only when the phosphatases were inhibited.

When using the phosphatase inhibitor okadaic acid (Bialojan and Takai 1988) we obtained a kinase activity from brain extract which was capable of conferring Alzheimer-like characteristics to tau protein. The kinase activity incorporated up to six Pi into tau, caused a mobility shift in brain tau as well as all recombinant tau isoforms, and induced an Alzheimer-like antibody reactivity with several antibodies (Biernat et al. 1992; Lichtenberg-Kraag et al. 1992). All antibodies that discriminated between PHF tau and normal tau were phosphorylation sensitive, either in a positive sense (i.e., reacting with phosphorylated epitopes of PHFs and of recombinant tau, for example AT8 and SMI31) or in a negative sense (i.e., reacting only with normal and unphosphorylated tau, for example TAU1 and SMI33; see Fig. 2). These results proved that the Alzheimer-like kinase activity was already present in normal brain and not generated by a pathological condition.

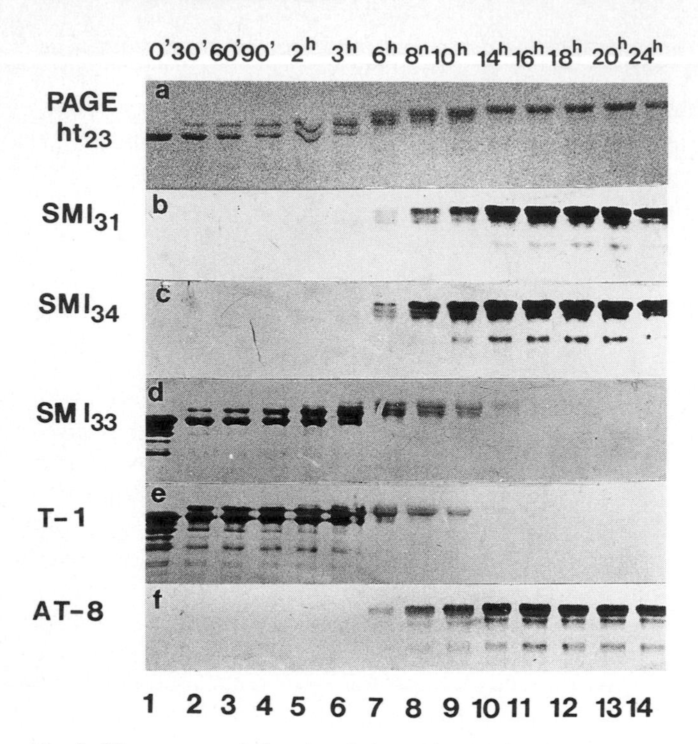

Fig. 2. Time course of phosphorylation of htau23 (ht_{23}) by kinase activity prepared from brain extract. **a** SDS-PAGE; **b-e** immunoblots with several antibodies (Lichtenberg-Kraag et al. 1992). There is a progressive shift in Mr and a change in reaction with antibodies that distinguish between normal and abnormal tau. Antibodies SMI31, SMI34, and AT-8 react with phosphorylated epitopes containing Ser-Pro motifs; antibodies SMI133 and Tau-1 react with unphosphorylated epitopes. Note that the kinase activity persists for up to 24 hours

The epitopes of the antibodies and the phosphorylation sites which controlled their reactivities were determined by a combination of phosphopeptide sequencing and site-directed mutagenesis. Most epitopes involved phosphorylatable serines followed by prolines (Fig. 3). An example is the epitope of the PHF-specific antibody AT8 (Mercken et al. 1992), which includes phosphorylated serines around residue 200 (Biernat et al. 1992; the details were a matter of debate; see Szendrei et al. 1993; Goedert et al. 1993, our recent re-investigation showed that the epitope of AT8 included phosphorylated Ser 202 and Thr 205). It turns out that this is roughly complementary to the epitope of the widely used antibody TAU1 (Binder et al. 1985; Kosik et al. 1988). We showed that its epitope is complementary to AT8 in that it requires these two serines in an unphosphorylated form (and therefore reacts with normal tau but not with PHF-tau). These two antibodies can therefore be used as positive and negative indicators of abnormal phosphorylation.

In addition to antibody epitopes, several other Ser-Pro motifs in recombinant tau have now been found to be phosphorylatable by the kinase

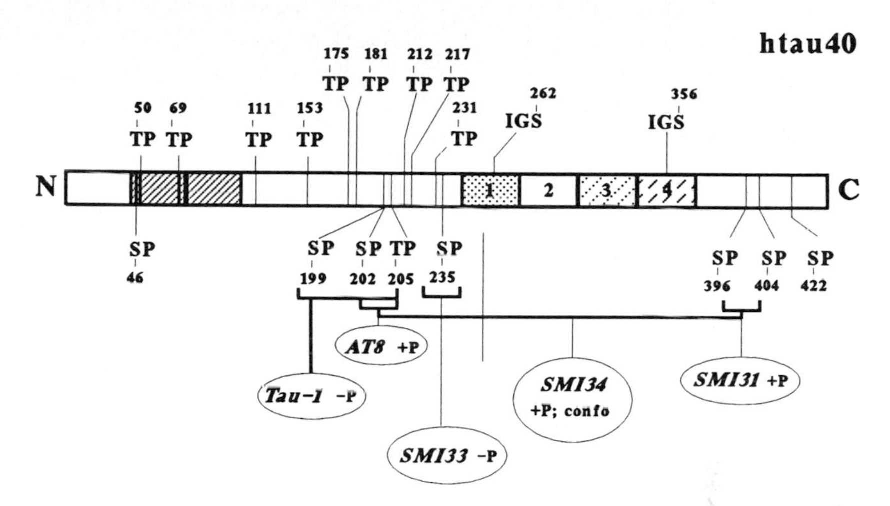

Fig. 3. Phosphorylatable sites and antibody epitopes in tau protein. Ser-Pro (SP) and Thr-Pro (TP) sites can be phosphorylated by proline-directed kinases, with MAP kinase being the most efficient, whereas the IGS motifs in the repeats are phosphorylated by a different kinase (Lichtenberg-Kraag et al. 1992; Biernat et al. 1993). Note that the Ser-Pro or Thr-Pro are clustered on both sides of the repeat domain

activity from brain, as well as several other residues (Gustke et al. 1992). Most sites are clustered in the vicinity of residue 200 or 400, i.e., flanking the region of internal repeats which bind to microtubules.

The phosphorylation causes a gel shift in several characteristic stages (Fig. 2). Starting from the original unphosphorylated state, we distinguish three main stages with several substages. The PHF-like antibody reaction is detectable from stage 2 onwards. Conversely, other antibodies reacting with normal tau lose their reactivity during stage 2. By these criteria, tau acquires its Alzheimer-like state during stage 2 of the phosphorylation (Lichtenberg-Kraag et al. 1992).

What Kinases Can Induce the Alzheimer-like Antibody Reactivity?

The results described above suggested that abnormal phosphorylation of tau occurs predominantly at Ser-Pro motifs (judging by phosphopeptide sequencing and PHF-specific antibodies), and that the corresponding kinase(s) are present in normal brain tissue. This meant that we had to search for a proline-directed kinase in brain tissue; it also explained why our earlier searches had failed (e.g., Steiner et al. 1990), since PKA, PKC, casein kinase and CaM kinase are not proline-directed kinases (see also Correas et al. 1992; Scott et al. 1993).

A number of proline-directed kinases are known (reviewed by Hunter 1991). We initially tested the cell cycle kinase p34(cdc2), which can be complexed with different regulatory subunits (cyclins), but this a low extent of phosphorylation or response with PHF-specific antibodies (Drewes et al. 1992). Our next attempt was MAP kinase. This enzyme was prepared from porcine brain and met the criteria required for abnormal phosphorylation: it induced the antibody reaction with PHF-specific antibodies, it incorporated between 12 and 15 phosphates into the tau molecule, and it phosphorylated Ser-Pro as well as Thr-Pro motifs in all recombinant isoforms tested (see Fig. 3). Moreover, when PHFs from Alzheimer brains were dephosphorylated with alkaline phosphatase, they lost their antibody reaction, but regained it when re-phosphorylated with MAP kinase.

All of these data would seem to be compatible with a model in which MAP kinase would be the main culprit for abnormal phosphorylation of tau. However, a further search shows that there are still other kinases in the brain extract. One of them is GSK-3, a kinase well known for its role in glycogen metabolism and activation of transcription factors (for review, see Woodgett 1991). When tau is phosphorylated with purified GSK-3 it shows the characteristic Mr shift in the SDS gel, and it acquires antibody reactivities similar to Alzheimer tau (similar to those shown in Fig. 2 for the brain extract kinase activity). However, the degree of phosphorylation is lower than with MAP kinase – only about 3–4 phosphates per tau molecule. The explanation is that Ser-Pro motifs that regulate the binding of PHF-specific antibodies (Fig. 3) were phosphorylated with similar kinetics and efficiency (Mandelkow et al. 1992; Hanger et al. 1992).

A search for further kinases showed that certain members of another class of kinases, the cyclin-dependent kinases, can induce abnormal phosphorylation as well. This includes the kinases cdk2 and cdk5 alias nclk (Baumann et al. 1993). The reaction with cdk2 may not be of physiological relevance because this kinase does not occur in brain tissue (Meyerson et al. 1992). However, cdk5 is abundant in brain and seems to play a role in neuronal development (Lew et al. 1992; Tsai et al. 1993; Shetty et al. 1993). In addition, other researchers have described proline-directed kinases phosphorylating tau in similar ways. Examples are a p40 neurofilament kinase, which is similar to ERK2, one of the isoforms of MAP kinase (Roder et al. 1993); proline-directed protein kinases (PDPK), which are members of the cdk family (Vulliet et al. 1992; Paudel et al. 1993); and two kinases originally termed tau kinases I and II (TK-I, TK-II), which now have been identified as GSK-3β (Ishiguro et al. 1993) and a cdk-like kinase (Hisanaga et al. 1993), respectively.

Summarizing these kinases, it seems that there are at least three types of proline-directed kinases which can transform tau into an abnormal state, represented by MAP kinase, GSK-3, and cdk5 (Fig. 4, upper left). They phosphorylate tau to different extents, but since most diagnostic antibodies recognize certain Ser-Pro motifs, the antibody reactivities are similar. It may be significant that these kinases are all involved in cellular signal transduction pathways. This suggests that an abnormal regulation of signal transduction may be important in the generation of the disease. Moreover, MAP kinase,

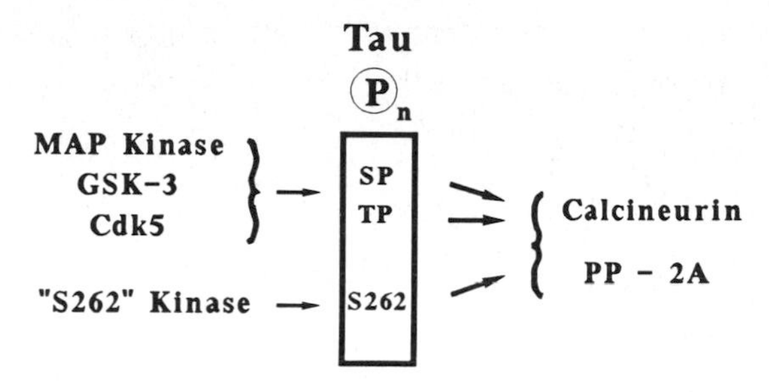

Fig. 4. Diagram of kinases and phosphatases acting on tau protein. **Middle box**, tau contains two classes of phosphorylation sites that appear to be important for AD. One class comprises the Ser-Pro (SP) and Thr-Pro motifs (TP) that determine the reactivity with diagnostic antibodies, and the second class contains (at least) Ser262 and has a strong effect on microtubule binding. *Left*, there are two corresponding classes of kinases that phosphorylate tau at these sites: MAP kinase, GSK-3, cdk5 phosphorylate Ser-Pro or Thr-Pro motifs; the Ser262 kinase phosphorylates Ser262 in the first repeat and the corresponding serines in the other repeats. *Right*, the phosphatases calcineurin and PP-2A can remove all of these phosphates on tau

GSK-3 and cdk5 appear to be physically associated with microtubules, as well as with PHFs from Alzheimer brain tissue. Thus any dysregulation of these kinases could affect tau protein directly. There is now growing evidence that some of these kinases may be unusually active during neuronal development, so that fetal tau shows similar phosphorylation characteristics as AD tau and reacts with similar diagnostic antibodies (Kanemaru et al. 1992; Watanabe et al. 1993; Bramblett et al. 1993; see also the chapter by Morishima-Kawashima and Ihara, this volume). This would lend support to the idea that, in AD, the neurons revert to a "fetal-like" state, trying to recover from some stress or toxic effect. Such effects are known to activate signal transduction pathways involving MAP kinase (see chapter by Kosik et al., this volume).

Does Phosphorylation Affect the Binding of Tau to Microtubules?

A working hypothesis shared by many in the field is that phosphorylation reduces the affinity of tau to microtubules which therefore break down so that cytoplasmic traffic becomes interrupted. There is substantial evidence that tau from PHFs is abnormally phosphorylated and not bound to microtubules (Lee et al. 1991; Bramblett et al. 1993; Yoshida and Ihara 1993; Lu and Wood 1993; Köpke et al. 1993), but it is less obvious what type of phosphorylation plays a role in the detachment of tau from microtubules.

It was therefore of interest to see if any of the kinases described so far had an effect on the binding of tau to microtubules in vitro (Biernat et al. 1993). In a typical binding experiment (Fig. 5), tau constructs are titrated against taxol-stabilized microtubules; this yields two parameters, the

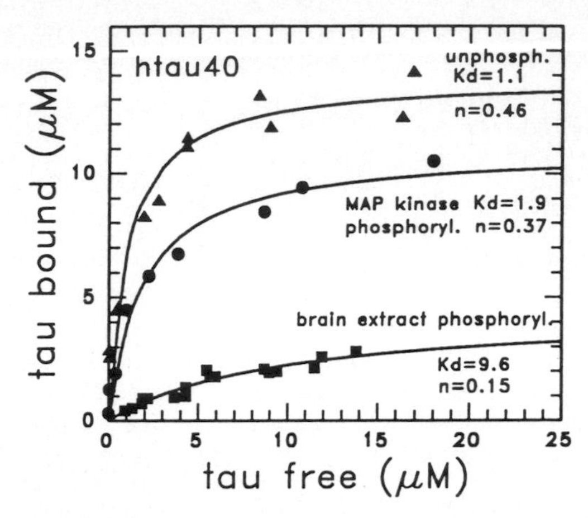

Fig. 5. Binding experiment of tau to 30 μM taxol-stabilized microtubules with or without phosphorylation by MAP kinase or the kinase activity from brain extract. Unphosphorylated tau (*top, triangles*) binds with a stoichiometry of nearly 0.46 (tau: tubulin dimer) and a dissociation constant of 1.1 μM. Phosphorylation by MAP kinase decreases the affinity $\approx$2-fold ($K_d \approx 1.9\,\mu$M) and reduces the stoichiometry to 0.33. Phosphorylation by the brain extract activity strongly reduces both the affinity ($K_d \approx 9.6\,\mu$M) and the stoichiometry (0.15). Note that although MAP kinase is much more efficient in terms of overall phosphorylation, the phosphorylation of Ser262 by the brain extract activity has a much more pronounced effect than all MAP kinase sites combined

stoichiometry and the dissociation constant where 50% of the tau is bound and 50% is free. Most tau constructs bind with a stoichiometry around n = 0.5 (one tau per two tubulin dimers) and Kd values in the μM range. Tau can be highly phosphorylated by MAP kinase, which incorporates up to 15 phosphates into tau, i.e., phosphorylates nearly all of the 17 Ser-Pro or Thr-Pro motifs in the largest human isoform. Surprisingly, however, this degree of phosphorylation has only a moderate effect on tau's binding to microbules ($\approx$20% decrease in stoichiometry). On the other hand, if one phosphorylates tau with the brain extract, the stoichiometry drops steeply down to 1/3 of the original value (to n $\approx$ 0.15) and the Kd increases several-fold, even when the degree of phosphorylation is only about 2–3 Pi per molecule. How can one explain this paradox?

As mentioned above, the brain extract contains a number of kinases. Analysis of the phosphopeptides shows that most sites phosphorylated in tau are part of Ser-Pro or Thr-Pro motifs but, in addition, there are sites that do not conform to this pattern. The most notable one is Ser262, followed by Ser356; these two serines are among the early ones to be phosphorylated by the brain extract (Gustke et al. 1992). We suspected that one of them was involved in microtubule binding. By generating a number of site-directed mutants, it was possible to show that the single site Ser262 phosphorylation had a dramatic effect on tau's affinity for microtubules, more than all other phosphorylation sites combined (Biernat et al. 1993). This explains why phosphorylation by the brain extract has the strong negative effect on microtubule binding, but not MAP kinase, GSK-3 or cdk5.

The significance of these findings can be appreciated when one considers the phosphorylation sites found directly in tau from AD brains, as determined by mass spectrometry (Hasegawa et al. 1992). Apart from several Ser-Pro or Thr-Pro motifs, there is a specific phosphorylation of Ser262. This does not occur in normal adult tau or in fetal tau, and thus appears to be characteristic of AD (Watanabe et al. 1993). The kinase responsible for this type of phosphorylation has not been identified so far, but the activity elutes as a pair of bands of Mr 35 and 41 kD (Biernat et al. 1993).

The results are summarized in the left half of Figure 4. Tau can be phosphorylated abnormally in two ways. One is phosphorylation by certain proline-dependent kinases; this conveys upon tau an altered Mr in SDS gels and an altered response with diagnostic antibodies that distinguish normal tau from PHF tau. The second way is phosphorylation at Ser262, which affects microtubule binding. Not included in this diagram are the other kinases which phosphorylate tau in vitro but show no clear changes with respect to the diagnostic antibodies or microtubule binding; these include PKA, PKC, CaM kinase, or casein kinase (Baudier and Cole 1987; Steiner et al. 1990; Correas et al. 1992; Scott et al. 1993).

What Phosphatases Protect Tau Against Abnormal Phosphorylation?

Normal neurons contain tau in a normal state of phosphorylation, and they contain kinases capable of inducing abnormal phosphorylation. These kinases must be active at least transiently as part of their role in signal transduction. So why is tau not always phosphorylated in an abnormal fashion? The likely answer is that there are phosphatases whose task it is to correct abnormal phosphorylation. Since tau is phosphorylated at Ser or Thr residues, the phosphatases must belong to the class of Ser/Thr phosphatases. They are generally classified as PP-1, PP-2a, PP-2b or PP-2c (for review see Cohen 1991); PP-2b is also called calcineurin because of its abundance in brain (reviewed by Stemmer and Klee 1991, and in a chapter by Kincaid, this volume). The phosphatases active towards tau have been identified by positive or negative approaches, involving either purified phosphatases or specific phosphatase inhibitors (Drewes et al. 1993).

The first approach is to phosphorylate tau with the brain extract in the presence of ATP (required for the kinases) and specific phosphatase inhibitors. As mentioned above, the brain extract usually has only minimal kinase activity because the kinases are overwhelmed by the phosphatases. However, when certain phosphatases are inactivated by inhibitors, the kinase reactions become visible and tau becomes abnormally phosphorylated. We found two phosphatase inhibitors to be effective: EGTA (a calcium chelator and thus an inhibitor of calcineurin) and okadaic acid, an inhibitor of PP-1 and PP-2a (Drewes et al. 1993).

A second approach is to pre-phosphorylate tau with a purified kinase and then use the brain extract (without ATP so that no kinases are active) for dephosphorylation in the presence of different phosphatase inhibitors. The advantage of this approach is that the pre-phosphorylation allows one to define the types of phosphorylation sites. We used purified MAP kinase because it phosphorylates tau to a high extent (almost every phosphorylatable motif, see above). The phosphatases present in the brain extract were capable of removing all phosphates from tau, but, as in the previous case, they could be inhibited by EGTA and okadaic acid.

As a positive control we used tau pre-phosphorylated by purified MAP kinase and dephosphorylated it with the purified phosphatases calcineurin and different forms of PP-2A or PP-1. Calcineurin (Fig. 6) and PP-2a were both capable of removing all phosphates from pre-phosphorylated tau; PP-1 was not. This finding agrees with the inhibitor studies in the first two types of experiments. (We note in passing that okadaic acid inhibits PP-2a and PP-

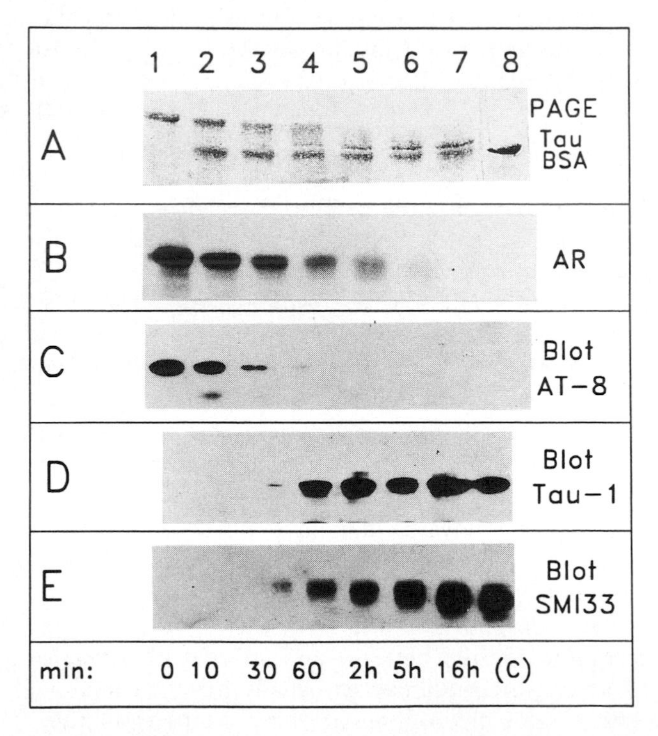

Fig. 6. Dephosphorylation of tau by calcineurin (Drewes et al. 1993). Isoform htau40 was first phosphorylated with MAP kinase using radioactive phosphate and then incubated with calcineurin (gift of C Klee, NIH). **A** The Mr shifts down; **B** autoradiography shows that phosphate disappears from the protein; **C** antibody AT-8 stops binding to the protein; **D** and **E** antibodies Tau-1 and SMI33 begin to bind. These changes are the opposite of those occurring during phosphorylation (compare to Fig. 2)

1 with different inhibition constants, ≈0.1 nM and 10 nM, respectively. Thus in principle PP-2a should be inhibitable by nM OA whereas PP-1 would require μM OA; in practice the concentration of PP-2A in cells approaches the μM range, so that several μM OA are required to saturate it. This effect can mask the intrinsic difference between the two phosphatases (for a discussion see Goris et al. 1989).

The same approaches can be applied to other phosphorylation sites. In particular we probed the phosphatases that remove phosphate from Ser262. This was again achieved both by calcineurin and PP-2a (Biernat et al. 1993; Drewes et al. 1993). These results are summarized in the right half of Figure 4. There are two classes of phosphorylation sites with different effects on tau's properties (Ser/Thr-Pro and Ser262), and there are two types of phosphatases, calcineurin and PP-2a, both of which are active towards all phosphorylation sites of tau.

How Does Tau Aggregate into Paired Helical Filaments?

The mechanism of aggregation is probably one of the most crucial problems to be solved because the deposition of PHFs in select neurons is one of the earliest signs of the disease. These PHFs are highly insoluble, which makes their extraction and analysis difficult, and they occur only in human AD brains, which means that the supply is limited to autopsies. The analysis of Alzheimer tau has improved as methods of isolation have been refined. One can distinguish several levels of pathological aggregation, all of which are based on PHFs (or variants thereof, straight filaments; Crowther 1991). The higher aggregates are neurofibrillary tangles or neuropil threads, which consist of coalesced PHFs with other material attached to them (from proteins to metal compounds, including aluminum). They are visible by light microscopy after silver staining. With regard to PHFs, one distinguishes between the SDS-insoluble and the SDS-soluble fraction (Wischik et al. 1988; Greenberg and Davies 1990). Both are more highly phosphorylated (6–8 Pi per tau) than normal adult tau (≈2 Pi; Ksiezak-Reding et al. 1992). The SDS-insoluble fraction probably represents a later and more extensive state of aggregation, it is more highly ubiquitinated, and it is resistant to proteolysis (pronase removes a fuzzy coat but leaves the core aggregated; Kondo et al. 1988; Wischik et al. 1988; Kziezak-Reding and Wall 1994). The SDS-soluble fraction can also be digested by pronase; analysis of this fraction has shown that PHFs consist mainly of tau and, moreover, all isoforms of tau (Brion et al. 1991; Goedert et al. 1992). The states of lower aggregation are more difficult to distinguish. Obviously even Alzheimer brains contain a fraction of normal tau which is highly soluble and in a state of low phosphorylation. However, there is also incipient aggregation in the form of soluble AD tau (Köpke et al. 1993).

One way to analyze PHFs is to induce the assembly of recombinant tau constructs, study the structures formed and compare them with PHFs from

Alzheimer brains. The advantages are that the starting material is soluble, it has a defined sequence and composition, and it can be modified by phosphorylation. Using this approach we discovered several features of tau (Wille et al. 1992): 1) Tau is a highly extended molecule, with lengths around 35–50 nm (depending on isoform). 2) Tau can associate into dimers which are antiparallel and roughly in register (Fig. 7). This can be shown by immuno-election microscopy using antibody lables that bind to the end of the tau molecule. 3) Certain tau constructs (comprising mainly the repeat region of tau) and their dimers can associate into PHFs that are very similar to those found in AD (Fig. 8). Such filaments are also formed from chemically cross-linked dimers of the repeat region of tau, suggesting that

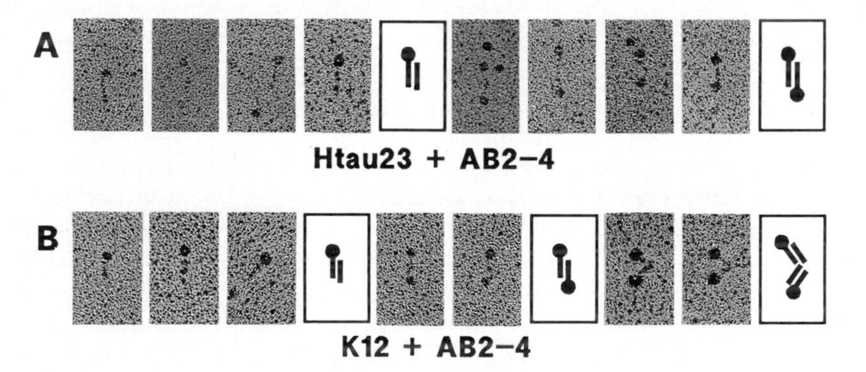

Fig. 7. Electron micrographs of tau isoform htau23 or construct K12. The rod-like particles are about 35 nm (htau23) or 25 nm (K12) long (Wille et al. 1992). They are labeled at one or both ends with antibody 2–4 (gift or R Vallee, Worcester Foundation), showing that the particles can form antiparallel dimers and more complex structures (see interpretative diagrams)

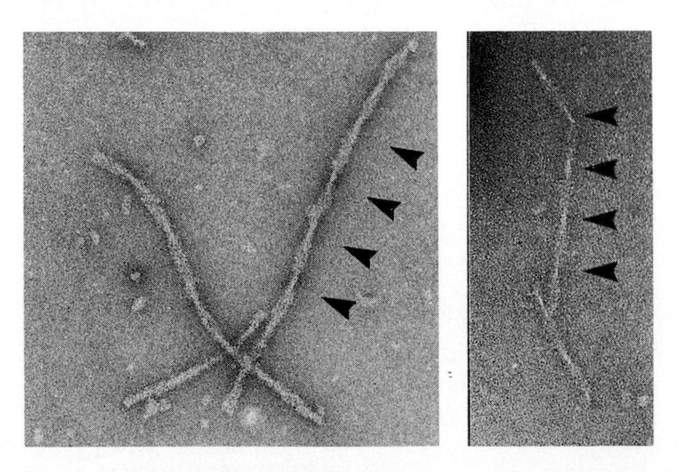

Fig. 8. Synthetic PHFs assembled from tau construct K12 (consisting of repeats 1, 3, 4 and a short tail). The cross-over periodicity is about 75–80 nm (*arrowheads*)

the dimer stage is an intermediate level of tau aggregation. By contrast it is quite difficult to form bona fide PHFs from intact tau (Crowther et al. 1994). Whether and how the aggregation of tau into synthetic PHFs depends on phosphorylation is not clear at present.

The observation of synthetic PHFs made from recombinant tau by Wille et al. (1992) was an important step towards elucidating the assembly mechanism, but it also illustrates our lack of detailed understanding of the control mechanisms. It is not yet possible to cycle between the states of assembly and disassembly (as is possible with other self-assembling polymers, such as actin filaments or microtubules). It is also not yet clear what the main principle behind the assembly is. Insoluble protein aggregates are commonly formed from denatured protein when the hydrophobic interior is exposed and then coalesces with other molecules. Such hydrophobic interactions are difficult to envisage with tau protein because its composition is unusually hydrophilic (Lee et al. 1988); this probably explains why it is heat stable and resistant to acid treatment (Cleveland et al. 1977; Fellous et al. 1977). Another possibility is a high extent of hydrogen-bonded β-structure (as proposed for "self-complementary" peptides by Zhang et al. 1993). Extensive β-structure indeed makes proteins very stable, and this is probably the basis for the aggregation of the β-amyloid into insoluble fibers (Inouye et al. 1993; Hilbich et al. 1992; see chapter by Selkoe, this volume). This principle probably does not apply to tau since tau has very little detectable secondary structure and behaves similar to a denatured protein (Schweers et al. 1994). A third mechanism would be covalent cross-linking via oxidized SH groups forming S-S linkages. This probably plays a role in the formation of tau dimers because they are generated most homogeneously from constructs having only one SH group which can be modified by S-S-crosslinking reagents (Wille et al. 1992). Whether this plays a role in the reducing environment of a cell is questionable, but it could take place once the redox state is perturbed. Finally, the cell has developed other aggregation mechanisms based on enzymatic cross-linking. One example is that of transglutaminase, which is capable of cross-linking tau (Dudek and Johnson 1993). Independently of which of these mechanisms apply, there remains the question of how tau's aggregation is influenced by phosphorylation, and how this is related to microtubule binding. In other words: Do microtubules decay because tau is hyperphosphorylated, detaches, and aggregates into PHFs, or do microtubules decay for some other reason, leaving tau free to interact with other proteins including itself? In fact, it is conceivable that the neuron could live without tau (because there are other MAPs that could take its place, Harada et al. 1994), but it is not conceivable that the neuron could live without microtubules, because they provide the tracks for intracellular transport. If one phrases the problem in this manner, it becomes apparent that it may be worth studying the other components of intracellular transport, from microtubules and their motor proteins to mechanisms of intracellular sorting, and how they become modified in AD.

Acknowledgements. We would like to thank Dr. M. Goedert for clones of human tau; Dr. W. Studier for the pET expression vector; Drs. J Dingus, R. Vallee, L. Binder, A. Vandevoorde and M. Mercken for antibodies. Phosphopeptide sequencing was done in collaboration with Dr. H. E. Meyer (Univ. Bochum). Brain tissue was generously provided by the Bryan Alzheimer Disease Research Center (Duke University Medical Center, Durham, NC), the Brain Tissue Resource Center (McLean Hospital/ Harvard Medical School, Belmont, MA), and the Alzheimer Research Center (Univ. Rochester Medical School, Rochester, NY). This work was supported by Bundesministerium für Forschung und Technologie (BMFT) and the Deutsche Forschungsgemeinschaft (DFG).

References

Anderton BH (1993) Expression and processing of pathological proteins in Alzheimer's disease. Hippocampus 3: 227–237

Bancher C, Grundke-Iqbal I, Iqbal K, Fried V, Smith H, Wisniewski H (1991) Abnormal phosphorylation of tau precedes ubiquitination in neurofibrillary pathology of Alzheimer disease. Brain Res 539: 11–18

Baudier J, Cole RD (1987) Phosphorylation of tau proteins to a state like that in Alzheimer's brain is catalyzed by a calcium/calmodulin dependent kinase and modulated by phospholipids. J Biol Chem 262: 17577–17583

Baumann K, Mandelkow E-M, Biernat J, Piwnica-Worms H, Mandelkow E (1993) Abnormal Alzheimer-like phosphorylation of tau protein by cyclin-dependent kinases cdk2 and cdk5. FEBS Lett 336: 417–424

Bialojan C, Takai A (1988) Inhibitory effect of a marine sponge toxin, okadaic acid, on protein phosphatases. Biochem J 256: 283–290

Biernat J, Mandelkow E-M, Schröter C, Lichtenberg-Kraag B, Steiner B, Berling B, Meyer HE, Mercken M, Vandermeeren A, Goedert M, Mandelkow E (1992) The switch of tau protein to an Alzheimer-like state includes the phosphorylation of two serine-proline motifs upstream of the microtubule binding region. EMBO J 11: 1593–1597

Biernat J, Gustke N, Drewes G, Mandelkow E-M, Mandelkow E (1993) Phosphorylation of serine 262 strongly reduces the binding of tau protein to microtubules: Distinction between PHF-like immunoreactivity and microtubule binding. Neuron 11: 153–163

Binder LI, Frankfurter A, Rebhun L (1985) The distribution of tau in the mammalian central nervous system. J Cell Biol 101: 1371–1378

Braak H, Braak E (1991) Neuropathological staging of Alzheimer-related changes. Acta Neuropathol 82: 239–259

Braak H, Braak E (1994) Pathology of Alzheimer's disease. In: Neurodegenerative diseases. Saunders, Calne D et al. (eds) Philadelphia, pp 585–613

Braak E, Braak H, Mandelkow E-M (1994) A sequence of cytoskeleton changes related to the formation of neurofibrillary tangles and neuropil threads. Acta Neuropathol 87: 554–567

Bramblett GT, Goedert M, Jakes R, Merrick SE, Trojanowski JQ, Lee VMY (1993) Abnormal tau phosphorylation at Ser(396) in Alzheimer's disease recapitulates development and contributes to reduced microtubule binding. Neuron 10: 1089–1099

Brandt R, Lee G (1993) Functional organization of microtubule-associated protein tau: Identification of regions which affect microtubule growth, nucleation, and bundle formation in vitro. J Biol Chem 268: 3414–3419

Brion J, Passareiro H, Nunez J, Flament-Durand J (1985) Mise en evidence immunologique de la proteine tau au niveau des lesions de degenerescence neurofibrillaire de la maladie d'Alzheimer. Arch Biol 95: 229–235

Brion JP, Hanger DP, Couck AM, Anderton BH (1991) A68 proteins in Alzheimer's disease are composed of several tau isoforms in a phosphorylated state which affects their electrophoretic mobilities. Biochem J 279: 831–836

Butner KA, Kirschner MW (1991) Tau-protein binds to microtubules through a flexible array of distributed weak sites. J Cell Biol 115: 717–730

Cleveland DW, Hwo S-Y, Kirschner MW (1977) Physical and chemical properties of purified tau factor and the role of tau in microtubule assembly. J Mol Biol 116: 227–247

Cohen P (1991) Classification of protein serine/threonine phosphatases: Identification and quantitation in cell extracts. Meth Enzym 201: 389–398

Correas I, Diaznido J, Avila J (1992) Microtubule associatedprotein tau is phosphorylated by protein kinase C on its tubulin binding domain. J Biol Chem 267: 15721–15728

Couchie D, Mavilia C, Georgieff I, Liem R, Shelanski M, Nunez J (1992) Primary structure of high molecular weight tau present in the peripheral nervous system. Proc Natl Acad Sci 89: 4378–4381

Crowther RA (1991) Straight and paired helical filaments in Alzheimer disease have a common structural unit. Proc Natl Acad Sci 88: 2288–2292

Crowther RA, Olesen OF, Smith MJ, Jakes R, Goedert M (1994) Assembly of Alzheimer-like filaments from full-length tau protein. FEBS Lett 337: 135–138

Drewes G, Lichtenberg-Kraag B, Döring F, Mandelkow E-M, Biernat J, Goris J, Doree M, Mandelkow E (1992) Mitogen-activated protein (MAP) kinase transforms tau protein into an Alzheimer-like state. EMBO J 11: 2131–2138

Drewes G, Mandelkow E-M, Baumann K, Goris J, Merlevede W, Mandelkow E (1993) Dephosphorylation of tau protein and Alzheimer paired helical filaments by calcineurin and phosphatase-2A. FEBS Lett 336: 425–432

Dudek SM, Johnson GVW (1993) Transglutaminase catalyzes the formation of SDS-insoluble, Alz50-reactive polymers of tau. J Neurochem 61: 1159–1162

Ennulat DJ, Liem RKH, Hashim GA, Shelanski ML (1989) Two separate 18-amino acid domains of tau promote the polymerization of tubulin. J Biol Chem 264: 5327–5330

Fellous A, Francon J, Lennon AM, Nunez J (1977) Microtubule assembly in vitro: Purification of assembly promoting factors. Eur J Biochem 78: 167–174

Goedert M (1993) Tau protein and the neurofibrillary pathology of Alzheimer S disease. Trends Neurosci 16: 460–465

Goedert M, Wischik C, Crowther R, Walker J, Klug A (1988) Cloning and sequencing of the cDNA encoding a core protein of the paired helical filament of Alzheimer disease: Identification as the microtubule-associated protein tau. Proc Natl Acad Sci 85: 4051–4055

Goedert M, Spillantini M, Jakes R, Rutherford D, Crowther RA (1989) Multiple isoforms of human microtubules-associated protein-tau: Sequences and localization in neurofibrillary tangles of Alzheimers-disease. Neuron 3: 519–526

Goedert M, Spillantini G, Cairns NJ, Crowther RA (1992) Tau proteins of Alzheimer paired helical filaments: Abnormal phosphorylation of all six brain isoforms. Neuron 8: 159–168

Goedert M, Jakes R, Crowther RA, Six J, Lübke U, Vandermeeren M, Cras P, Trojanowski JQ, Lee VMY (1993) The abnormal phosphorylation of tau protein at Ser202 in Alzheimer's disease recapitulates phosphorylation during development. Proc Natl Acad Sci 90: 5066–5070

Goris J, Hermann J, Hendrix P, Ozon R, Merlevede W (1989) Okadaic acid, a sepcific protein phosphatase inhibitor, induces maturation and MPF formation in Xenopus laevis oocytes. FEBS Letters 245: 91–94

Greenberg SG, Davies P (1990) A preparation of Alzheimer paired helical filaments that displays distinct tau-proteins by polyacrylamide-gel electrophoresis. Proc Natl Acad Sci 87: 5827–5831

Grundke-Iqbal I, Iqbal K, Tung Y, Quinlan M, Wisniewski H, Binder L (1986) Abnormal phosphorylation of the microtubule-associated protein tau in Alzheimer cytoskeletal pathology. Proc Natl Acad Sci 83: 4913–4917

Gustke N, Steiner B, Mandelkow E-M, Biernat J, Meyer HE, Goedert M, Mandelkow E (1992) The Alzheimer-like phosphorylation of tau protein reduces microtubule binding and involves Ser-Pro and Thr-Pro motifs. FEBS Lett 307: 199–205

Gustke N, Trinczek B, Biernat J, Mandelkow E-M, Mandelkow E (1994) Domains of tau protein and interactions with microtubules. Biochemistry 33: 9511–9522

Hanger D, Hughes K, Woodgett J, Brion J, Anderton B (1992) Glycogen-synthase kinase-3 induces Alzheimer's disease-like phosphorylation of tau: Generation of paired helical filament epitopes and neuronal localization of the kinase. Neurosci Lett 147: 58–62

Harada A, Oguchi K, Okabe S, Kuno J, Terada S, Ohshima T, Sato-Yoshitake R, Takei Y, Noda T, Hirokawa N (1994) Altered microtubule organization in small-caliber axons of mice lacking tau protein. Nature 369: 488–491

Hasegawa M, Morishima-Kawashima M, Takio K, Suzuki M, Titani K, Ihara Y (1992) Protein sequence and mass spectrometric analyses of tau in the Alzheimer's disease brain. J Biol Chem 26: 17047–17054

Hilbich C, Kisters-Woike B, Reed J, Masters C, Beyreuther K (1992) Substitutions of hydrophobic amino-acids reduce the amyloidogenicity of Alzheimer's disease βA4 peptides. J Mol Biol 228: 460–473

Himmler A, Drechsel D, Kirschner M, Martin D (1989) Tau consists of a set of proteins with repeated C-terminal microtubule-binding domains and variable N-terminal domains. Molec Cell Biol 9: 1381–1388

Hisanaga S, Ishiguro K, Uchida T, Okumura E, Okano T, Kishimoto T (1993) Tau-protein kinase II has a similar characteristic to cdc2 kinase for phosphorylating neurofilament proteins. J Biol Chem 268: 15056–15060

Hunter T (1991) Protein kinase classification. Meth Enzym 200: 3–37

Inouye H, Fraser PE, Kirschner DA (1993) Structure of beta-crystallite assemblies formed by Alzheimer beta-amyloid protein analogues: analysis by X-ray diffraction. Biophys J 64: 502–519

Ishiguro K, Shiratsuchi A, Sato S, Omori A, Arioka M, Kobayashi S, Uchida T, Imahori K (1993) Glycogen-synthase kinase 3-beta is identical to tau protein kinase I generating several epitopes of paired helical filaments. FEBS Lett 325: 167–172

Kanai Y, Chen J, Hirokawa N (1992) Microtubule bundling by tau proteins in vivo: Analysis of functional domains. EMBO J 11: 3953–3961

Kanemaru K, Takio K, Miura R, Titani K, Ihara Y (1992) Fetal-type phosphorylation of the tau in paired helical filaments. J Neurochem 58: 1667–1675

Knops J, Kosik K, Lee G, Pardee J, Cohengould L, McConlogue L (1991) Overexpression of tau in a nonneuronal cell induces long cellular processes. J Cell Biol 114: 725–733

Kondo J, Honda T, Mori H, Hamada Y, Miura R, Ogawara M, Ihara Y (1988) The carboxyl third of tau is tightly bound to paired helical filaments. Neuron 1: 827–834

Köpke E, Tung Y, Shaikh S, Alonso A, Iqbal K, Grundke-Iqbal I (1993) Microtubule-associated protein tau: Abnormal phosphorylation of a non-paired helical filament pool in Alzheimer's disease. J Biol Chem 268: 24374–24384

Kosik K, Orecchio L, Binder L, Trojanowski J, Lee V, Lee G (1988) Epitopes that span the tau molecule are shared with paired helical filaments. Neuron 1: 817–825

Ksiezak-Reding H, Yen SH (1991) Structural stability of paired helical filaments requires microtubule-binding domains of tau: A model for self-association. Neuron 6: 717–728

Ksiezak-Reding H, Wall JS (1994) Mass and physical dimensions of 2 distinct populations of paired helical filaments. Neurobiol Aging 15: 11–19

Ksiezak-Reding H, Liu WK, Yen SH (1992) Phosphate analysis and dephosphorylation of modified tau associated with paired helical filaments. Brain Res 597: 209–219

Lee G, Cowan N, Kirschner M (1988) The primary structure and heterogeneity of tau protein from mouse brain. Sicence 239: 285–288

Lee VMY, Balin BJ, Otvos L, Trojanowski JQ (1991) A68 – a major subunit of paired helical filaments and derivatized forms of normal tau. Science (Wash.) 251: 675–678

Lew J, Winkfein RJ, Paudel HK, Wang JH (1992) Brain proline-directed protein kinase is a neurofilament kinase which displays high sequence homology to p34(cdc2). J Biol Chem 267: 25922–25926

Lichtenberg-Kraag B, Mandelkow E-M, Biernat J, Steiner B, Schröter C, Gustke N, Meyer HE, Mandelkow E (1992) Phosphorylation dependent interaction of neurofilament antibodies with tau protein: Epitopes, phosphorylation sites, and relationship with Alzheimer tau. Proc Natl Acad Sci 89: 5384–5388

Lo MMS, Fieles AW, Norris TE, Dargis PG, Caputo CB, Scott CW, Lee VMY, Goedert M (1993) Human tau isoforms confer distinct morphological and functional properties to stably transfected fibroblasts. Mol Brain Res 20: 209–220

Lu Q, Wood JG (1993) Functional studies of Alzheimers disease tau protein. J Neurosci 13: 508–515

Mandelkow E-M, Mandelkow E (1993) Tau as a marker for Alzheimer's disease. TIBS 18: 480–483

Mandelkow E-M, Drewes G, Biernat J, Gustke N, Van Lint J, Vandenheede JR, Mandelkow E (1992) Glycogen synthase kinase-3 and the Alzheimer-like state of microtubule-associated protein tau. FEBS Lett 314: 315–321

Mercken M, Vandermeeren M, Lübke U, Six J, Boons J, Van de Voorde A, Martin J-J, Gheuens J (1992) Monoclonal antibodies with selective specificity for Alzheimer tau are directed against phosphatase-sensitive epitopes. Acta Neuropathol 84: 265–272

Meyerson M, Enders GH, Wu CL, Su LK, Gorka C, Nelson C, Harlow E, Tsai LH (1992) A family of human cdc2-related protein-kinases. EMBO J 11: 2909–2917

Morishima-Kawashima M, Hasegawa M, Takio K, Suzuki M, Titani K, Ihara Y (1993) Ubiquitin is conjugated with amino-terminally processed tau in paired helical filaments. Neuron 10: 1151–1160

Novak M, Kabat J, Wischik CM (1993) Molecular characterization of the minimal protease resistant tau-unit of the Alzheimer's-disease paired helical filament. EMBO J 12: 365–370

Paudel H, Lew J, Ali Z, Wang J (1993) Brain proline-directed protein kinase phosphorylates tau on sites that are abnormally phosphorylated in tau associated with Alzheimer's paired helical filaments. J Biol Chem 268: 23512–23518

Roder HM, Eden PA, Ingram VM (1993) Brain protein kinase pk40(erk) converts tau into a PHF-like form as found in Alzheimer's disease. Biochem Biophys Res Comm 193: 639–647

Schweers O, Schönbrunn-Hanebeck E, Marx A, Mandelkow E (1994) Structural studies of tau protein and Alzheimer paired helical filaments show no evidence for β structure. J Biol Chem 269: 24290–24297

Scott C, Spreen R, Herman J, Chow F, Davison M, Young J, Caputo C (1993) Phosphorylation of recombinant tau by cAMP-dependent protein kinase: Identification of phosphorylation sites and effect on microtubule assembly. J Biol Chem 268: 1166–1173

Shetty KT, Link WT, Pant HC (1993) Cdc2-like kinase from rat spinal-cord specifically phosphorylates KSPXK motifs in neurofilament proteins: Isolation and characterization. Proc Natl Acad Sci 90: 6844–6848

Steiner B, Mandelkow E-M, Biernat J, Gustke N, Meyer HE, Schmidt B, Mieskes G, Söling HD, Drechsel D, Kirschner MW, Goedert M, Mandelkow E (1990) Phosphorylation of microtubule-associated protein tau: Identification of the site for Ca^{++}-calmodulin dependent kinase and relationship with tau phosphorylation in Alzheimer tangles. EMBO J 9: 3539–3544

Stemmer P, Klee C (1991) Serine/threonine phosphatases in the nervous system. Curr Opin Neurobiol 1: 53–64

Szendrei GI, Lee VM-Y, Otvos L (1993) Recognition of the minimal epitope of monoclonal antibody Tau-1 depends upon the presence of a phosphate group but not its location. J Neurosci Res 34: 243–249

Tsai LH, Takahashi T, Caviness V, Harlow E (1993) Activity and expression pattern of cyclin-dependent kinase 5 in the embryonic mouse nervous system. Development 119: 1029–1040

Vulliet R, Halloran S, Braun R, Smith A, Lee G (1992) Proline-directed phosphorylation of human tau protein. J Biol Chem 267: 22570–22574

Watanabe A, Hasegawa M, Suzuki M, Takio K, Morishima-Kawashima M, Titani K, Arai T, Kosik KS, Ihara Y (1993) In-vivo phosphorylation sites in fetal and adult rat tau. J Biol Chem 268: 25712–25717

Wille H, Drewes G, Biernat J, Mandelkow E-M, Mandelkow E (1992) Alzheimer-like paired helical filaments and antiparallel dimers formed from microtubule-associated protein tau in vitro. J Cell Biol 118: 573–584

Wischik C, Novak M, Thogersen H, Edwards P, Runswick M, Jakes R, Walker J, Milstein C, Roth M, Klug A (1988) Isolation of a fragment of tau derived from the core of the paired helical filament of Alzheimer disease. Proc Natl Acad Sci 85: 4506–4510

Woodgett JR (1991) A common denominator linking glycogen metabolism, nuclear oncogenes, and development. TIBS 16: 177–181

Yoshida H, Ihara Y (1993) Tau in paired helical filaments is functionally distinct from fetal tau: Assembly incompetence of paired helical filament tau. J Neurochem 61: 1183–1186

Zhang S, Holmes T, Lockshin C, Rich A (1993) Spontaneous assembly of a self-complementary oligopeptide to form a stable macroscopic membrane. Proc Natl Acad Sci 90: 3334–3338

Posttranslational Modifications of the Tau in PHF: Phosphorylation and Ubiquitination

*M. Morishima-Kawashima** and *Y. Ihara*

Summary

Paired helical filaments (PHF) are a unit fibril of neurofibrillary tangles and their accumulation appears to be correlated with the degree of dementia. To learn more about PHF genesis, we investigated the posttranslational modifications of tau in PHF by direct protein chemical analysis.

Highly phosphorylated full-length tau, PHF-tau, is the only building block of PHF at the early stage. The exact phosphorylation sites of PHF-tau have been determined; PHF-tau is phosphorylated on no more than 19 Ser/Thr sites. All the sites except Ser-262 are localized to both amino- and carboxy-terminal flanking regions of the microtubule-binding domain. Half of them are proline-directed and the other half are non-proline-directed. Overall, the phosphorylation of PHF-tau can be considered to consist of fetal-type phosphorylation and additional proline-directed and non-proline-directed phosphorylation. This extraphosphorylation may provide PHF-tau with its unusual properties, including assembly incompetence.

The other major modification of PHF is ubiquitination. From analysis of the PHF smear that presumably represents more modified PHF, the ubiquitin-targeted protein was identified as tau and the conjugated sites were localized to the microtubule-binding domain. Since the PHF smear consists largely of the carboxy-half of tau and ubiquitin, ubiquitin is conjugated with amino-terminally processed tau in PHF.

Introduction

Alzheimer's disease (AD) is characterized by innumerable β-amyloid plaques and neurofibrillary tangles throughout the cortex (for review see Selkoe 1991). Although β-amyloid is now believed to be involved in AD pathogenesis (Mullan and Crawford 1993), neurofibrillary tangles are considered to be closely related to neuronal death in AD brain because these tangle-bearing

* Department of Neuropathology, Institute for Brain Research, Faculty of Medicine, University of Tokyo, 7-3-1 Hongo, Bunkyoku, Tokyo 113, Japan

K.S. Kosik et al. (Eds.)
Alzheimer's Disease: Lessons from Cell Biology

neurons are lost during the progression of AD (Terry et al. 1981). Recently, the best correlate with the degree of dementia has been shown to be the loss of synapses (Terry et al. 1991), which in turn is correlated with tangle formation (Arriagada et al. 1992).

Neurofibrillary tangles exist in subsets of neurons and in their small neuronal, presumably dendritic, processes (neuropil threads or curly fibers; Braak et al. 1986; Kowall and Kosik 1987; Ihara 1988; Iwatsubo et al. 1992). Paired helical filaments (PHF), a unit fibril of the neurofibrillary tangles, have an unusual morphology; they are apparently composed of two 10 nm filaments wound around each other with a half periodicity of 80 nm. PHF also show unusual biochemical characteristics, insolubility and remarkable resistance to proteases (Selkoe et al. 1982), which prevented their extensive biochemical analysis for a long time. It was only six years ago that SDS-insoluble PHF (SDS-PHF) were clearly shown to be composed of tau, a microtubule-associated phosphoprotein (Kondo et al. 1988; Wischik et al. 1988) and ubiquitin, an essential element in the cytosolic ATP-dependent protein degradation system (Mori et al. 1987). In those PHF, tau appeared to be truncated because only the carboxyl third of tau was detected by protein chemical and immunochemical analyses (Kondo et al. 1988).

Recent studies have shown that hyperphosphorylated full-length tau, called A68 or PHF-tau, composes solely SDS-soluble, Sarkosyl-insoluble PHF (Greenberg and Davies 1990; Lee et al. 1991), which presumably represent their early stage. PHF-tau is characterized by an unusually slow mobility on SDS-PAGE and loss of tubulin assembly-promoting activity (Greenberg et al. 1992; Yoshida and Ihara 1993). Both parameters can be normalized by dephosphorylation. Consistent with this finding, we were unable to detect any abnormal modifications in PHF-tau other than phosphorylation by extensive protein chemical analysis (Hasegawa et al. 1992). Thus, hyperphosphorylation of tau seems to be the first critical step toward PHF formation.

In contrast to the phosphorylation, ubiquitination of PHF appears to be a late event (Bancher et al. 1991). The affected neurons respond by the ubiquitin pathway to abnormal cytoplasmic aggregates, but are unable to remove them effectively.

Over the last three years, we have been analyzing PHF protein chemically by means of ion-spray mass spectrometry (ISMS) and protein sequencing. Here we report on details of phosphorylation and ubiquitination found in tau in PHF.

Abnormal Phosphorylation of Tau in PHF

PHF-tau Shares Phosphorylation Characteristics with Fetal Tau

In the course of our work on the phosphorylation of tau in PHF, we found that its phosphorylation is very similar to that normally found in fetal tau;

two PHF monoclonal antibodies and two types of PHF polyclonal antibodies, all of which are phosphorylation-dependent and have different specificities, preferentially labeled fetal and juvenile tau, but not much adult tau (Kanemaru et al. 1992; Hasegawa et al. 1993). This indicates that PHF-tau shares several phosphorylated epitopes with fetal tau.

As a first step, we determined *in vivo* phosphorylation sites of fetal tau by ISMS and protein sequencing (Watanabe et al. 1993). Phosphoserine was identified as S-ethylcysteine by sequencing of ethanethiol-modified peptides (Meyer et al. 1991). Fetal tau was found to be phosphorylated on 10 Ser/Thr residues at most, including Ser-198, -199, -202, -235, -396, -400, and -404, and Thr-181, -217, and -231 (the numbering follows a 441-amino acid human tau isoform; Goedert et al. 1989; Fig. 1). All of the phosphorylation sites are localized to both amino- and carboxy-terminal flanking regions of the microtubule-binding domain. It should be noted that remarkable heterogeneity was present; the phosphorylation extent at a given site varied 20 to 80%. Most importently, eight of 10 sites are in Ser-Pro or Thr-Pro motif (Fig. 1). This finding indicates that fetal tau is an *in vivo* substrate for proline-directed protein kinases (PDPKs), including cdc2 or cdc2-related kinases (the most abundant species in brain is cyclin-dependent protein

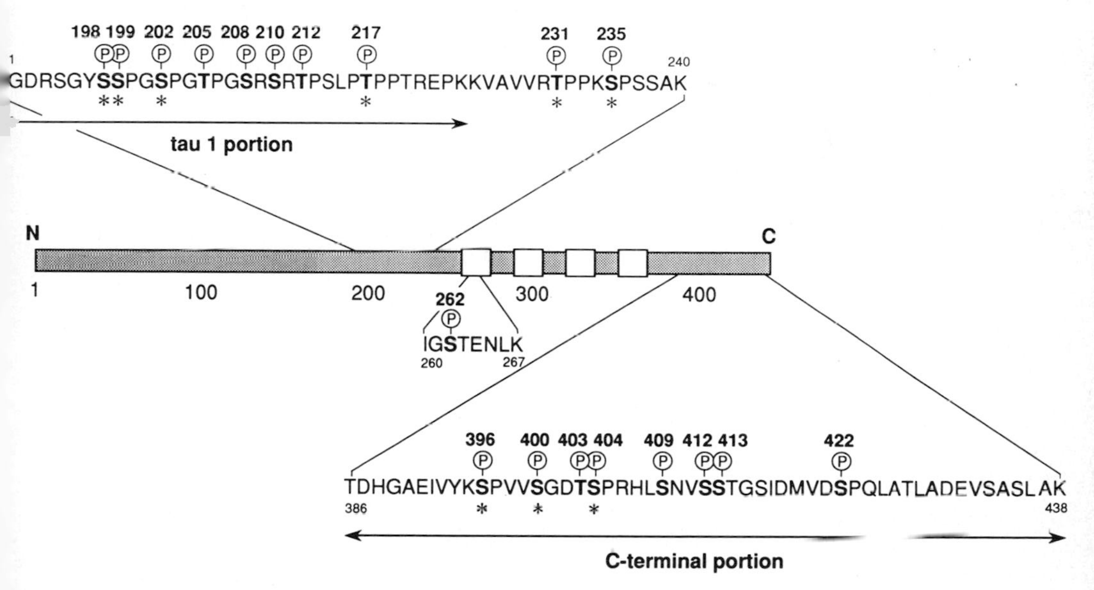

Fig. 1. Phosphorylation sites in PHF-tau and fetal tau. Phosphorylation sites in PHF-tau and fetal tau are marked by *P* and *asterisks*, respectively. Amino acids are numbered according to a 441-amino acid human tau isoform (Goedert et al. 1989). *Open boxes* indicate 18-amino acid microtubule-binding domains. Fetal tau is a three-repeat isoform which lacks the second repeat. Note that all the phosphorylation sites except Ser-262 are localized to both amino- and carboxyl-terminal flanking portions of the microtubule-binding region. Thr-181 is also phosphorylated in fetal tau (not shown)

kinase (cdk) 5; Meyerson et al. 1992; Lew et al. 1992) and mitogen-activated protein (MAP) kinases (Drewes et al. 1992). Glycogen synthase kinase 3β(GSK3β)/TPK I appears to be only partly proline-directed, because it also phosphorylates non-proline-directed sites in the proline-rich region as well (Woodgett 1990; Ishiguro et al. 1992, 1993). Interestingly, the two non-proline-directed sites, Ser-198 and 400, conform to the recognition motif of GSK3 once Ser-202 and Ser-404, respectively, are phosphorylated (Kemp and Pearson 1990).

PHF-tau: Proline-directed and Non-proline-directed Phosphorylation

The immunochemical approach leaves some ambiguity regarding the exact phosphorylation sites and entirely depends on the available phosphorylation-dependent antibodies. Thus, we have determined the exact phosphorylation sites in PHF-tau in the same manner as in fetal tau (Hasegawa et al. 1992; Morishima-Kawashima et al. 1995a).

AP I (*Achromobacter lyticus* protease I) peptide mapping of PHF-tau showed three anomalously eluted peaks which became normal by dephosphorylation. The analysis of the peptides indicated that Thr-231, Ser-235, and Ser-262 are the abnormal phosphorylation sites, with the last being only partially phosphorylated (Hasegawa et al. 1992; Fig. 1). We also found that PHF-tau is highly phosphorylated in the tau 1 portion and the carboxyl-terminal portion outside the binding domain. It was very difficult to determine the exact phosphorylation sites on these portions, because 1) both relatively large peptides provide no significant mass signals, unless dephosphorylated, and 2) both portions are unusually resistant to proteinases and generated peptides either are not or are only poorly recovered from a HPLC column. After trial and error, we finally established the proper conditions to digest and separate these peptides and analyzed all the generated peptides (Morishima-Kawashima et al. 1995a).

Our results showed that Ser-198, -199, -202, Thr-205, Ser-208, -210, Thr-212 and -217 in the tau 1 portion and Ser-396, -400, Thr-403, Ser-404, -409, -412, -413, and -422 in the carboxyl-terminal portion are phosphorylated (Fig. 1). Thus, 19 Ser/Thr sites are identified; all the sites except for Ser-262 are clustered in both flanking regions of the microtubule-binding domain. Most of the phosphorylation sites found in fetal tau are indeed phosphorylated in PHF-tau. This provides a rationale for the immunochemical similarities between the two molecules. However, PHF-tau is much more phosphorylated than fetal tau in both the tau 1 and carboxyl-terminal portions. Unexpectedly, half of the sites are non-proline-directed, whereas the other half are proline-directed (Fig. 1). Most of the PHF-tau-specific phosphorylation sites that are not found in fetal tau are non-proline-directed. Thus, the abnormal phosphorylation of PHF-tau can be considered to consist of fetal-type phosphorylation and additional proline-directed and non-proline-directed phosphorylation.

The similarities in the phosphorylation of PHF-tau and fetal tau may reflect an unsuccessful regenerative attempt of tangle-bearing neurons in AD brain, because degeneration always accompanies regeneration. Besides fetal sites, there are 10 phosphorylation sites apparently specific for PHF-tau. A list of consensus recognition motifs by various kinases suggests that non-fetal-type sites in PHF-tau may be phosphorylated by GSK3 (Ishiguro et al. 1992; Yang et al. 1993a,b), MAP kinase (Drewes et al. 1992), cAMP-dependent protein kinase (PKA; Scott et al. 1993), and/or 35/41 kd kinase (Biernat et al. 1993). As in the case of fetal tau, Ser-208 and -409 become the GSK recognition motif when Thr-212 and Ser-413 are phosphorylated (Kemp and Pearson 1990). Thr-212 and Ser-413 in turn can be phosphorylated by MAP kinase and GSK3β, respectively (Drewes et al. 1992; Ishiguro et al. 1992). Thus, it is quite possible that multiple kinases are involved in the hyperphosphorylation of PHF-tau in a synergistic way.

A remarkable functional characteristic of PHF-tau is its tubulin assembly incompetence (Yoshida and Ihara 1993). This contrasts with fetal tau. Despite its high degree of phosphorylation, fetal tau retains a lower, but still significant, promoting activity of tubulin assembly (Yoshida and Ihara 1993). PHF-tau is distinct from fetal tau in its phosphorylation on many non-proline-directed sites. One such site, Ser-409, is known to be preferentially phosphorylated by PKA in vitro (Scott et al. 1993). Therefore, we examined how PKA phosphorylation affects the promoting activity of fetal tau (Morishima-Kawashima et al. 1995b). Although PKA phosphorylation of dephosphorylated fetal tau suppressed its promoting activity only slightly, the phosphorylation of nontreated fetal tau almost eliminated the activity. This finding suggests that the non-proline-directed extraphosphorylation may provide PHF-tau with its unusual properties, including assembly incompetence.

A most important question remains: why does excess proline-directed and non-proline-directed phosphorylation occur on PHF-tau? A possible explanation would be activation or upregulation of protein kinases and/or inactivation or downregulation of counteracting protein phosphatases. Regarding protein phosphatases, at least three are known – protein phosphatase 1, protein phosphatase 2A and protein phosphatase 2B – all of which are reported to dephosphorylate tau on proline-directed sites *in vitro* (Goedert et al. 1992; Drewes et al. 1993; Gong et al. 1993, 1994). In this context, it should be noted that microtubules appear to be lost where PHF are being formed (Perry et al. 1991). This finding raises the possibility that the phosphorylation specific for PHF-tau may be related to the loss of microtubules. It is possible that free tau molecules are good substrates for several protein kinases, creating PHF-tau-specific phosphorylated epitopes. In any case, the phosphorylation that distinguishes PHF-tau from fetal tau should be investigated in depth. This should lead to a better understanding of the mechanism of PHF formation and the underlying neuritic dystrophy in AD brain.

Ubiquitin Is Conjugated with Amino-terminally Processed Tau in PHF

Ubiquitin, a highly conservative protein among eukaryotes, is an essential element of the ATP-dependent protein degradation system. Ubiquitin marks a target protein through an isopeptide bond between its carboxy-terminal Gly and ε-amino group of Lys in a target protein, and its ubiquitinated form is rapidly degraded by a 26S proteasome complex (for reviews see Finley and Chau 1991; Hershko and Ciechanover 1992). Ubiquitination of protein exists in a monoubiquitinated form or a multiubiquitin chain: the former may not be a degradation signal, but may modulate the function of a target protein, whereas the latter, Lys-48-linked ubiquitin-ubiquitin conjugate, works as a strong degradation signal when joined to Lys in a target protein.

Since we showed that ubiquitin is a component of PHF (Mori et al. 1987), ubiquitin immunoreactivities have been demonstrated in various cytoplasmic inclusion bodies in degenerative diseases, which include Pick bodies in Pick's disease, Lewy bodies in Parkinson's disease and diffuse Lewy body disease, Lewy-like bodies in amyotrophic lateral sclerosis, and Mallory bodies in alcoholic liver disease (Mayer et al. 1991). In addition, ubiquitin-immunoreactive aggregates have recently been shown to emerge even in normal human brain in an age-dependent fashion (Pappolla et al. 1989). Thus, the ubiquitin pathway may also have an important role in removing such abnormal cytoplasmic aggregates under pathological conditions. However, these abnormal inclusions appear not to be effectively degraded and removed despite ubiquitination.

We have undertaken protein chemical analysis of PHF to address the following issues: whether ubiquitin in PHF occurs in a free form or a conjugated form; if conjugated, what the target protein is in PHF; and why the ubiquitin pathway is not effective in removing PHF. Immunoblot analysis of AD Sarkosyl-insoluble pellets with various tau antibodies displayed diffuse smearing substances in addition to PHF-tau. This smearing on the blot, presumably representing more modified tau, is highly characteristic of PHF-enriched fraction (Ihara et al. 1983). On the other hand, a monoclonal antibody to ubiquitin, DF2, which was originally raised against SDS-PHF (Mori et al. 1987), labeled the smear but not PHF-tau or other distinct bands. This suggests that ubiquitin is not attached to PHF-tau but to more processed smearing tau. We therefore purified the smear having both tau- and ubiquitin-immunoreactivities, which is referred to as the PHF smear, and identified the ubiquitin-conjugated protein by direct protein chemical analysis (Morishima-Kawashima et al. 1993).

In the course of purification of the PHF smear, we noticed the presence of two types of PHF smear: one is only tau-immunoreactive (the Ub(−) PHF smear) and the other is both tau- and ubiquitin-reactive (the Ub(+) PHF smear). AP I peptide mapping of the Ub(+) PHF smear, PHF-tau and ubiquitin revealed that the profile of the Ub(+) PHF smear was very similar to those of PHF-tau and ubiquitin put together, but that the most carboxy-terminal peptide of ubiquitin was undetectable in the PHF smear map. This suggests that its carboxy-terminal portion was conjugated with a certain peptide and eluted elsewhere. Several peaks specific for the PHF smear were subjected to protein sequence and ISMS analyses. In this way, we identified five ubiquitin-ligated tau peptides (Morishima-Kawashima et al. 1993; Fig. 2). Thus, ubiquitin targets tau in PHF at multiple sites; Lys-254, 257, 311, and 317 in the tau serve as acceptors for ubiquitin. These ubiquitin-conjugated sites are all localized to the microtubule-binding region of tau, which is in sharp contrast to the phosphorylation sites found in PHF-tau. Further, we found a ubiquitin-ubiquitin conjugate derived from a multiubiquitin chain. Although a multiubiquitin chain exists, its proportion seems to be very small and the monoubiquitinated form is predominant in PHF.

In the PHF smear, the amounts of AP I peptides from the amino-terminal portion of tau were greatly reduced as compared to those from the carboxyl-terminal portion (Morishima-Kawashima et al. 1993; Fig. 2). This implies that the amino-terminal portion of tau is cleaved off in the PHF smear. This finding was also confirmed by immunoblots of the PHF smear probed with various tau antibodies whose epitopes spanned the tau molecule. The extent of the processing is considerable in the Ub(−) PHF smear,

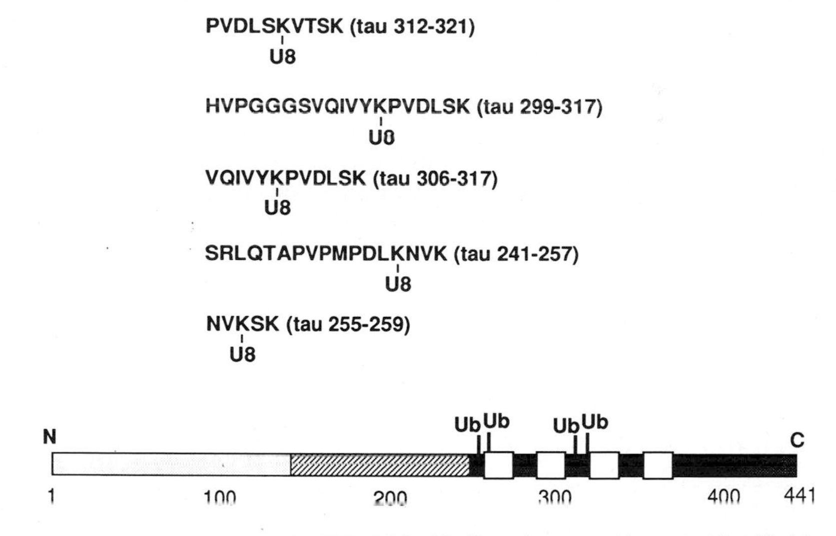

Fig. 2. Ubiquitination sites in the tau in PHF. Ubiquitin-ligated tau peptides were identified in the PHF smear (*top*). The numbering is according to a 441-residue human tau isoform (Goedert et al. 1989). U8 is the most carboxy-terminal peptide generated from ubiquitin by AP I digestion. These ubiquitin conjugation sites are illustrated by Ub along the tau molecule (*bottom*). *Open boxes* indicate 18-amino acid microtubule-binding domains. *Dotted and shadowed boxes* indicate greatly and significantly reduced portions in the tau from the PHF smear as compared with PHF-tau, respectively. Note that all the ubiquitin conjugation sites are localized to the microtubule-binding region

and much more so in the Ub(+) PHF smear. The most straightforward explanation for the findings is that the amino-terminal processing of tau in PHF precedes its ubiquitination; PHF-tau, hyperphosphorylated full-length tau, accumulates as PHF, which is then gradually cleaved off at its amino-terminus, followed by ubiquitination. Only after considerable amino-terminal processing does the tau in PHF become ubiquitinated. This indicates that the formation of PHF (aggregation of insoluble protein) alone does not trigger ubiquitination. It might be due to the blocked (acetylated) amino terminus of PHF-tau (Hasegawa et al. 1992), since unmodified amino-terminus is necessary for ubiquitination and subsequent degradation (see also Mayer et al. 1989).

Why are PHF not degraded despite ubiquitination? There are several possible factors which should be taken into account. First, the multiubiquitin chain, a strong degradation signal, comprises only a small proportion of ubiquitin in PHF (Morishima-Kawashima et al. 1993). Most PHF may not be recognized by the 26S proteasome complex as the target to be degraded even after ubiquitination, and they are allowed to accumulate. Second, the newly assembled PHF in affected neurons presumably take a long time to be ubiquitinated and attacked by the ubiquitin degradation system. This assumption is also consistent with our recent confocal microscopic observation that a large number of curly fibers remain non-ubiquitinated (Iwatsubo et al. 1992). Once dispersed PHF get together and form tightly packed PHF bundles (neurofibrillary tangles), the large proteasome complex may not be able to reach the ubiquitination sites, which are found abundantly deep inside the tangle but not on the superficial area. Related to this, it is noteworthy that antibodies to the 26S proteasome complex (Kwak et al. 1991) and ubiquitin carboxyl-terminal hydrolase (Lowe et al. 1990) do not label mature tangles, whereas Lewy bodies are intensely labeled. Third, the inability to degrade may be attributable to the unusual properties of the PHF core (Selkoe et al. 1985), such as the resistance of its rigid structure (antiparallel β pleated sheet) to various proteases, including the 26S proteasome. Finally, we suggest the possibility that the ubiquitin-dependent degradation system is affected in the degenerating neuron of AD brain, because it is an energy-dependent metabolic pathway vulnerable to degeneration (see Morishima-Kawashima et al. 1993).

References

Arriagada PV, Growdon JH, Hedley-Whyte ET, Hyman BT (1992) Neurofibrillary tangles but not senile plaques parallel duration and severity of Alzheimer's disease. Neurology 42: 631–639

Bancher C, Grundke-Iqbal I, Iqbal K, Fried VA, Smith HT, Wisniewski HM (1991) Abnormal phosphorylation of tau precedes ubiquitination in neurofibrillary pathology of Alzheimer disease. Brain Res 539: 11–18

Biernat J, Gustke N, Drewes G, Mandelkow E-M, Mandelkow E (1993) Phosphorylation of Ser^{262} strongly reduces binding of tau to microtubules: Distinction between PHF-like immunoreactivity and microtubule binding. Neuron 11: 153–163

Braak H, Braak E, Grundke-Iqbal I, Iqbal K (1986) Occurrence of neuropil threads in the senile human brain and in Alzheimer's disease: a third location of paired helical filaments outside of neurofibrillary tangles and neuritic plaques. Neurosci Lett 65: 351–355

Drewes G, Lichtenberg-Kraag B, Döring F, Mandelkow E-M, Biernat J, Goris J, Dorée M, Mandelkow E (1992) Mitogen activated protein (MAP) kinase transforms tau protein into an Alzheimer-like state. EMBO J 11: 2131–2138

Drewes G, Mandelkow E-M, Baumann K, Goris J, Merlevede W, Mandelkow E (1993) Dephosphorylation of tau protein and Alzheimer paired helical filaments by calcineurin and phosphatase-2A. FEBS Lett 336: 425–432

Finley D, Chau V (1991) Ubiquitination. Ann Rev Cell Biol 7: 25–69

Goedert M, Spillantini MG, Jakes R, Rutherford D, Crowther RA (1989) Multiple isoforms of human microtubule-associated protein tau: Sequences and localization in neurofibrillary tangles of Alzheimer's disease. Neuron 3: 519–526

Goedert M, Cohen ES, Jakes R, Cohen P (1992) p 42 MAP kinase phosphorylation sites in microtubule-associated protein tau are dephosphorylated by protein phosphatase 2A1. Implications for Alzheimer's disease. FEBS Lett 312: 95–99

Gong C-X, Singh TJ, Grundke-Iqbal I, Iqbal K (1993) Phosphoprotein phosphatase activities in Alzheimer disease brain. J Neurochem 61: 921–927

Gong C-X, Grundke-Iqbal I, Damuni Z, Iqbal K (1994) Dephosphorylation of microtubule-associated protein tau by protein phosphatase-1 and -2C and its implication in Alzheimer disease. FEBS Lett 341: 94–98

Greenberg SG, Davies P (1990) A preparation of Alzheimer paired helical filaments that displays distinct τ proteins by polyacrylamide gel electrophoresis. Proc Natl Acad Sci USA 87: 5827–5831

Greenberg SG, Davies P, Schein JD, Binder LI (1992) Hydrofluoric acid-treated τPHF proteins display the same biochemical properties as normal τ. J Biol Chem 267: 564–569

Hasegawa M, Morishima-Kawashima M, Takio K, Suzuki M, Titani K, Ihara Y (1992) Protein sequence and mass spectrometric analyses of tau in the Alzheimer's disease brain. J Biol Chem 267: 17047–17054

Hasegawa M, Watanabe A, Takio K, Suzuki M, Arai T, Titani K, Ihara Y (1993) Characterization of two distinct monoclonal antibodies to paired helical filaments: Further evidence for fetal-type phosphorylation of the τ in paired helical filaments J Neurochem. 60: 2068–2077

Hershko A, Ciechanover A (1992) The ubiquitin system for protein degradation. Annu Rev Biochem 61: 761–807

Ihara Y (1988) Massive somatodendritic sprouting of cortical neurons in Alzheimer's disease. Brain Res 459: 138–144

Ihara Y, Abraham C, Selkoe DJ (1983) Antibodies to paired helical filaments in Alzheimer's disease do not recognize normal brain proteins. Nature 304: 727–730

Ishiguro K, Omori A, Takamatsu M, Sato K, Arioka M, Uchida T, Imahori K (1992) Phosphorylation sites on tau by tau protein kinase I, a bovine derived kinase generating an epitope of paired helical filaments. Neurosci Lett 148: 202–206

Ishiguro K, Shiratsuchi A, Sato S, Omori A, Arioka M, Kobayashi S, Uchida T, Imahori K (1993) Glycogen synthase kinase 3β is identical to tau protein kinase I generating several epitopes of paired helical filaments. FEBS Lett 325: 167–172

Iwatsubo T, Hasegawa M, Esaki Y, Ihara Y (1992) Lack of ubiquitin immunoreactivities at both ends of neuropil threads. Possible bidirectional growth of neuropil threads. Am J Pathol 140: 277–282

Kanemaru K, Takio K, Miura R, Titani K, Ihara Y (1992) Fetal-type phosphorylation of the τ in paired helical filaments. J Neurochem 58: 1667–1675

Kemp BE, Pearson RB (1990) Protein kinase recognition sequence motifs. Trends Biochem Sci 15: 342–346

Kondo J, Honda T, Mori H, Hamada Y, Miura R, Ogawara M, Ihara Y (1988) The carboxyl third of tau is tightly bound to paired helical filaments. Neuron 1: 827–834
Kowall NW, Kosik SK (1987) Axonal disruption and aberrant localization of tau protein characterize the neuropil pathology of Alzheimer's disease. Ann Neurol 22: 639–643
Kwak S, Masaki T, Ishiura S, Sugita H (1991) Multicatalytic proteinase is present in Lewy bodies and neurofibrillary tangles in diffuse Lewy body disease brains. Neurosci Lett 128: 21–24
Lee VM-Y, Balin BJ, Otvos L Jr, Trojanowski JQ (1991) A68: a major subunit of paired helical filaments and derivatized forms of normal tau. Science 251: 675–678
Lew J, Winkfein RJ, Paudel HK, Wang JH (1992) Brain proline-directed protein kinase is a neurofilament kinase which displays high sequence homology to $p34^{cdc2}$. J Biol Chem 267: 25922–25926
Lowe J, McDermott H, Landon M, Mayer RJ, Wilkinson KD (1990) Ubiquitin carboxyl-terminal hydrolase (PGP 9.5) is selectively present in ubiquitinated inclusion bodies characteristic of human neurodegenerative diseases. J Pathol 161: 153–160
Mayer A, Siegel NR, Schwartz AL, Ciechanover A (1989) Degradation of proteins with acetylated amino termini by the ubiquitin system. Science 244: 1480–1483
Mayer RJ, Arnold J, László L, Landon M, Lowe J (1991) Ubiquitin in health and disease. Biochim Biophys Acta 1089: 141–157
Meyer HE, Hottmann-Posorske E, Heilmeyer LMG Jr (1991) Determination and localization of phosphoserine in proteins and peptides by conversion to S-ethylcysteine. Meth Enzymol 201: 169–185
Meyerson M, Enders GH, Wu C-L, Su L-K, Gorka C, Nelson C, Harlow E, Tsai L-H (1992) A family of human cdc2-related protein kinases. EMBO J 11: 2909–2917
Mori H, Kondo J, Ihara Y (1987) Ubiquitin is a component of paired helical filaments in Alzheimer's disease. Science 235: 1641–1644
Morishima-Kawashima M, Hasegawa M, Takio K, Suzuki M, Titani K, Ihara, Y (1993) Ubiquitin is conjugated with amino-terminally processed tau in paired helical filaments. Neuron 10: 1151–1160
Morishima-Kawashima M, Hasegawa M, Takio K, Suzuki M, Yoshida H, Titani K, Ihara Y (1995a) Proline-directed and non-proline-directed phosphorylation of PHF-tau. J Biol Chem (in press)
Morishima-Kawashima M, Hasegawa M, Takio K, Suzuki M, Yoshida H, Watanabe A, Titani K, Ihara Y (1995b) Hyperphosphorylation of tau in PHF. Neurobiol Aging (in press)
Mullan M, Crawford F (1993) Genetic and molecular advances in Alzheimer's disease. Trends Neurosci 16: 978–979
Pappolla MA, Omar R, Saran B (1989) The "normal" brain: "Abnormal" ubiquitinlated deposits highlight an age-related protein change. Am J Pathol 135: 585–591
Perry G, Kawai M, Tabaton M, Onorato M, Mulvihill P, Richey P, Monandi A, Connolly JA, Gambetti P (1991) Neuropil threads of Alzheimer's disease show a marked alteration of the normal cytoskeleton. J Neurosci 11: 1748–1755
Scott CW, Spreen RC, Herman JL, Chow FP, Davison MD, Young J, Caputo CB (1993) Phosphorylation of recombinant tau by cAMP-dependent protein kinase. Indentification of phosphorylation sites and effects on microtubule assembly. J Biol Chem 268: 1166–1173
Selkoe DJ (1991) The molecular pathology of Alzheimer's disease. Neuron 6: 487–498
Selkoe DJ, Ihara Y, Salazar FJ (1982) Alzheimer's disease: insolubility of partially purified paired helical filaments in sodium dodecyl sulfate and urea. Science 215: 1243–1245
Selkoe DJ, Ihara Y, Abraham C, Rasool CG, McCluskey AH (1985) Biochemical and immunocytochemical studies of Alzheimer paired helical filaments. In: Katzman R (ed) Biological aspects of Alzheimer's disease (Banbury Report 15). Cold Spring Harbor Laboratory, New York, pp 125–134
Terry RD, Peck A, De Teresa R, Schechter R, Horoupian DS (1981) Some morphometric aspects of the brain in senile dementia of the Alzheimer type. Ann Neurol 10: 184–192

Terry RD, Masliah E, Salmon DP, Butters N, DeTeresa R, Hill R, Hansen LA, Katzman R (1991) Physical basis of cognitive alterations in Alzheimer's disease: synapse loss is the major correlate of cognitive impairment. Ann Neurol 30: 572–580

Watanabe A, Hasegawa M, Suzuki M, Takio K, Morishima-Kawashima M, Titani K, Arai T, Kosik KS, Ihara Y (1993) *In vivo* phosphorylation sites in fetal and adult rat tau. J Biol Chem 268: 25712–25717

Wischik C, Novak M, Thøgersen HC, Edwards PC, Runswick MJ, Jakes R, Walker JE, Milstein C, Roth M, Klug A (1988) Isolation of a fragment of tau derived from the core of the paired helical filament of Alzheimer disease. Proc Natl Acad Sci USA 85: 4506–4510

Woodgett JR (1990) Molecular cloning and expression of glycogen synthase kinase-3/Factor A. EMBO J 9: 2431–2438

Yang S-D, Song J-S, Yu J-S, Shiah S-G (1993a) Protein kinase FA/GSK-3 phosphorylates τ on Ser^{235}-Pro and Ser^{404}-Pro that are abnormally phosphorylated in Alzheimer's disease brain. J Neurochem 61: 1742–1747

Yang S-D, Song J-S, Liu W-K, Yen S-H (1993b) Synergistic control mechanism for abnormal site phosphorylation of Alzheimer's diseased brain tau by kinase FA/GSK-3 α. Biochem Biophys Res Commun 197: 400–406

Yoshida H, Ihara Y (1993) τ in paired helical filaments is functionally distinct from fetal-τ: Assembly incompetence of paired helical filament-τ. J Neurochem 61: 1183–1186

Tau Variants in Aging and Neurodegenerative Disorders

V. Buée-Scherrer, L. Buée, P.R. Hof, P. Vermersch, B. Leveugle, A. Wattez, C. Bouras, D.P. Perl, and *A. Delacourte**

Introduction

One of the main features of many neurodegenerative disorders is the presence of neurofibrillary tangles (NFT) in the cerebral cortex. They are commonly found in Alzheimer's disease (AD), amyotrophic lateral sclerosis/parkinsonism-dementia complex of Guam (ALS/PDC), dementia pugilistica, head injury, Hallervorden-Spatz disease, post-encephalitic parkinsonism, and progressive supranuclear palsy (PSP) and are frequently observed in Pick's disease (Hof et al. 1991, 1992a, 1994a; Rewcastle 1991). They are also seen in normal aging (Bouras et al. 1994). In all of these disorders, NFT share similar antigenic properties (Yen et al. 1983; Dickson et al. 1985; Goldman and Yen 1986; Vickers et al. 1992, 1994). Abnormally phosphorylated tau proteins are the main components of the aggregated filaments found in NFT in AD (Grundke-Iqbal et al. 1986; Delacourte et al. 1990; Greenberg et al. 1992). Similar tau immunoreactivity is observed in NFT of most of these neurodegenerative disorders (Pollock et al. 1986; Tabaton et al. 1988; Shankar et al. 1989; Cammarata et al. 1990; Trojanowski et al. 1993). Tau proteins belong to the microtubule-associated proteins family (Weingarten et al. 1975). In the human brain, six isoforms are produced from a single gene by alternative mRNA splicing. They differ from each other by either the presence of three or four microtubule-binding regions in the carboxy-terminal (C-terminal) part of the molecule and one or two inserts in the amino-terminal (N-terminal) part (Goedert et al. 1989; Himmler et al. 1989; Kosik et al. 1989; Lee et al. 1989). The molecular weight of the six isoforms ranges from 45 to 62 kDa when run on polyacrylamide gel electrophoresis in the presence of sodium dodecyl sulfate.

Among these disorders, we recently described differences in the tau electrophoretic profile. We report here recent data on the characterization of tau proteins in different neurodegenerative disorders as well as in normal aging.

* INSERM U422, Place de Verdun, 59045 Lille Cédex, France

K.S. Kosik et al. (Eds.)
Alzheimer's Disease: Lessons from Cell Biology

Tau Proteins in Alzheimer's Disease

Alzheimer's disease (AD) is a neurodegenerative disorder characterized by a progressive and irreversible loss of cortical neurons and by the presence of two neuropathological lesions: senile plaques (SP) and NFT. SP result from the extracellular accumulation of straight filaments made of βA4 peptide (Glenner and Wong 1984), whereas NFT correspond to the intraneuronal aggregation of paired helical filaments (PHF; Kidd 1963). At the microscopic level, NFT are preferentially observed in the large pyramidal cells of the infragranular layers (V-VI) of the associative cortical areas (they are widespread in frontal, parietal and temporal regions, fewer in the occipital lobe) and to a lesser extent in pyramidal cells of supragranular layers (II-III), and also in pyramidal cells of the entorhinal cortex and hippocampus (Fig. 1A,B). Many other cortical and subcortical areas, such as nucleus basalis of Meynert, amygdala, locus coeruleus and raphe, are also affected by this degenerating process (Braak and Braak 1985; Wisniewski et al. 1989; Rewcastle 1991).

The main components of PHF are abnormally phosphorylated tau proteins (Brion et al. 1985; Delacourte and Défossez 1986; Grundke-Iqbal et al. 1986; Kosik et al. 1986). In fact, immunoblotting has shown that PHF are composed of three pathological tau proteins called tau 55, 64 and 69, because of their molecular weight, or A68 (Flament and Delacourte 1989, 1990; Delacourte et al. 1990; Lee et al. 1991). According to Goedert et al. (1992a), tau 55 results from the hyperphosphorylation of the two tau variants with no N-terminal insert and with or without C-terminal microtubule binding domain, tau 64 from the variants with one N-terminal insert and with or without the additional C-terminal domain, and tau 69 from the variants with two N-terminal inserts and with or without the C-terminal domain. Using specific monoclonal antibodies against pathological tau proteins and peptide mapping, it has been demonstrated that 6 to 8 sites are abnormally phosphorylated (Hasegawa et al. 1992). Several kinases have been implicated in this abnormal phosphorylation, all directed against serine-proline or threonine-proline motives, like mitogen-activated protein kinases, glycogen synthase kinase-3, or proline-directed protein kinases (Ishiguro et al. 1991; Drewes et al. 1992; Goedert et al. 1992b; Gustke et al. 1992; Hanger et al. 1992; Ishiguro et al. 1992; Mandelkow et al. 1992; Vulliet et al. 1992; Hisanaga et al. 1993; Ishiguro et al. 1993).

We have also developed immunological probes specific for these sites to investigate the specificity of the tau triplet. By immunoblotting with these antibodies, we detected the abnormal tau proteins in all AD brains but never in the isocortical areas from normal adult brains. Immunoblotting also indicated a strong correlation between the immunohistochemical detection of NFT and the presence of the tau triplet, showing that it is a reliable and early marker of the degenerating process. Therefore, these pathological tau proteins can be used to quantify the neurofibrillary degeneration (Flament

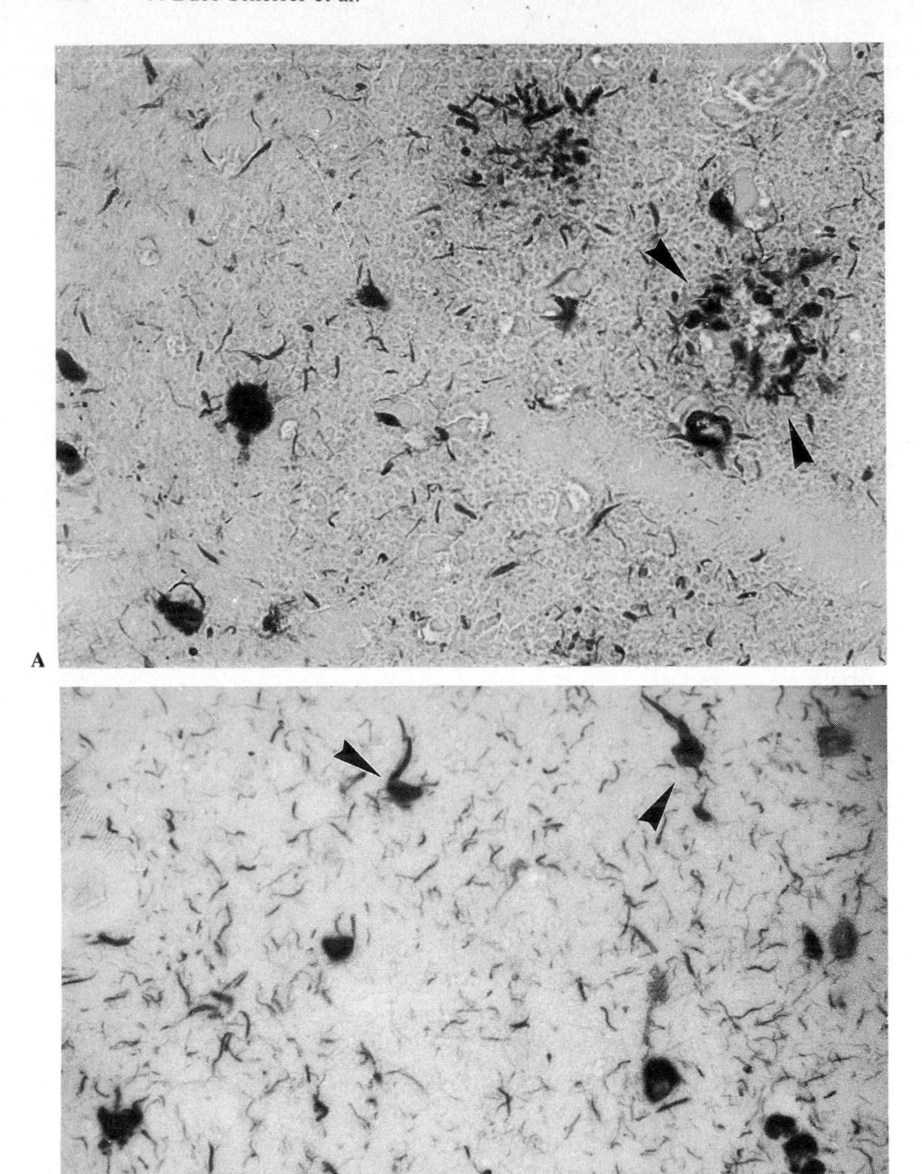

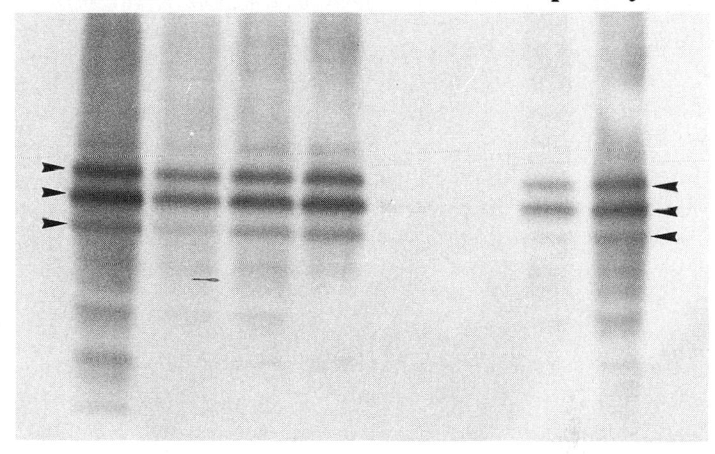

Fig. 1. A Distribution of NFT in the cerebral cortex of an AD case. CA1 field of the hippocampus. Note the presence of clusters of dystrophic neurites (*arrowheads*). **B** Layers II-III of Brodmann area 9 (superior frontal cortex). Note the presence of NFT and curly fibers (dystrophic neurites). Materials were stained with the antibody against PHF. **C** Immunoblot detection of pathological tau proteins with the absorbed anti-PHF. Proteins from brain homogenates were resolved onto 10–20% SDS-PAGE. After transfer, only tau 55, 64 and 69 (*arrowheads*) were detected, especially in the cortical associative areas (A 20, A9, and A11), amygdala (Amy) and hippocampus (Hip). Normal tau proteins ranging from 45 to 62 kDa were not labeled. However, note that in this particular case, tau immunoreactivity was weaker in the hippocampus as compared to the other areas. Note that no immunoreactivity is seen in primary motor cortex (A4) or subcortical structures, including basal ganglia (BG) and thalamus (Tha). The labeling of the abnormal tau triplet was surprisingly intense in the primary visual cortex (A17), but still weaker than in temporal neocortical and other association areas

et al. 1990a). A biochemical mapping was performed in several cortical areas of the brain from patients with senile dementia of the Alzheimer type. This analysis revealed that the detection of the pathological tau triplet was positive in all areas studied (Fig. 1C), with the exception of the primary motor and visual cortex (areas 4 and 17) in some cases, and that the detection was especially strong in temporal neocortical and limbic areas. The scores were higher in associative cortex than in primary sensory cortex but differed from one patient to another (Vermersch et al. 1992a).

It is interesting to note that tau pathology is also found in brain cortical areas from patients presenting with Down syndrome (Flament et al. 1990b) and, to a much lesser extent, in the entorhinal cortex and hippocampus of non-demented elderly individuals (Vermersch et al. 1992b; Bouras et al. 1993).

Tau Proteins in Non-demented Elderly Individuals

NFT are also encountered in the aging brain (Bouras et al. 1994; Giannakopoulos et al. 1994). The presence of these lesions and amyloid deposits raises the problem of "normal" and "pathological aging" (Dickson et al. 1991). Indeed, there are no clinical criteria to differentiate between these two types of aging, since these individuals do not demonstrate any significant cognitive changes raising the question of "incidental" NFT. Non-demented individuals demonstrate a consistent pattern of progressive changes with increasing age: NFT in layer II of the entorhinal cortex and in the CA1 field of the hippocampus (Mann et al. 1987; Braak and Braak 1991; Bouras et al. 1993, 1994; Giannakopoulos et al. 1994). In some cases, cortical involvement of NFT in association areas is also seen but it is likely to be related to changes other than "normal aging" (Hof et al. 1992b).

Using immunoblotting, we obtained results that correlated with these neuropathological data. Indeed, the abnormal tau triplet referred to as tau 55, 64 and 69 is also found in the same restricted areas of all non-demented elderly individuals in entorhinal cortex, and sometimes in the hippocampus (Vermersch et al. 1992b). However, the intensity of tau proteins immunodetection differs quantitatively as compared to AD patients, suggesting that the amount of tau proteins is lower (Fig. 2). The immunodetection of the tau triplet is present in the hippocampal formation of all of the patients over 75 years of age, and these findings do not correlate with most of the previous immunohistochemical data (Braak and Braak 1991; Dickson et al. 1991). Only one group reported the presence of NFT in all of the investigated

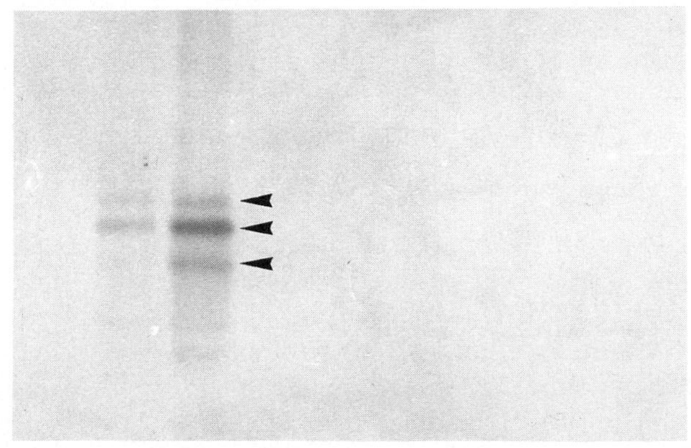

Fig. 2. Immunoblot detection of pathologic tau proteins with the abs. anti-PHF on brain homogenates from different areas of a typical non-demented individual. Tau 55, 64 and 69 indicated by arrowheads. They are only found in amygdala (Amy) and the hippocampal formation (Hip) but never in the other areas studied (BG, basal ganglia; Tha, thalamus; A4, A9, A11, A17 and A20, Brodmann areas 4, 9, 11, 17 and 20, respectively)

non-demented elderly cases (Bouras et al. 1993, 1994). These discrepancies are likely to be related to the high sensitivity of our biochemical method and the specificity of our immunological probes.

In conclusion, we demonstrated that, in aging, neurofibrillary degeneration is restricted to the hippocampal formation. Furthermore, this neurodegeneration is of the Alzheimer type, since it is characterized by the presence of the abnormal tau triplet (tau 55, 64 and 69). The specificity in the regional distribution in aging is completely different from the pathology encountered in AD. Indeed, NFT appear first in the same region, namely the hippocampal formation, but they rapidly extend to the close associative isocortical areas, such as inferior temporal cortex. Such data do not support the concepts of normal and pathological aging.

Biochemical features of tau proteins from NFT found in other neurodegenerative disorders than AD, including progressive supranuclear palsy (PSP), amyotrophic lateral sclerosis/parkinsonism dementia complex of Guam (ALS/PDC) and Parkinson's disease with dementia, were also investigated.

Tau Proteins in Progressive Supranuclear Palsy

PSP is a sporadic condition that was first described by Steele et al. in 1964. This neurodegenerative disorder is characterized by paralysis of downward and lateral gaze, gait ataxia, and nuchal and facial dystonia. Dementia is also a common feature at the end-stage of the disease. Neuropathologically, PSP is characterized by neuronal loss, gliosis and NFT. NFT were first described in basal ganglia, brain stem, and cerebellum (Steele et al. 1964). The subcortical localization of the neuropathological lesions has led to the definition of PSP as a model of "subcortical dementias" (Albert et al. 1974; Cummings and Benson 1984). More recently, a cortical involvement of the degenerating process has been described, with the same features as subcortical NFT (Takahashi et al. 1989; Hauw et al. 1990; Hof et al. 1992a). These studies demonstrated that the primary motor cortex was more severely affected than neocortical association areas, as compared to AD (Hauw et al. 1990; Hof et al. 1992a). Immunohistochemical studies (Fig. 3A) have revealed that NFT are denser in supragranular layers (III) than in infragranular layers (V) in PSP, whereas in AD, NFT are denser in layer V than in layer III (Hauw et al. 1990; Hof et al. 1992a). Ultrastructural analyses further support the difference between AD and PSP, since PHF are found in AD whereas straight filaments are observed in PSP (Kidd 1963; Tellez-Nagel and Wisniewski 1973; Tomonaga 1977; Montpetit et al. 1985).

Despite structural differences, straight filaments and PHF share common epitopes regarding cytoskeletal proteins like tau proteins (Pollock et al. 1986; Bancher et al. 1987; Tabaton et al. 1988; Flament et al. 1991).

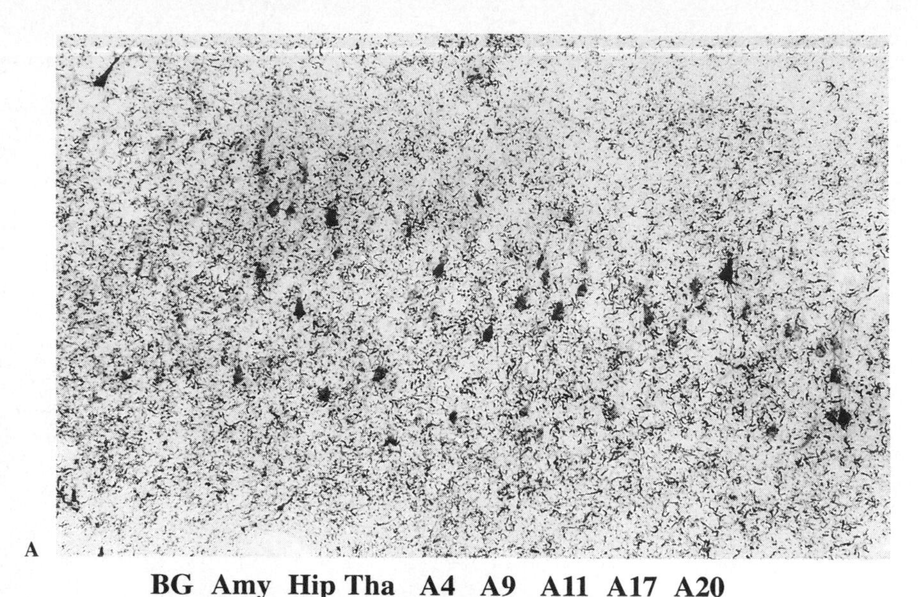

A

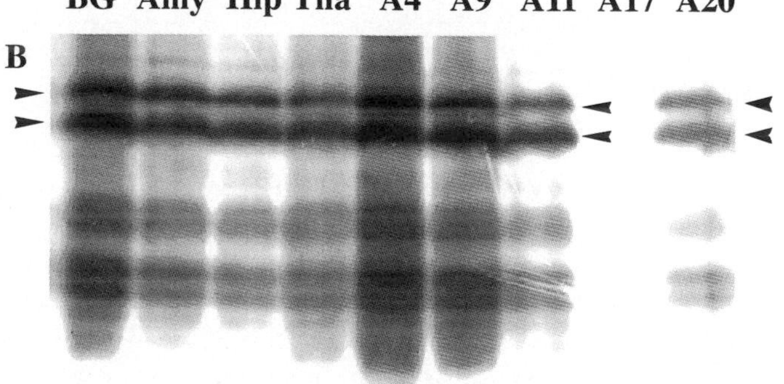

B

Fig. 3. A Distribution of NFT in the CA1 field of one case presenting with PSP and signs of dementia. Note that NFT are not in large numbers, but dystrophic neurites are numerous. Materials were labeled with the anti-PHF. **B** Immunoblot detection of pathologic tau proteins with the abs. anti-PHF on brain homogenates from different areas of a typical PSP case. Tau 55, 64 and 69 are indicated by arrowheads. They are found in all areas studied (BG, basal ganglia; Amy, amygdala; Hip, hippocampus; Tha, thalamus; A4, A9, A11, A20, Brodmann areas), with the exception of the primary visual cortex (A17). Note the strong immunoreactivity in primary motor cortex (A4) as compared to inferior temporal cortex (A20)

Biochemical approaches have shown that, in PSP brains, a doublet of abnormally phosphorylated tau proteins was detected (tau 64 and 69), instead of an abnormal tau triplet (55, 64 and 69) as in AD cases (Flament et al. 1991). Moreover, two-dimensional analysis revealed that the isoelectric properties of the doublet in PSP were different from those obtained in AD. Indeed, while tau 64 is the most acidic component in AD, tau 69 is the most acidic one in PSP (Flament et al. 1991). Recently, a biochemical mapping has been done on several cortical and subcortical areas from one PSP brain. This study revealed first, that the doublet of tau 64 and 69 is detected in the subcortical regions where NFT are found, and second, that neocortical areas are also affected (Vermersch et al. 1994). These results were in good agreement with previous neuropathological results that showed a cortical involvement (Hauw et al. 1990; Hof et al. 1992a). It is interesting to note that the presence of NFT in cortical areas is always correlated to dementia. We had the opportunity to analyze materials from a non-demented young PSP patient (33 years of age). In this case, abnormally phosphorylated tau proteins were found in both basal ganglia and thalamus whereas they were absent in all of the other areas studied, including amygdala, hippocampus, and Brodmann's areas 4, 9, 11, 17/18, 20 and 24. Conversely, the other PSP patients with dementia who were studied contained large amounts of pathological tau proteins in the neocortex, especially in Brodmann areas 4 and 6 and in subcortical structures (Fig. 3B). These findings suggest that the cortical pathology may play a role in the cognitive changes observed in PSP.

Tau Proteins in Parkinson's Disease

Parkinson's disease (PD) is usually sporadic. The most characteristic clinical features include tremor, expressionless face, rigidity, and slowness in initiating and performing voluntary movements. Approximatively 15 to 20% of patients with PD develop dementia (Mayeux et al. 1988). However, the nature and neuropathological basis of the dementia remain controversial (Cummings 1988). This dementia may be related either to cell loss in subcortical nuclei (Albert 1978; McGeer et al. 1984) or to AD pathology, since the incidence of pathologically proven AD in patients with PD appears to be higher than in the general population (Hakim and Mathieson 1979). The widespread distribution of Lewy bodies in the central nervous system in some demented PD cases may be an alternate explanation (Kosaka et al. 1984). Since patients with Lewy bodies may be a separate entity (Kosaka et al. 1984), only pathologic changes from PD cases with rare or no cortical Lewy bodies are described. All of these cases showed neuronal loss, and depigmentation of the substantia nigra and the locus coeruleus as well as gliosis. In PD cases with dementia, NFT were found in cortical areas.

We compared the electrophoretic tau profile in PD patients with or without dementia (data not shown). The pathological tau triplet referred to as tau 55, 64 and 69 was found in all cases. In cases without dementia, its presence was restricted to the hippocampal formation as described in non-demented elderly individuals. In patients with dementia, it was mainly detected in frontal and temporal areas, whereas cingulate and occipital

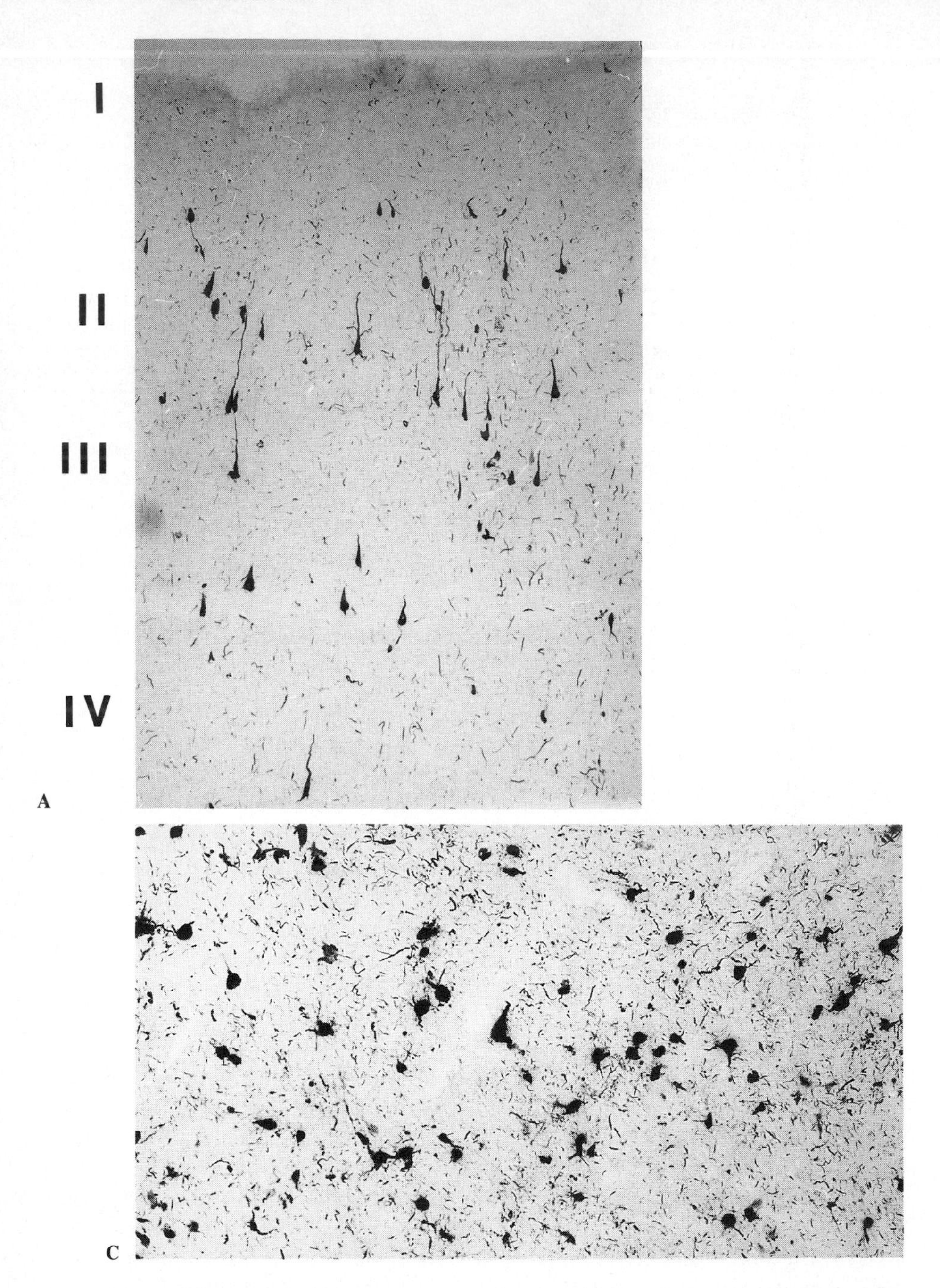
I
II
III
IV
A
C

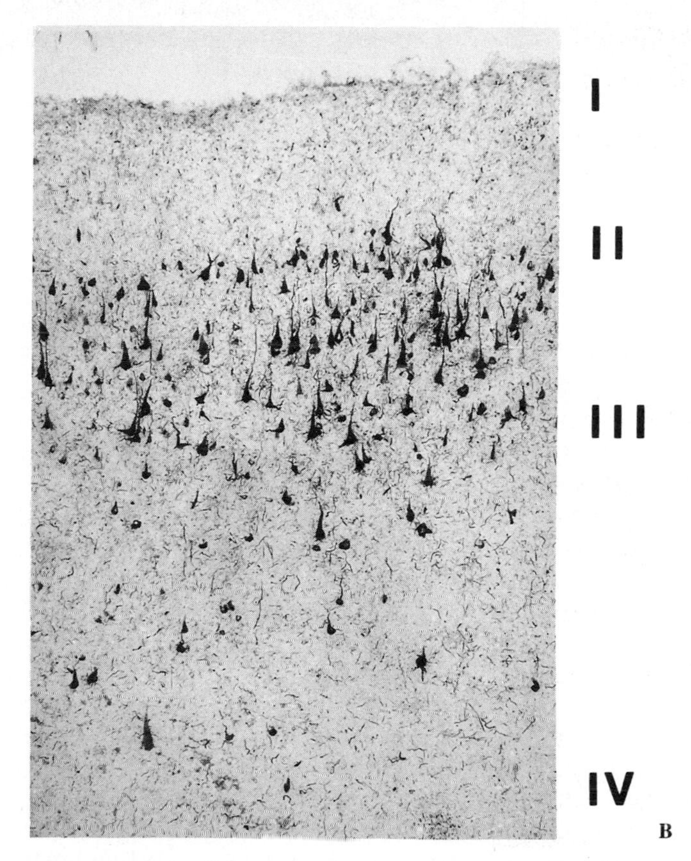

Fig. 4. Comparison of the NFT distribution between area 20 (**A**) and area 11 (**B**) in a ALS/PDC case (G1). Note that the inferior temporal cortex (**A**) is more affected than the orbito frontal cortex (**B**). Also note the laminar NFT distribution typical of Guamanian ALS/PDC, with very high densities in the supragranular layers (II-III) in both regions. **C**, Presence of NFT in the same Guamanian patient in subcortical structures such as the nucleus reuniens of the thalamus. Note that the subcortical areas are also severely affected by the degenerating process. Materials were stained with the anti-PHF antibody

cortex were relatively spared. It is interesting to note that, in these latter cases, the prefrontal cortex was always more affected than temporal areas. This is a striking difference from the pathologic changes seen in AD, which are always more severe in the temporal lobe. This prefrontal localization of NFT may be correlated to the clinical signs observed in PD with dementia. Indeed, cognitive changes observed in PD evoke a frontal lobe dysfunction (Pillon et al. 1986) that may be a primary event in the disease (Taylor et al. 1986). Our findings suggest that NFT in the frontal lobe may contribute to the onset of cognitive changes (Vermersch et al. 1993).

Tau Proteins in Amyotrophic Lateral Sclerosis/Parkinsonism Dementia Complex of Guam

Amyotrophic lateral sclerosis/parkinsonism-dementia complex (ALS/PDC) is a chronic degenerative disorder highly prevalent in the native Chamorro population of Guam island in the western Pacific Ocean (Hirano et al. 1961; Brody et al. 1975; Garruto and Yase 1986). Clinically, Guamanian ALS is undistinguishable from sporadic continental ALS and presents with fasciculations and lower and upper motor neuron signs. Parkinsonism-dementia is characterized by an insidious progressive mental decline and extrapyramidal signs (Hirano et al. 1961; Chen and Chase 1986). Guamanian ALS/PDC is characterized by a severe cortical atrophy and neuronal loss. The neuropathological hallmark of ALS/PDC is the extensive development of NFT, especially in the neocortex (Fig. 4A,B) and hippocampal formation (Hirano et al. 1961, 1966, 1967, 1968) but also in subcortical structures (Fig. 4C). Although NFT are present in large numbers in both AD and ALS/PDC, these two conditions do not display a similar laminar distribution in neocortex. NFT are preferentially distributed within layers II-III in the neocortical areas of Guamanian ALS/PDC cases (Fig. 4A,B) and are relatively sparser in layers V-VI. In AD, NFT are denser in layers V-VI than in layers II-III (Hof et al. 1991, 1994b). Immunohistochemical studies have also revealed that several cytoskeletal proteins (neurofilaments, microtubule-associated protein 2, tau and ubiquitin) are present in NFT of ALS/PDC patients (Shankar et al. 1989). The ultrastructure of NFT in Guamanian ALS/PDC consists of straight filaments and PHF (Rewcastle 1991), and PHF have been shown to be essentially similar to those observed in AD (Hirano et al. 1968).

Pathological tau proteins were consistently found in Guamanian ALS/PDC but their biochemical profile differed quantitatively among cases, showing that it is highly heterogeneous. For instance, five cases exhibited three distinct tau protein profiles, depending on their regional distribution (Buée-Scherrer et al. 1993). First, the abnormal tau triplet (tau 55, 64 and 69) was found in all of the brain regions (in both cortical and subcortical regions; Fig. 5A), in contrast to AD patients, where the triplet is mainly located in neocortical association areas (Vermersch et al. 1992a). Second, one elderly Guamanian control case and one case with early PD displayed the same abnormal tau triplet distribution restricted to the hippocampal formation and amygdala as observed in non-demented elderly Caucasian individuals (Vermersch et al. 1992b; Fig 2), demonstrating a pattern comparable to "normal aging" within the Chamorro population. Third, in a case with rapidly advancing ALS/PDC for one year prior to death, degenerative changes occurred first in limbic (amygdala, hippocampal formation) and polymodal association areas (Brodmann area 20; Fig. 5B), as observed in mild AD. The biochemical data always closely correlated with the clinical and neuropathological examination of these cases (Buée-Scherrer et al. 1993).

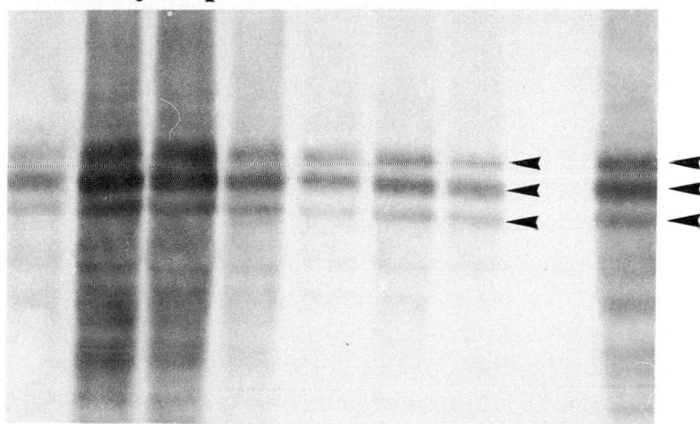

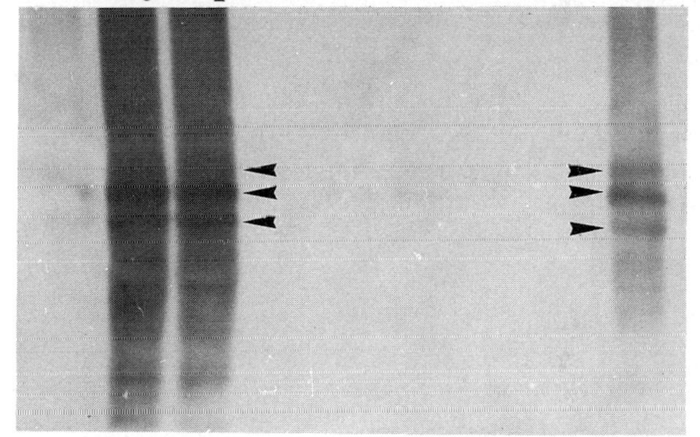

Fig. 5. A Immunoblotting of brain homogenates from case G1 with the abs. anti-PHF antibody. All regions biochemically studied from case G1 contain the abnormal tau triplet (*arrowheads*), except area 17/18. Amygdala (Amy) and hippocampal formation (Hip) demonstrate the strongest labeling, followed by area 20, indicating that this association area is affected later than Hip and Amy. Note that the other association areas (A4, A9 and A11) are less affected than the temporal cortex since the abnormal tau triplet staining is weaker. The abnormal tau triplet is also present in the subcortical areas such as basal ganglia (BG) and thalamus (Tha). **B** Immunoblotting of brain homogenates from a patient of Guam with an early onset of ALS/PDC, using the abs. anti-PHF antibody. Tau 55, 64 and 69 are strongly immunoreactive in the hippocampal formation (Hip), amygdala (Amy) and area 20. Note the presence of the abnormal tau triplet with a much lower immunoreactivity in area 4, area 9, and also in basal ganglia (BG) (*arrowheads*). The abnormal tau triplet is not found in primary and secondary visual cortex (A 17/18) and in other remaining association areas such as area 11 and in thalamus (Tha)

Conclusion

The present data indicate that there are at least two types of electrophoretic tau profile: the AD-type triplet and the PSP-type doublet. Moreover, the differences in the tau electrophoretic profiles might be related to the type of filaments (PHF for the triplet; straight filaments for the doublet) found in these NFT (Greenberg 1993). Among the different neurodegenerative disorders described, differences in the regional tau distribution were also observed. Since laminar and regional distributions of NFT are different among the dementing conditions, the presence of a pathologic tau triplet or doublet may be specific to a subtype of neurons. The differences in tau profile among these different neurodegenerative disorders probably reflect different etiopathogenic pathways involving either different sets of enzymes (kinases, phosphatases, etc.) or different types of neurons with their own pool of tau proteins. A better understanding of the molecular mechanisms leading to neuronal degeneration in these dementing illnesses will be of crucial importance to developing strategies aimed at the therapeutic protection of the vulnerable neuronal populations. Since a number of investigators have been able to induce abnormal phosphorylation of tau proteins in a cell model by using okadaic acid, a phosphatase inhibitor (Vandermeeren et al. 1993; Sautière et al. 1994; Vincent et al. 1994), these strategies might be developed in the near future.

Acknowledgements. Supported by the French Research Ministry, the France Alzheimer Association, NIH Grants AG05138 and AG08802, the Brookdale Foundation and the Florence J. Gould Association.

References

Albert ML, Feldman RG, Willis AL (1974) The "subcortical dementia" of progressive supranuclear palsy. J Neurol Neurosurg Psychiat 37: 121–130

Bancher C, Lassman H, Budka H, Grundke-Iqbal I, Iqbal K, Wiche G, Seitelberg F, Wiesniewski HM (1987) Neurofibrillary tangles in Alzheimer's disease and progressive supranuclear palsy: antigenic similarities and differences. Microtubule-associated tau protein is prominent in all types of tangles. Acta Neuropathol 74: 39–46

Bouras C, Hof PR, Morrison JH (1993) Neurofibrillary tangle densities in the hippocampal formation in a non-demented population define subgroups of patients with differential early pathologic changes. Neurosci Lett 153: 131–135

Bouras C, Hof PR, Giannakopoulos P, Michel J-P, Morrison JH (1994) Regional distribution of neurofibrillary tangles and senile plaques in the cerebral cortex of elderly patients: a quantitative evaluation of a one-year autopsy population from a geriatric hospital. Cerebral Cortex 4: 138–150

Braak H, Braak E (1985) On areas of transition between entorhinal allocortex and temporal isocortex in the human brain. Normal morphology and lamina-specific pathology in Alzheimer's disease. Acta Neuropathol 68: 325–332

Braak H, Braak E (1991) Neuropathological stageing of Alzheimer-related changes. Acta Neuropathol 82: 239–259

Brion JP, Passareiro H, Nunez J, Flament-Durand J (1985) Immunological detection of tau protein in neurofibrillary tangles of Alzheimer's disease. Arch Biol 95: 229–235

Brody JA, Stanhope JM, Kurland LT (1975) Patterns of amyotrophic lateral sclerosis and parkinsonism-dementia on Guam. In: Hornabrook RW, Davis FA (eds) Topics on tropical neurology, contemporary neurology series, vol 12. Philadelphia, Davis, pp 45–70

Buée-Scherrer V, Buée L, Vermersch P, Hof PR, Leveugle B, Loerzel A, Steele JC, Delacourte A, Perl DP (1993) Biochemical characterization of the Tau proteins in amyotrophic lateral sclerosis/parkinsonism dementia complex (ALS/PDC) of Guam. Soc Neurosci Abstr 19: 195

Cammarata S, Mancardi G, Tabaton M (1990) Formic acid treatment exposes hidden neurofilament and tau epitopes in abnormal cytoskeletal filaments from patients with progressive supranuclear palsy and Alzheimer's disease. Neurosci Lett 115: 351–355

Chen KM, Chase TN (1986) Parkinsonism-dementia. In: Vinken PJ, Bruyn GW, Klawans HL (eds) Extrapyramidal disorders. Handbook of clinical neurology, vol 49. Elsevier, Amsterdam, pp 167–183

Cummings JL (1988) The dementias of Parkinson's disease: prevalence, characteristics, neurobiology, and comparison with dementia of the Alzheimer type. Eur Neurol 28: 15–23

Cummings JL, Benson DF (1984) Subcortical dementia: review of an emerging concept. Arch Neurol 41: 874–879

Delacourte A, Défossez A (1986) Alzheimer's disease: tau proteins, the promoting factors of microtubule assembly are major components of paired helical filaments. J Neurol Sci 76: 173–186

Delacourte A, Flament S, Dibe EM, Hublau P, Sablonnière B, Hémon B, Scherrer V, Défossez A (1990) Pathological proteins tau 64 and 69 are specifically expressed in the somatodendritic domain of the degenerating cortical neurons during Alzheimer's disease: demonstration with a panel of antibodies against tau proteins. Acta Neuropathol 80: 111–117

Dickson WD, Kress Y, Crowe A, Yen SH (1985) Monoclonal antibodies to Alzheimer neurofibrillary tangles II: Demonstration of a common antigenic determinant between ANT and neurofibrillary degeneration in progressive supranuclear palsy. Am J Pathol 120: 292–303

Dickson WD, Crystal HA, Mattiace LA, Masur DM, Blau AD, Davies P, Yen SH, Aronson MK (1991) Identification of normal and pathological aging in prospectively studied nondemented elderly humans. Neurobiol Aging 13: 179–189

Drewes G, Lichtenberg-Kraag B, Döring F, Mandelkow EM, Biernat J, Dorée M, Mandelkow E (1992) Mitogen activated protein (MAP) kinase transforms tau protein into an Alzheimer-like state. EMBO J 11: 2131–2138

Flament S, Delacourte A (1989) Abnormal tau species are produced during Alzheimer's disease neurodegenerating process. FEBS Lett 247: 213–216

Flament S, Delacourte A (1990) Tau marker? Nature 346: 22

Flament S, Delacourte A, Delaère P, Duyckaerts C, Hauw JJ (1990a) Correlation between microscopical changes and tau 64 and 69 biochemical detection in senile dementia of the Alzheimer-type. Tau 64 and 69 are reliable markers of the neurofibrillary degeneration. Acta Neuropathol 80: 212–215

Flament S, Delacourte A, Mann DMA (1990b) Phosphorylation of tau proteins: a major event during the process of neurofibrillary degeneration. A comparative study between Alzheimer's disease and Down's syndrome. Brain Res 516: 15–19

Flament S, Delacourte A, Verny M, Hauw JJ, Javoy-Agid F (1991) Abnormal tau proteins in progressive supranuclear palsy. Similarities and differences with the neurofibrillary degeneration of the Alzheimer type. Acta Neuropathol 81: 591–596

Garruto RM, Yase Y (1986) Neurodegenerative disorders of the Western Pacific: the search for mechanisms of pathogenesis. Trends Neurosci 9: 368–371

Giannakopoulos P, Hof PR, Mottier S, Michel JP, Bouras C (1994) Neuropathological changes in the cerebral cortex of 1258 cases from a geriatric hospital: retrospective clinicopathological evaluation of a 10-year autopsy evaluation. Acta Neuropathol 87: 456–468

Glenner G, Wong CW (1984) Alzheimer's disease: initial report of the purification and characterization of a novel cerebrovascular amyloid protein. Biochem Biophys Res Commun 120: 885–890

Goedert M, Spillantini MG, Jakes R, Rutherford D, Crowther RA (1989) Multiple isoforms of human microtubule-associated protein tau: sequences and localization in neurofibrillary tangles of Alzheimer's disease. Neuron 3: 519–526

Goedert M, Spillantini MG, Cairns NJ, Crowther RA (1992a) Tau proteins of Alzheimer paired helical filaments: abnormal phosphorylation of all six brain isoforms. Neuron 8: 159–168

Goedert M, Cohen ES, Jakes R, Cohen P (1992b) p42 MAP kinase phosphorylation sites in microtubule associated protein tau are dephosphorylated by protein phosphatase $2A_1$. FEBS Lett 312: 95–99

Goldman JE, Yen S-H (1986) Cytoskeletal protein abnormalities in neurodegenerative diseases. Ann Neurol 19: 209–223

Greenberg SG (1993) Filament-associated tau proteins in neurodegenerative diseases. Soc Neurosci Abstr 19: 196

Greenberg SG, Davies P, Scheim JD, Binder LI (1992) Hydrofluoric acid-treated tau-PHF proteins display the same biochemical properties as normal tau. J Biol Chem 267: 564–569

Grundke-Iqbal I, Iqbal K, Tung Y-C, Zaidi MS, Wisniewski HM, Binder LI (1986) Abnormal phosphorylation of the microtubule-associated protein tau in Alzheimer cytoskeletal pathology. Proc Natl Acad Sci USA 83: 4913–4917

Gustke N, Steiner B, Mandelkow EM, Biernat J, Meyer HE, Goedert M, Mandelkow E (1992) The Alzheimer-like phosphorylation of tau protein reduces microtubule binding and involves Ser-Pro and Thr-Pro motifs. FEBS Lett 307: 199–205

Hakim AM, Mathieson G (1979) Dementia in Parkinson's disease: a neuropathological study. Neurology 29: 1209–1214

Hanger DP, Hughes K, Woodgett JR, Brion JP, Anderton BH (1992) Glycogen synthase kinase-3 induces Alzheimer's disease-like phosphorylation of tau: generation of paired helical filament epitopes and neuronal localization of the kinase. Neurosci Lett 147: 58–62

Hasegawa M, Morishima-Kawashima M, Takio K, Suzuki M, Titani K, Ihara Y (1992) Protein sequence and mass spectrometric analyses of tau in the Alzheimer's disease brain. J Biol Chem 267: 17047–17054

Hauw JJ, Verny M, Delaère P, Cervera P, He Y, Duyckaerts C (1990) Constant neurofibrillary changes in the neocortex in progressive supranuclear palsy. Basic differences with Alzheimer's disease and aging. Neurosci Lett 119: 182–186

Himmler A, Drechsel D, Kirschner MW, Martin DW Jr (1989) Tau consists of a set of proteins with repeated C-terminal microtubule-binding domains and variable N-terminal domains. Mol Cell Biol 9: 1381–1388

Hirano A, Kurland LT, Krooth RS, Lessel S (1961) Parkinsonism-dementia complex, and endemic disease on the island of Guam: I. Clinical features. Brain 84: 642–661

Hirano A, Malamud N, Elizan TS, Kurland LT (1966) Amyotrophic lateral sclerosis and parkinsonism-dementia complex on Guam: further pathological studies. Arch Neurol 15: 35–51

Hirano A, Arumugasamy N, Zimmerman HM (1967) Amyotrophic lateral sclerosis: a comparison of Guam and classical cases. Arch Neurol 16: 357–363

Hirano A, Dembitzer HM, Kurland LT, Zimmerman HM (1968) The fine structure of some intraganglionic alterations: neurofibrillary tangles, granulovacuolar bodies and "rod-like" structures in Guam amyotrophic lateral sclerosis and parkinsonism-dementia complex. J Neuropathol Exp Neurol 27: 167–182

Hisanaga S-I, Ishiguro K, Uchida T, Okumura E, Okano T, Kishimoto T (1993) Tau protein kinase II has similar characteristic to cdc2 kinase for phosphorylating neurofilament proteins. J Biol Chem 268: 15056–15060

Hof PR, Perl D, Loerzel AJ, Morrison JH (1991) Neurofibrillary tangle distribution in the cerebral cortex of parkinsonism-dementia cases from Guam: differences with Alzheimer's disease. Brain Res 564: 306–313

Hof PR, Delacourte A, Bouras C (1992a) Distribution of cortical neurofibrillary tangles in progressive supranuclear palsy: a quantitative analysis of six cases. Acta Neuropathol 84: 45–51

Hof PR, Bierer LM, Perl DP, Delacourte A, Buée L, Bouras C, Morrison JH (1992b) Evidence for early vulnerability of the medial and inferior temporal lobe in a 82-year-old patient with possible preclinical signs of dementia: regional and laminar distribution of neurofibrillary tangles and senile plaques. Arch Neurol 49: 946–953

Hof PR, Bouras C, Perl DP, Morrison JH (1994a) Quantitative neuropathology analysis of Pick's disease cases: cortical distribution of Pick bodies and coexistence with Alzheimer's disease. Acta Neuropathol 87: 115–124

Hof PR, Perl DP, Loerzel J, Steele JC, Morrison JH (1994b) Amyotrophic lateral sclerosis and parkinsonism-dementia from Guam: differences in neurofibrillary tangles distribution and density in the hippocampal formation and neocortex. Brain Res 650: 107–116

Ishiguro K, Omori A, Kato K, Tomizawa K, Imahori K, Uchida T (1991) A serine/threonine proline kinase activity is included in the tau protein kinase fraction forming a paired helical filament epitope. Neurosci Lett 128: 195–198

Ishiguro K, Takamatsu M, Tomizawa K, Omori A, Kato K, Takahashi M, Arioka M, Uchida T, Imahori K (1992) Tau protein Kinase I converts normal tau protein into A68-like component of paired helical filaments. J Biol Chem 267: 10897–10901

Ishiguro K, Shiratsuchi A, Sato S, Omori A, Arioka M, Kobayashi SU (1993) Glycogen synthase kinase 3β is identical to tau protein kinase-I generating several epitopes of paired helical filaments. FEBS Lett 325: 167–172

Kidd M (1963) Paired helical filaments in electron microscopy of Alzheimer's disease. Nature 197: 192–193

Kosaka K, Yoshimura M, Ikeda K, Budka H (1984) Diffuse type of Lewy Body disease: progressive dementia with abundant cortical Lewy bodies and senile changes from varying degree – a new disease? Clin Neuropathol 3: 185

Kosik KS, Joachim CL, Selkoe DJ (1986) Microtubule associated protein tau is a major antigenic component of paired helical filaments in Alzheimer's disease. Proc Natl Acad Sci USA 83: 4044–4048

Kosik KS, Orecchio LD, Bakalis S, Neve RL (1989) Developmentally regulated expression of specific tau sequences. Neuron 2: 1389–1397

Lee G, Neve RL, Kosik KS (1989) The microtubule binding domain of tau protein. Neuron 2: 1615–1624

Lee VM-Y, Balin BJ, Otvos Jr L, Trojanowski JQ (1991) A68: a major subunit of paired helical filaments and derivatized forms of normal tau. Science 251: 675–678

McGeer PL, McGeer EG, Suzuki J, Dolman CE, Nagi T (1984) Aging, Alzheimer's disease, and the cholinergic system of the basal forebrain. Neurology 34: 741

Mandelkow EM, Drewes G, Biernat J, Gustke N, Van Lint J, Vandenheede JR, Mandelkow E (1992) Glycogen synthase kinase-3 and the Alzheimer-like state of microtubule-associated protein tau. FEBS Lett 314: 315–321

Mann DMA, Tuckert CM, Yates PO (1987) The topographic distribution of senile plaques and neurofibrillary tangles in the brains of non-demented persons at different ages. Neuropathol Appl Neurobiol 13: 123–139

Mayeux R, Stern Y, Rosenstein R, Marder K, Hanser A, Cote L, Fahn S (1988) An estimate of the prevalence of dementia in idiopathic Parkinson's disease. Arch Neurol 45: 260–262

Montpetit V, Clapin DF, Guberman A (1985) Substructure of 20-nm filaments of progressive supranuclear palsy. Acta Neuropathol 68: 311–318

Pillon B, Dubois B, Lhermitte F, Agid Y (1986) Heterogeneity of cognitive impairment in progressive supranuclear palsy, Parkinson's disease and Alzheimer's disease. Neurology 36: 1179–1185

Pollock NJ, Mirra SS, Binder LI, Hansen LA, Wood JG (1986) Filamentous aggregates in Pick's disease, progressive supranuclear palsy, and Alzheimer's disease show antigenic determinants with microtubule-associated protein tau. Lancet 2: 1211

Rewcastle NB (1991) Degenerative diseases of the central nervous system. In: Davis RL, Robertson DM (eds) Textbook of neuropathology, 2nd edn. Williams and Wilkins, Baltimore, pp 903–961

Sautière PE, Caillet-Boudin ML, Wattez A, Delacourte A (1994) Detection of Alzheimer-type tau proteins in okadaic acid-treated SKNSH-SY5Y neuroblastoma cells. Neurodegeneration 3: 53–60

Shankar S, Yanagihara R, Garruto RM, Grundke-Iqbal I, Kosik KS, Gajdusek DC (1989) Immunocytochemical characterization of neurofibrillary tangles in amyotrophic lateral sclerosis and parkinsonism-dementia on Guam. Ann Neurol 25: 146–151

Steele JC, Richardson JC, Olszewski J (1964) Progressive supranuclear palsy. A heterogeneous degeneration involving brain stem, basal ganglia and cerebellum with vertical gaze and pseudobulbar palsy, nuchal dystonia and dementia. Arch Neurol 10: 333–359

Tabaton M, Whitehouse PJ, Perry G, Davies P, Autilio-Gambetti L, Gambetti P (1988) Alz-50 recognizes abnormal filaments in Alzheimer's disease and progressive supranuclear palsy. Ann Neurol 24: 407–413

Takahashi H, Oyanagi K, Takeda S, Hinkuma K, Ikuta F (1989) Occurrence of 15-nm-wide straight tubules in neocortical neurons in progressive supranuclear palsy. Acta Neuropathol 79: 233–239

Taylor AE, Saint-Cyr JA, Lang AE (1986) Frontal lobe dysfunction in Parkinson's disease. The cortical focus of neostriatal outflow. Brain 109: 845–883

Tellez-Nagel I, Wisniewski HM (1973) Ultrastructure of neurofibrillary tangles in Steele-Richardson-Olszewski syndrome. Arch Neurol 29: 324–327

Tomonaga M (1977) Ultrastructure of neurofibrillary tangles in progressive supranuclear palsy. Acta Neuropathol 37: 177–181

Trojanowski JQ, Schmidt ML, Shin R-W, Bramblett GT, Rao D, Lee VMY (1993) Altered tau and neurofilament proteins in neurodegenerative diseases: diagnostic implications for Alzheimer's disease and Lewy body dementias. Brain Pathol 3: 45–54

Vandermeeren M, Lübke U, Six J, Cras P (1993) The phosphate inhibitor okadaic acid induces a phosphorylated paired helical filament tau epitope in human LA-N-5 neuroblastoma cells. Neurosci Lett 153: 57–60

Vermersch P, Frigard B, Delacourte A (1992a) Mapping of neurofibrillary degeneration in Alzheimer's disease: evaluation of heterogeneity using the quantification of abnormal tau proteins. Acta Neuropathol 85: 48–54

Vermersch P, Frigard B, David JP, Fallet-Bianco C, Delacourte A (1992b) Presence of abnormally phosphorylated tau proteins in the entorhinal cortex of aged non-demented subjects. Neurosci Lett 144: 143–146

Vermersch P, Delacourte A, Javoy-Agid F, Hauw JJ, Agid Y (1993) Dementia in Parkinson's disease: biochemical evidence for cortical involvement using the immunodetection of abnormal tau proteins. Ann Neurol 33: 445–450

Vermersch P, Robitaille Y, Bernier L, Wattez A, Gauvreau D, Delacourte A (1994) Biochemical mapping of neurofibrillary degeneration in a case of progressive supranuclear palsy: evidence for general cortical involvement. Acta Neuropathol 87: 572–577

Vickers JC, Delacourte A, Morrison JH (1992) Progressive transformation of the cytoskeleton associated with normal aging and Alzheimer's disease. Brain Res 594: 273–278

Vickers JC, Riederer BM, Marugg MA, Buée-Scherrer V, Buée L, Delacourte A, Morrison JH (1994) Alterations in neurofilament protein immunoreactivity in human hippocampal neurons related to normal aging and Alzheimer's disease. Neuroscience 62: 1–13

Vincent I, Rosado M, Kim E, Davies P (1994) Increased production of paired helical filament epitopes in a cell culture system reduces the turnover of tau. J Neurochem 62: 715–723

Vulliet R, Halbran SM, Braun RK, Smith AJ, Lee G (1992) Proline-directed phosphorylation of human tau protein. J Biol Chem 267: 22570–22574

Weingarten MD, Lockwood AH, Hwo S-H, Kirschner MW (1975) A protein factor essential for microtubule assembly. Proc Natl Acad Sci USA 72: 1858–1862

Wisniewski HM, Rabe A, Zigman W, Silverman W (1989) Neuropathological diagnosis of Alzheimer's disease. J Neuropathol Exp Neurol 48: 606–609
Yen SH, Horoupian DS, Terry RD (1983) Immunocytochemical comparison of neurofibrillary tangles in senile dementia of Alzheimer type, progressive supranuclear palsy and post-encephalitic parkinsonism. Ann Neurol 13: 172–175

A Phosphorylated Tau Species Is Transiently Present in Developing Cortical Neurons and Is Not Associated with Stable Microtubules

*J.P. Brion**, *A.M. Couck, J.L. Conreur,* and *J.N. Octave*

Summary

During brain development, the microtubule-associated protein tau presents a transient state of high phosphorylation, similar to the phosphorylation status of paired helical filaments-tau in Alzheimer's disease. We have investigated the developmental distribution of this phosphorylated foetal-type tau in the developing rat cortex and in cultures of embryonic cortical neurons using antibodies that react with tau in a phosphorylation-dependent manner. The phosphorylated foetal-type tau was present in the developing cortex at 20 days but not at 18 days of embryonic life and was not detected before 4–5 days in neuronal culture. The cyclin-dependent kinase $p34^{cdc2}$ was expressed only in germinal layers in the embryonic brain and was not co-localized with phosphorylated tau. After 10 days of postnatal life, this phosphorylated tau progressively disappeared from cortical neurons, disappearing first from the deepest cortical layers where neurons are ontogenetically the oldest. This phosphorylated tau was found in axons and dendrites of cortical neurons at all developmental stages, whereas unphosphorylated tau tended to disappear from dendrites during development. The timing of appearance of phosphorylated tau in the cortex, by comparison with the expression of other developmental markers, indicates that phosphorylated tau is present at a high level only during the period of intense neuritic outgrowth and that it disappears during the period of neurite stabilization and synaptogenesis, concomitant with the expression of adult tau isoforms. In control cultures and in cultures treated with colchicine, this phosphorylated tau was not associated with cold-stable and colchicine-resistant microtubules.

These in vivo results suggest that the high expression of this phosphorylated tau species is correlated with the presence of a dynamic microtubule network during a period of high plasticity in the developing brain.

---

* Laboratory of Pathology and Electron Microscopy, Université Libre de Bruxelles, 808 route Lennik, Bldg C-10, 1070 Brussels, Belgium

K.S. Kosik et al. (Eds.)
Alzheimer's Disease: Lessons from Cell Biology

Introduction

The morphological differenciation of neurons is a complex phenomenon encompassing different processes, i.e., growth of cell bodies, extension of dendrites and axons, and synaptogenesis. Microtubules play a crucial role during neuronal differenciation since the outgrowth of neurites is known to be strongly dependent on microtubule assembly (Yamada et al. 1970). Brain microtubules, in addition to dimers of α- and β-tubulins, contain several groups of microtubule-associated proteins (MAPs). Both the level of expression of these proteins and the composition of microtubules show extensive changes during neuronal development (for review, see Tucker 1990; Matus 1991). Among MAPs, tau proteins are thought to play a role in axonal development (Caceres and Kosik 1990; Baas et al. 1991). Tau proteins can also promote microtubule assembly in vitro (Cleveland et al. 1977), bind to microtubules and stabilize them in vivo (Drubin and Kirschner 1986), and induce bundling of microtubules (Kanai et al. 1989; Scott et al. 1992). Tau proteins can be phosphorylated in vitro by several protein kinases and this phosphorylation is thought to modulate their function in vivo. For instance, phosphorylated tau has been shown to be less efficient than unphosphorylated tau in promoting microtubule assembly (Lindwall and Cole 1984).

Tau proteins have been found to be integral components of the paired helical filaments (PHF) which compose the neurofibrillary tangles, a characteristic neuronal lesion of Alzheimer's disease (Brion et al. 1985; Delacourte and Defossez 1986; Grundke-Iqbal et al. 1986; Kosik et al. 1986; Nukina and Ihara 1986; Wood et al. 1986; Goedert et al. 1988; Wischik et al. 1988; Lee et al. 1991). It has recently become evident that tau is highly phosphorylated in the developing brain in a manner similar to the phosphorylation state of tau proteins in Alzheimer's disease (Kanemaru et al. 1992; Brion et al. 1993; Bramblett et al. 1993; Goedert et al. 1993; Kenessey and Yen 1993; Watanabe et al. 1993). This phosphorylation is strongly reduced after the first 2–3 weeks of postnatal development in the rat (Brion et al. 1993). These findings suggest that the developmental-type regulation of tau phosphorylation may reappear in some neurones in Alzheimer's disease.

In the adult central nervous system, tau consists of several isoforms generated from a single gene by alternative splicing. These isoforms, ranging in size from 352 to 441 amino acids in the human species (Goedert et al. 1989), differ in the presence towards the carboxyl terminus of the three or four tandem repeat regions and the presence or absence of insertions in the amino-terminal half of the molecule. The repeat regions constitute the microtubule-binding domains of tau. Tau proteins show a striking developmental evolution (Francon et al. 1982). The shortest tau isoform containing three tandem repeats and no insertion towards the amino domain is the only isoform present in the embryonic brain and its expression decreases sharply at the end of the first postnatal week (Kosik et al. 1989; Brion et al. 1993), whereas adult tau isoforms begin to appear during the second postnatal week. This foetal-type tau isoform has been found to be less efficient than adult isoforms in promoting microtubules assembly in vitro (Francon et al. 1982; Goedert and Jakes 1990). This property might account for the presence of a less stable and more "plastic" microtubule network in developing neurones, which is necessary during the extensive remodeling and the complex neuritic growth that these neurones perform during development.

In this study we have investigated the differential distribution of phosphorylated and unphosphorylated tau during cortical development in the rat from embryonic to adult stages and during the differenciation of neurones in primary cultures. The timing of appearance and the distribution of a phosphorylated tau species were compared to the distribution of other molecules showing a developmental evolution to investigate to what extent the changes in tau phosphorylation are correlated with steps in the differenciation of neurones.

Experimental Procedures

Antibodies

The B19 rabbit polyclonal antiserum, raised to adult bovine tau proteins (Brion et al. 1991a), reacts with all known adult and foetal tau isoforms, and this recognition is insensitive to the phosphorylation status of tau proteins. The tau-1 mouse monoclonal antibody (Binder et al. 1985) was purchased from Boehringer. The reactivity of tau-1 antibody with tau requires the presence of an unphosphorylated serine in position 199 and/or 202 of the tau molecule (Biernat et al. 1992; according to the numbering of the longest of the low molecular weight human tau isoform), i.e., in a domain common to all tau isoforms. The AT8 mouse monoclonal antibody (kindly supplied by Innogenetics) reacts strongly with the phosphorylated PHF-tau found in Alzheimer's disease but not with the tau isoforms found in the adult brain (Mercken et al. 1992; Biernat et al. 1992). It also reacts with phosphorylated foetal tau (Brion et al. 1993; Goedert et al. 1993). This antibody was found to recognize the phosphorylated form of the tau-1 epitope, i.e., the recognition of tau by AT8 depends of the phosphorylation of Ser^{202} (Mercken et al. 1992; Goedert et al. 1993). The B9 antibody is a rabbit polyclonal antiserum raised to rat high molecular weight MAP2 proteins (Brion et al. 1988). The mouse monoclonal antibodies to α-tubulin, MAP5, and MAP1A were purchased from Sigma. The other antibodies used in this study were a mouse monoclonal antibody to GAP-43 (Boehringer), a mouse monoclonal antibody to synaptophysin (BioTest) and a mouse monoclonal antibody to $p34^{cdc2}$ kinase (Santa Cruz).

Preparation of Tissue Samples at Different Developmental Stages

Five Pregnant female rats were sacrificed at 18 or 20 days of gestation and the brains of the rat foetuses (E18 and E20) immediately dissected. Rat brains (five rats from each stage) were also dissected from pups at three days post-natal (P3), 10 days post-natal (P10), 13 days post-natal (P13), and 20 days post-natal (P20) and from adult rats (3 months old). These brains were fixed in methacarn, dehydrated, and embedded in parrafin. Tissue sections with a thickness of 7 μm were cut from the tissue blocks and processed for immunocytochemistry.

Neuronal Cultures

Primary cultures of embryonic cortical neurones were performed as reported elsewhere (Vanisberg et al. 1991). Briefly, the forebrains of rat embryos were taken at 18 days of gestation and mechanically dissociated, and the cells were seeded on Petri dishes previously coated with poly-L-lysine (10 μg/ml). The cells were cultured in a mixture of DMEM and Ham's F12 (3:1) supplemented with 10% fetal calf serum, 0.6% D-glucose, 100 I.U. penicillin and 100 μg/ml of streptomycin. Cytosine arabinoside (10 μM) was added to the culture medium after three days. Neuronal cultures were maintained at 37°C in the presence of 5% CO2.

Some cultures were treated at six to seven days with 10 μm colchicine (Janssens) for six hours. Control and treated cultures were fixed in 4% (w/v) paraformaldehyde in phosphate buffer (0.1 M, pH 7.4). Other cultures were kept frozen at −20°C for Western blot analysis (see below). In some experiments, cultures were extracted with Triton X-100 before fixation, as reported elsewhere (Ferreira et al. 1989). These detergent-extracted cultures contained an insoluble cytoskeletal fraction.

Immunocytochemistry

The immunolabeling on tissue sections was performed with the peroxidase-antiperoxidase (PAP) method. Briefly, tissue sections were treated with H_2O_2 to inhibit endogenous peroxidase, incubated with the blocking solution (20% normal goat serum in TBS: 0.01 M Tris, 0.15 M NaCl, pH 7.4), and then sequentially with the diluted primary antibody, a goat anti-rabbit or anti-mouse antibody (Nordic) followed by a rabbit or a mouse PAP complex (Nordic). The peroxidase activity was revealed using diamininobenzidine as chromogenic substrate. Prior to immunolabeling, another set of tissue sections was incubated for 20 hours at 37°C with alkaline phosphatase (30 U/ml, Sigma, from calf intestine) diluted in 0.1 M Tris-HCL, pH 8.0, 0.01 M.

The neuronal cultures were labeled with the peroxidase-antiperoxidase (PAP) method as described above. Before immunolabeling, fixed cultures were treated for 30 minutes with 0.2% (v/v) Triton X-100 in TBS.

Western Blot Analysis

For Western blot analysis, treated and untreated cultures were resuspended in a homogenisation buffer (50 mM Tris-HCl, pH 7.4, 0.1 M NaCl, 10 mM EDTA, 1 mM phenylmethylsulfonylfluoride, 25 μg leupeptin/ml, 25 μg pepstatin/ml) and kept frozen at −20°C. The homogenates were centrifugated at 20,000 g for 30 min to separate a soluble cytosolic fraction (S1) from an insoluble material (P1). This pellet P1 was resuspended in the homogenisation buffer containing 1% Triton X-100 and was again centrifugated; the supernatant (S2) and the pellet (P2) were kept at −20°C. The pellet P2 constitutes a cold-stable and detergent-extracted insoluble cytoskeletal fraction.

The different fractions were run on 10% polyacrylamide gels (SDS-PAGE) and electrophoretically transferred to nitrocellulose membrane. Nitrocellulose membranes were treated as previously described (Brion et al. 1991b): after blocking in semi-fat dried mile (10% (w/v) in TBS), they were incubated with the primary antibodies followed by goat anti-rabbit or anti-mouse immunoglobulins conjugated to alkaline phosphatase (Sigma).

Results and Discussion

We have detailed the cellular distribution of a phosphorylated tau species during cortical development and in cultures of embryonic cortical neurons, using in parallel several tau antibodies which recognize tau in a phosphorylation-dependent manner, i.e., their reactivity requires either unphosphorylated (antibody tau-1) or phosphorylated (antibody AT8) serine residues in position 199 and/or 202. A preferential phosphorylation of ser^{202} in the foetal brain and in Alzheimer's disease has been reported both on the basis of immunochemical evidence (Biernat et al. 1992; Mercken et al. 1992; Brion et al. 1993; Goedert et al. 1993) and protein chemistry (Watanabe et al. 1993).

⟶

Fig. 1. Immunolabeling on sagittal tissue sections of rat forebrain at 18 days of embryonic life. **A** Tau-1 antibody. **B** AT8 antibody to phosphorylated tau. **C** MAP2 antiserum. **D** $p34^{cdc2}$ antibody. **E** GAP-43 antibody. **F** MAP5 antibody. CP, cortical plate; IZ, intermediary zone; H, hippocampus; VZ, ventricular zone CPu, caudate-putamen; V, ventricle. Scale bar, 500 μm

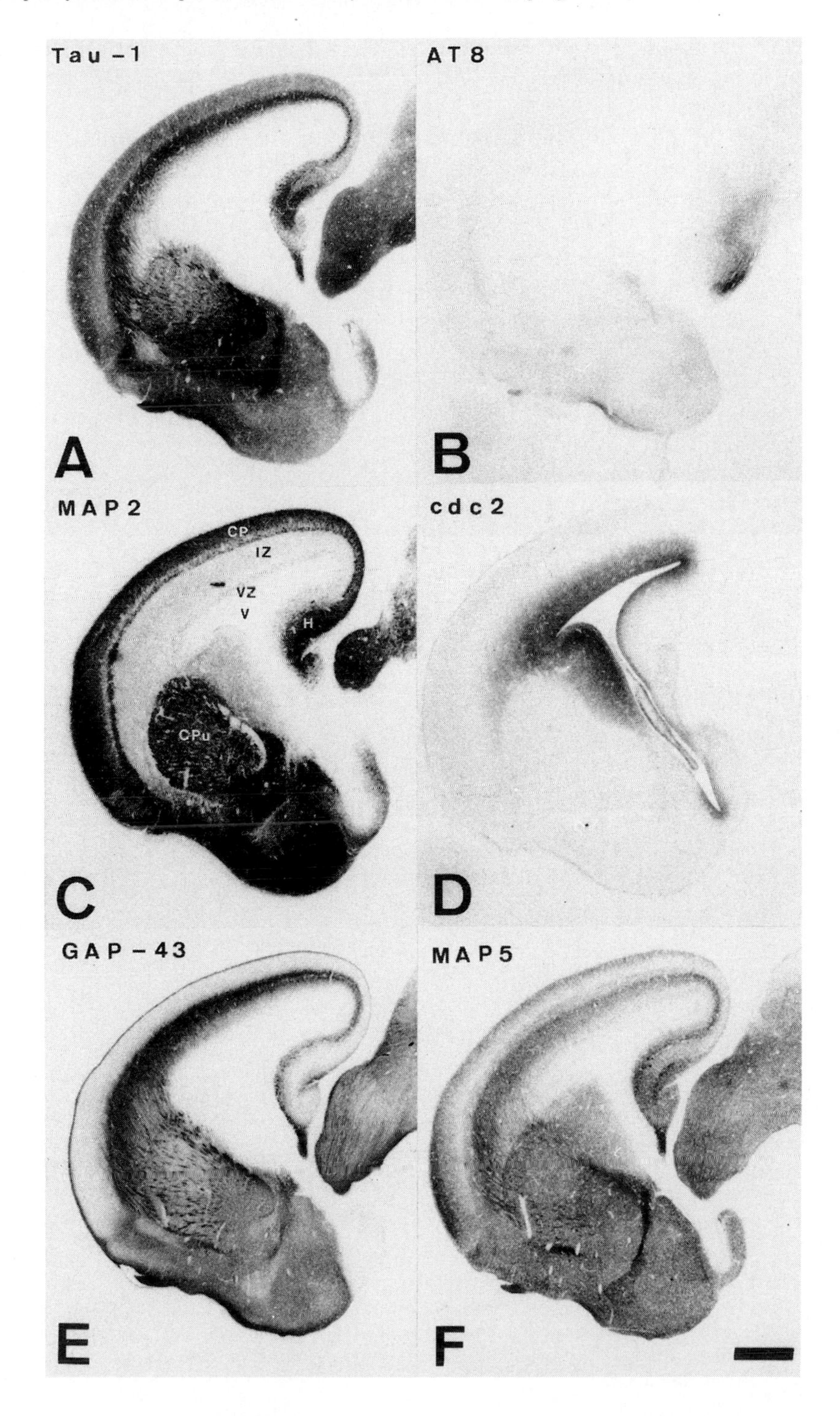
Tau -1
AT8
A
B
MAP2
CP
IZ
VZ
V
H
CPu
cdc2
C
D
GAP-43
MAP5
E
F

AT8-immunoreactive Phosphorylated Tau Is Absent from Cortical Neurons at Early Stages of Differenciation

Using the phosphorylation-independent antiserum B19 and the tau-1 antibody, tau immunoreactivity was clearly observed in all neurones localized in the marginal zone, the cortical plate and the subplate at E18 (Fig. 1) and at E20 (Fig. 2). The cells localized in the ventricular zone, the subventricular zone and the intermediate zone were unlabeled. A strong tau immunoreactivity was also observed in the fibre tracts present in the intermediate zone. In contrast with the B19 and tau-1 antibodies, the AT8 antibody did not label any neurone in the developing forebrain at E18, including the cortex, the caudate-putamen, and the hippocampus (Fig. 1B). However, some fibre tracts were labeled by the AT8 antibody in the developing brainstem.

At E20, the AT8 antibody was found to label numerous neurones in the developing cortex, i.e., in the cortical plate and the subplate (Fig. 2B). AT8 also labeled fibres in the intermediate zone but not in the ventricular and subventricular zones.

A strong AT8 immunoreactivity was still present at 3 days postnatal and was observed up to 10 days postnatal but started to decrease after this date;

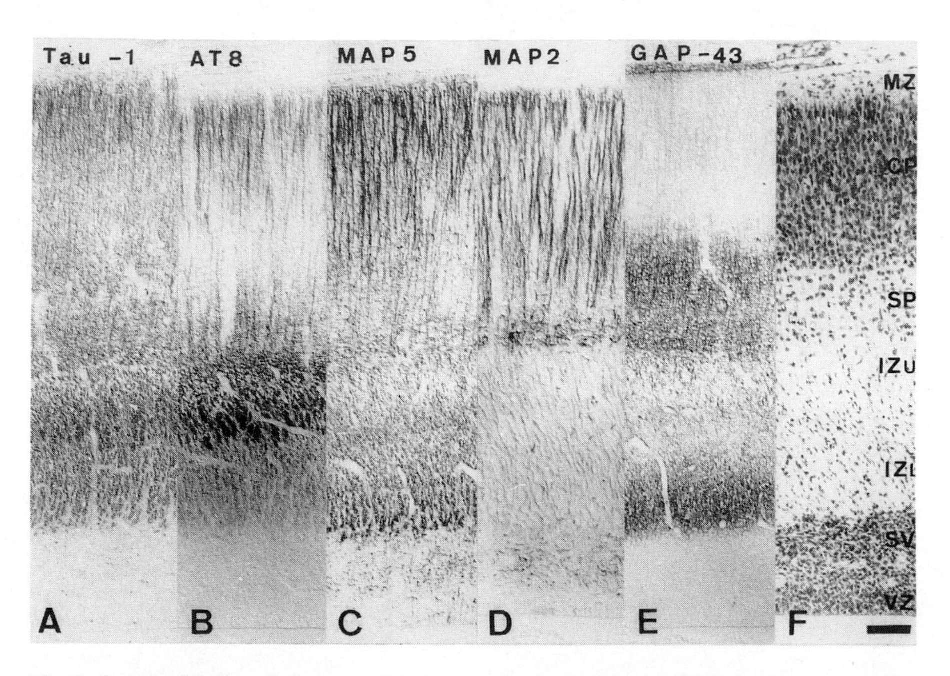

Fig. 2. Immunolabeling of the rat occipital cortex (sagittal section) at 20 days of embryonic life. **A** Tau-1 antibody; **B** AT8 antibody; **C** MAP5 antibody; **D** MAP2 antiserum; **E** GAP-43 antibody; **F** Nissl staining. MZ, marginal zone; CP, cortical plate; SP, subplate; IZU, upper intermediary zone; IZL, lower intermediary zone; SV, subventricular zone; VZ, ventricular zone. Scale bar, 25 μm

at P13, the AT8 immunoreactivity had disappeared from the deep cortical layers but was still present in the superficial layers (Fig. 3B). At 20 days postnatal, the AT8 immunoreactivity was almost undetectable. The tau-1 immunoreactivity became relatively weaker during the first postnatal week but increased during the second and third postnatal weeks and was present at all ages in the whole thickness of the cortex.

The observations confirm that tau is expressed in the developing cortex at E18, as reported previously by immunocytochemistry (Ferreira et al. 1987) and by in situ hybridization (Takemura et al. 1991). The expression of tau in the embryonic nervous system had also been described by Western blot (Couchie and Nunez 1985), Northern blot (Kosik et al. 1989; Mangin et al. 1989) or PCR analysis (Brion et al. 1993). We detected tau in post-migratory neurones localized in the cortical plate and the subplate but not in precursor cells in the ventricular zone or in neuroblasts in the subventricular zone or in the intermediary layer. Previous studies also indicated that tau was excluded from immature neuroepithelial cells in the cortex (Ferreira et al. 1987; Takemura et al. 1991) and in the cerebellum (Brion et al. 1988; Couchie et al. 1990). This suggests that tau is not expressed in actively dividing precursors or in immature neurones migrating from the subventricular zone to the cortical palte. Similar observations have been made for the expression of MAP2 (Crandall et al. 1986).

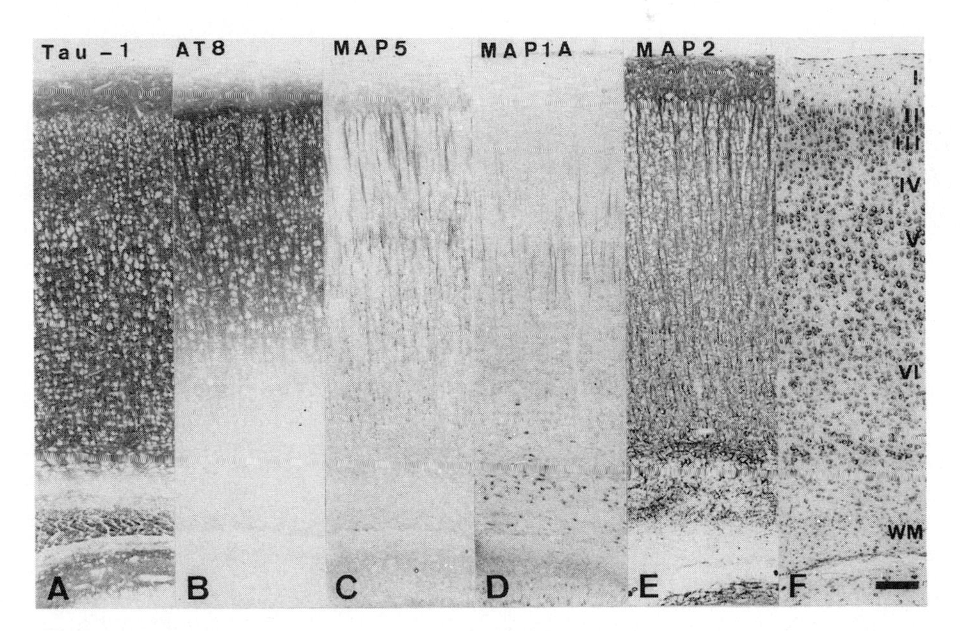

Fig. 3. Immunolabeling of the rat occipital cortex (sagittal section) at 13 days of postnatal life. **A** Tau-1 antibody; **B** AT8 antibody; **C** MAP5 antibody; **D** MAP1A antibody; **E** MAP2 antiserum; **F** Nissl staining. I-VI, cortical layers; WM, white matter. Scale bar, 50 μm

Although tau is expressed in the developing cortex at E18, the phosphorylated tau species detected by the AT8 antibody was only present at E20 and for a transient period during postnatal development. These immunocytochemical results are in agreement with our previous analysis by Western blotting of the developmental evolution of phosphorylated tau (Brion et al. 1993). The presence of a highly phosphorylated foetal-type tau in the developing brain has also been documented in other recent studies (Kanemaru et al. 1992; Bramblett et al. 1993; Goedert et al. 1993; Kenessey and Yen 1993; Watanabe et al. 1993).

The timing of appearance of AT8-immunoreactive tau was also investigated in primary cultures of embryonic cortical neurones, both by immunocytochemistry and immunoblotting. As already reported (Couchie et al. 1986; Kosik and Finch 1987; Dotti et al. 1987), tau immunoreactivity was readily detectable in neurones a few hours after plating, when a phosphorylation-independent antibody such as B19 was used, and was uniformly distributed in rounded cell bodies without neurites or with a few small neurites after the first hours (not shown). Tau immunoreactivity was observed in dendrites, pericarya and axons during the first week after plating (Fig. 4A). Thereafter, the tau immunoreactivity became progressively weaker in dendrites. The tau-1 antibody weakly labeled pericarya and dendrites after a few days of culture. However, this antibody strongly labeled some long axons (Fig. 4C). By contrast, AT8 immunoreactivity did not appear in cultured neurones before 4–5 days of culture. As with the B19 antibody, this labeling was observed in dendrites, pericarya and axons (Fig. 4B)

The delayed appearance of this phosphorylated tau species was confirmed by Western blotting analysis: a tau species of 48 kDa was detected by the B19 phosphorylation-independent antibody in the soluble fraction (S1) of cultures of different ages (Fig. 5A; at 13 days of culture another tau species of 60 kDa became apparent) but the tau species present at the early stage of the culture was not detected by the AT8 antibody (Fig. 5B).

These observations in culture parallel the observations made in vivo during brain development, i.e., the tau species expressed initially in developing cortical neurons are mainly unphosphorylated at Ser^{202}. Thus, although tau expressed during foetal and postnatal development has been shown to be transiently highly phosphorylated, we observed that there is a lag period of several days in vivo and in culture before the appearance of AT8-immunoreactive phosphorylated tau, suggesting that very early neurite outgrowth does not reuire this phosphorylated tau species.

Phosphorylated Tau, But Not Unphosphorylated Tau, Remains in Dendrites of Cortical Neurones During Development

During embryonic development, the tau immunoreactivity detected at E18 and E20 by the B19 antiserum and tau-1 and at E20 by AT8 was present in axons, pericarya and dendrites of postmigratory neurones in the developing

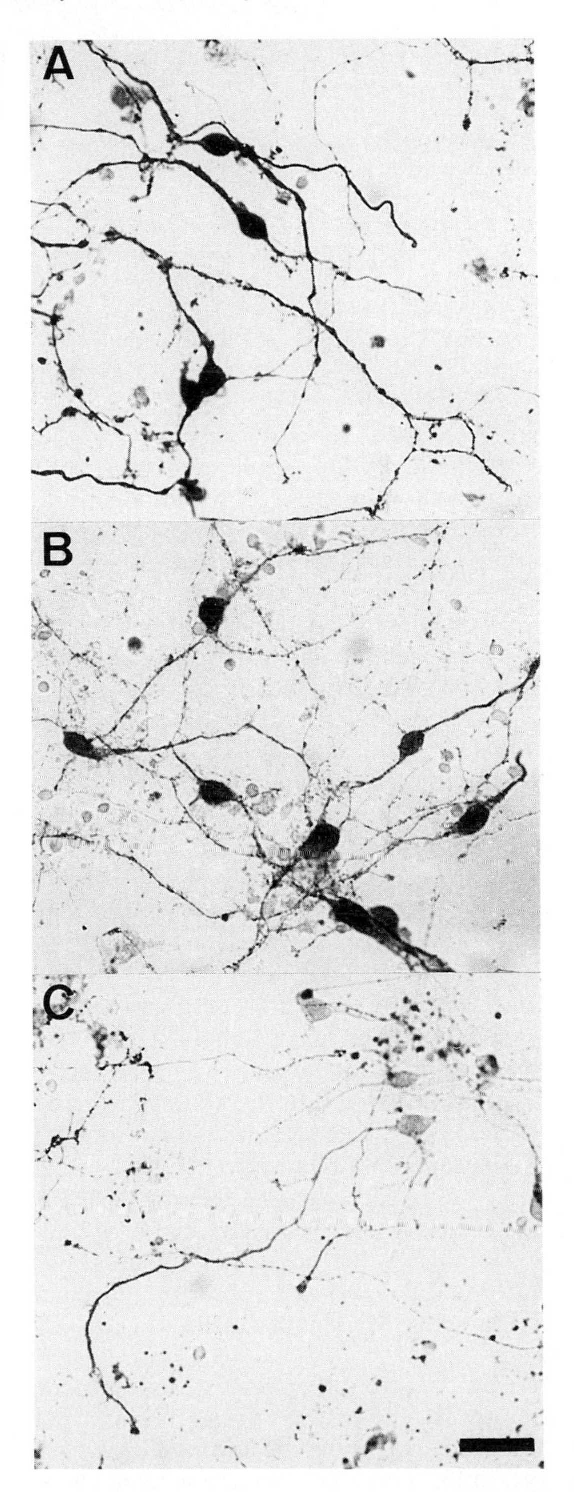

Fig. 4. Immunolabeling of cortical embryonic neurones cultured for seven days. **A** B19 antiserum to tau; **B** AT8 antibody; **C** Tau-1 antibody. Scale bar, 25 μm

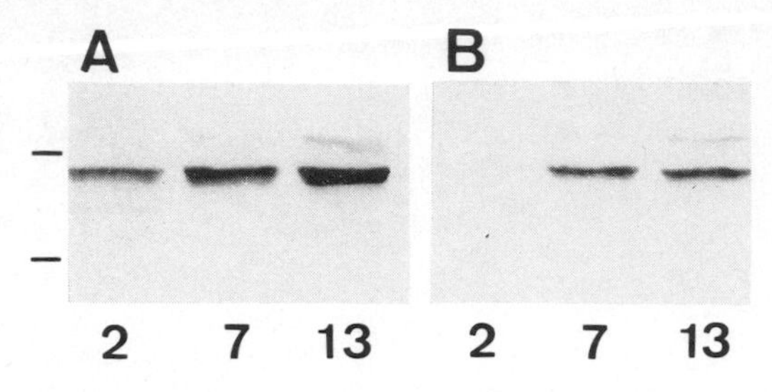

Fig. 5. Western blots of the soluble fraction of 2-, 7- and 13-day-old neuronal cultures. The blots were incubated with the B19 antiserum to tau (**A**) or the monoclonal antibody AT8 (**B**). Bars on the left indicate the position of the following molecular mass markers: 58.1 kDa (catalase, *top*), 39.8 kDa (alcohol dehydrogenase, *bottom*)

cortex (Fig. 2). During the postnatal period, different patterns of tau compartmentalization were observed when different tau antibodies were used. During the first postnatal days, AT8 immunoreactivity was still observed in the whole thickness of the cortex and was present in axons (e.g., in the corpus callosum), in pericarya and in dendrites. As described above, the AT8 immunoreactivity progressively disappeared from deep cortical layers but remained in dendrites of labeled cortical neurones at all ages. The pericaryal labeling of cortical neurons by AT8 decreased after the first postnatal days. Prior treatment of tissue sections with alkaline phosphatase completely abolished the labeling with AT8 antibody.

A different situation emerged for tau-1 immunoreactivity. During the first postnatal days, tau-1 weakly labeled axons in the white and grey matter and short segments of apical dendrites of pyramidal neurones mainly in cortical layer V (not shown). During these first postnatal days a treatment of tissue sections with alkaline phosphatase increased this dendritic labeling by tau-1 in superficial cortical layers and also induced a labeling of the cell bodies of pyramidal neurons. At 10–13 days postnatal, the tau-1-IR was considerably decreased in pericarya and dendrites but was strong in axons. At this age, the dephosphorylation induced only a marginal increase in dendritic labeling by tau-1. This axonal labeling by tau-1 was still more obvious at P20 and in the adult. At these latter ages, alkaline phosphatase treatment of tissue sections did not induce a dendritic labeling by tau-1.

The phosphorylation-independent B19 tau antibody labeled axons and dendrites in cortex during the first postnatal days, but this dendritic labeling progressively disappeared and, after 10 days, the labeling with this antibody was similar to that observed with tau-1, i.e., strongly concentrated in axons. Tau-1 and B19 were also observed to label the interfascicular oligodendroglia in the adult (Migheli et al. 1988). The AT8 antibody did not label the latter cells.

In the cat visual cortex, tau proteins detected by tau-1 have been reported to be present during the first postnatal week in axons, dendrites and pericarya, and to be localized exclusively in axons at later ages (Riederer and Innocenti 1991). In previous studies, it was also reported that tau-1 and other tau antibodies strongly label axons in adult rat and that the somatodendritic domain of neurones was not labeled (Binder et al. 1985) or was weakly labeled (Migheli et al. 1988). Dephosphorylation was reported to induce a labeling of dendrites and pericarya by tau-1 in rat (Papasozomenos and Binder 1987) and in cat (Riederer and Innocenti 1991); however, this phosphorylation-dependent somatodendritic staining was observed to decrease during postnatal development of cat cortex, concerning only a few neurones and their dendrites in cortical layers II and III in the adult cat (Riederer and Innocenti 1991). We similarly observed that dephosphorylation led to a somatodendritic labeling of cortical neurons by tau-1, but only during the first postnatal week. At later ages, the tau immunoreactivity was concentrated in axons and dephosphorylation did not lead to a significant labeling of the somatodendritic domain of cortical neurones. The relative inability of dephosphorylation to unmask somatodendritic tau under our conditions seems improbable, since the AT8 immunoreactivity was systematically abolished by the same treatment. In addition, the phosphorylation-independent B19 antibody also localized tau in axons in the adult rat.

Our results thus strongly suggest that, in the adult rat, tau is concentrated in axons of cortical neurones and that, relative to axons, only trace amount of tau can be present in the somatodendritic domain. The targeting of tau to axons in the adult might involve different mechanisms, such as the preferential localization of its mRNA at the basis of the axon (Litman et al. 1993). However, it should be stressed that the evolution of tau compartmentalization during development might differ according to species, neuronal type and brain area. For instance, during most of the development of the rat cerebellar cortex, tau is absent from dendrites of Purkinje cells (Brion et al. 1988; Ferreira et al. 1987). Differences have also been observed in cultures: tau is not excluded from dendrites and cell bodies of cultured hippocampal neurons (Dotti et al. 1987) but the segregation of tau into axons was reported to be more or less completely realized in cultured cortical neurones (Kosik and Finch 1987; Litman et al. 1993).

It thus appears that tau detected in the somatodendritic or axonal domain during development of cortical neurones cannot be unequivocally considered as either phosphorylated or unphosphorylated. For example, at E18, tau appears to be unphosphorylated on Ser^{202} in axons and dendrites but soon after, at E20, both axons and dendrites contain a mixture of phosphorylated and unphosphorylated tau species. This finding suggests that this phosphorylation does not play a direct role in the establishment of neuronal polarity.

AT8-immunoreactive Phosphorylated Tau Disappears First from Cortical Neurones That Are Ontogenetically the Oldest

The disappearance of the AT8 immunoreactivity in the cortical plate followed an inside-out gradient, i.e., the AT8 immunoreactivity disappeared first from white matter tracts and from the deepest cortical layers. At 13 days postnatal, only the superficial cortical layers were still AT8-immunoreactive. This gradient of disappearance of AT8 immunoreactivity might reflect the inside-to-outside sequence of cortical development: neurones in the deeper layers develop before neurones in superficial layers. Thus the persistence of AT8 immunoreactivity in superficial layers might be related to the more immature state of these neurones.

The expression of foetal tau decreases sharply *in vivo* at the end of the first postnatal week (Francon et al. 1982; Kosik et al. 1989), whereas adult tau isoforms begin to appear during the second postnatal week. In embryonic neurones cultured for 10–13 days, we did not observe a disappearance of AT8 immunoreactivity and this was associated with an absence of expression of the full pattern of adult-type tau isoforms. This absence of tau evolution in culture has also been previously observed in some other (Couchie et al. 1986; Dotti et al. 1987) but not all (Litman et al. 1993) studies. Our data suggest, however, that there is a temporal relationship between the disappearance of AT8-immunoreactive phosphorylated tau and the expression of adult-type tau isoforms.

AT8-immunoreactive Phosphorylated Tau Is Found Only During the Period of Intense Dendritic Outgrowth

For the sake of comparison with AT8-immunoreactive phosphorylated tau, the distribution in the cortex of several proteins known to show a developmental evolution was studied in parallel on adjacent sections. The timing of disappearance of AT8 immunoreactivity paralleled to some extent the decrease of MAP5-IR during cortical development. MAP5 was strongly expressed at E18 and E20 in axonal tracts running in the intermediate zone in the developing cortex (Fig. 1F) and was also present in dendrites in the cortical plate (Fig. 2C). MAP5 (also termed MAP1B or MAP1X; Calvert and Anderton 1985; Riederer et al. 1986; Bloom et al. 1985) is strongly expressed during the period of active neuritic outgrowth in the embryonic brain and during early postnatal days, but MAP5 levels drop rapidly between postnatal days 7 and 10 (Riederer et al. 1986; Schoenfeld et al. 1989). The extension of dendrites and the increase in the complexity of dendritic trees occurs mainly during postnatal weeks 1–3; the first 10 days is the period when dendritic growth is most active (Miller 1988). This suggests that AT8-immunoreactive tau is mainly expressed during this period of active dendritic growth. The AT8 immunoreactivity had a different timing of appearance and a different distribution than that of GAP43, known to be a marker of developmental axonal outgrowth (Jacobson et al. 1986; Oestreicher and Gispen 1986; Goslin et al. 1988). In addition GAP-43, in contrast to AT8-immunoreactive tau, is still significantly expressed in the adult brain (Benowitz et al. 1988).

Converse observations were made when comparing AT8 immunoreactivity with MAP1A-IR and synaptophysin-IR. MAP1A and synaptophysin are expressed later than MAP5 during brain development and are considered to be molecules involved in differenciation processes-subsequent to active neuritic outgrowth. MAP1A is thought to be involved in neurite stabilization and, as reported (Riederer and Matus 1985; Schoenfeld et al. 1989), a strong MAP1A-IR was not observed before P10, i.e., at a period when AT8 immunoreactivity started to decrease. In addition, MAP1A is strongly expressed mainly in the basal portion of primary dendrites of pyramidal cells in layer V (Tucker et al. 1989), i.e., neurones that are systematically unlabeled by AT8 at this period. Synaptophysin is an integral component of the membrane of synaptic vesicles and, as reported, a significant synaptophysin-IR was not observed in the cortex before P10, i.e., before the period of intense synaptogenesis (Knaus et al. 1986).

AT8-immunoreactive Phosphorylated tau and $p34^{cdc2}$ Kinase Do Not Co-localize in the Embryonic Brain

The protein kinases involved in the generation of the AT8 epitope (i.e., phosphorylation of Ser^{202}) in developing cortical neurones are still unknown. Previous studies performed in vitro have demonstrated that tau can be phosphorylated by several proline-directed protein kinases, giving rise to a paired helical filament-like phosphorylation status similar to the phosphorylation status of foetal tau. One such candidate kinase is the cyclin-dependent kinase $p34^{cdc2}$ (Ledesma et al. 1992; Scott et al. 1993). The $p34^{cdc2}$ kinase has also been reported to phosphorylate the tau-1 epitope in vitro (Liu et al. 1993). The antibody to $p34^{cdc2}$ kinase strongly labeled neurones in the ventricular zone and in the intermediate zone in the embryonic brain (Fig. 1D). During postnatal development, a $p34^{cdc2}$-IR was still present in the external granular layer in the cerebellum (not shown), i.e., in a layer still containing immature neurones during the second and third postnatal weeks. The expression of $p34^{cdc2}$ kinase was thus limited to germinal layers and was not observed in the cortical plate where a strong AT8 immunoreactivity appears at E20. This absence of co-localization suggests that the $p34^{cdc2}$ kinase is not involved in the generation of the AT8 epitope in the developing brain. However, other proline-directed protein kinases might be responsible for the appearance of AT8-immunoreactive tau; for example, tau phosphorylation on Ser^{202} can also be achieved by the ERK2 kinase (Drewes et al. 1992) and a $p34^{cdc2}$-related kinase, the cyclin-dependent kinase 5 (cdk5; Kobayashi et al. 1993).

Tau Phosphorylated on Ser^{202} Is Not Associated with Cold-stable Microtubules

A subset of neuronal microtubules is known to be cold-stable, i.e., not recovered in the soluble fraction after brain homogenisation at cold temperatures (Brady et al. 1984). In our culture system, an important proportion of the total tubulin present in seven-day-old culture is associated with these old-stable microtubules. The proportion of stable microtubules has been previously shown to increase during neuronal development in vivo and also in cerebellar macroneurones developing in culture (Ferreira et al. 1989).

To assess the association of phosphorylated and unphosphorylated tau with cold-stable microtubules, the partition of tau with cold-soluble and insoluble fractions was studied in neuronal cultures. Tau-immunoreactive species were detected in the soluble fraction (see above) but also in the insoluble, detergent-extracted cytoskeletal fraction (P2; Fig. 6), whereas only a very weak tau immunoreactivity was observed in the S2 fraction, i.e., the fraction solubilized by the detergent treatment (not shown). At seven days in culture, these insoluble tau species were estimated to represent from 10 to 20% of the amount of soluble tau, as judged from the volume of sample loadings necessary to give similar signals on Western blots probed with the B19 polyclonal antiserum to tau. These insoluble tau species consisted of three closely adjacent bands migrating with a higher electrophoretic mobility than the tau species detected in the soluble fraction. These tau species associated with the insoluble, cold-stable cytoskeletal fraction were labeled by the B19 antiserum and the tau-1 antibody but not by the AT8 antibody, even when an AT8-immunoreactive tau species was present in the soluble fraction (Fig. 6). Thus soluble tau, but not tau associated to cold-stable microtubules, is phosphorylated on Ser^{202}. The tau species found in the insoluble fraction show quicker electrophoretic mobility, again suggesting that they are not phosphorylated on sites which, when they are phosphorylated, are known to reduce the electrophoretic mobility of tau. The presence of several tau bands in the insoluble fraction does not indicate the presence of several tau isoforms, since a single tau isoform is expressed in the embryonic brain and in neurones during the first week of culture. Rather, they probably correspond to tau species differentially phosphorylated on other sites than

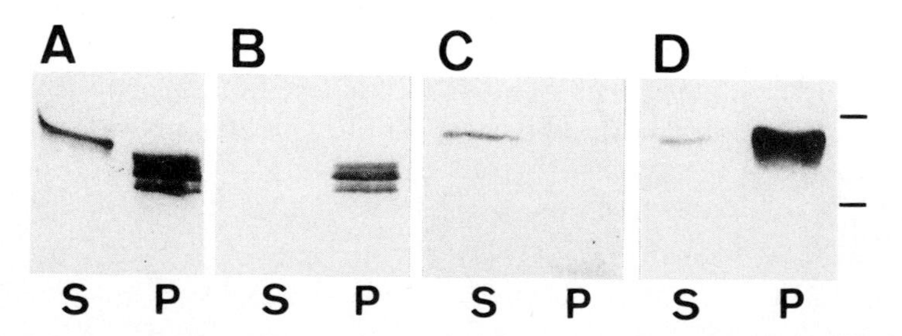

Fig. 6. Western blots of the soluble (S) or insoluble (P) fraction of 7-day-old neuronal cultures. The blots were incubated with the B19 antiserum to tau (**A**), the monoclonal antibody tau-1 (**B**), the monoclonal antibody AT8 (**C**), and the α-tubulin monoclonal antibody (**D**). Bars on the right indicate the position of the following molecular mass markers: 58.1 kDa (catalase, *top*), 39.8 kDa (alcohol dehydrogenase, *bottom*)

Ser^{202}. In CHO (Gallo et al. 1992) and 3T3 (Sygowski et al. 1993) cells transfected by a single recombinant tau isoform, tau was similarly observed to migrate as a set of several bands by one-dimensional SDS-PAGE as a consequence of its phosphorylation in the cell.

After extraction in situ of neuronal cultures with Triton X-100, most of the AT8 immunoreactivity was lost, again suggesting that this phosphorylated tau is poorly associated with the insoluble cytoskeletal fraction.

Tau Phosphorylated on Ser^{202} Is Not Associated with Colchicine-resistant Microtubules

To further investigate the relationship between microtubule stability and tau phosphorylation, six- to seven-day-old culture were treated with colchicine. After this treatment, we still recovered significant amounts of tubulin in the cold-extracted, insoluble fraction. This tubulin most probably belongs to the subset of neuronal microtubules known to be resistant to the depolymerising action of colchicine; this particular population of insoluble microtubules would thus combine cold and colchicine resistance, although our data do not indicate if there is a complete overlap between cold-stable and colchicine-resistant microtubules.

The colchicine treatment induced marked changes in the phosphorylation status of soluble tau and in the immunoreactivity of tau in cultured neurones. Many neurones treated with colchicine displayed a significant increase in pericaryal tau immunoreactivity with the phosphorylation-independent B19 tau antiserum (not shown). With the tau-1 antibody, this increase in pericaryal tau immunoreactivity was strong and observed in most neurones (Mattson 1992). By contrast, the colchicine treatment led to an important decrease of neuronal labeling by the AT8 antibody. This increase in tau immunoreactivity corresponds to the accumulation of an unphosphorylated tau species (Davis et al., manuscript submitted). Also, this dephosphorylation concerns the tau proteins present in the soluble fraction. This finding suggests that, following the depolymerisation action of colchicine, a soluble pool of dephosphorylated tau proteins accumulates in the cytoplasm.

On Western blots, the soluble fraction (S1) of colchicine-treated cultures showed two to three tau bands labeled by the B19 antibody (Fig. 7B,C). These tau species showed slightly quicker electrophoretic mobility than in the soluble fraction of control cultures. The cytoskeletal fraction (P2) of colchicine-treated cultures did contain three closely adjacent tau bands migrating at the same level as the tau bands found in the same fraction of control cultures (not shown).

The Western blot analysis also showed that, for a similar intensity of B19 labeling, the tau species present in the soluble fractions were more intensely labeled by the tau-1 antibody in colchicine-treated cultures than in control cultures (Fig. 7A). Conversely, these tau species were less intensely

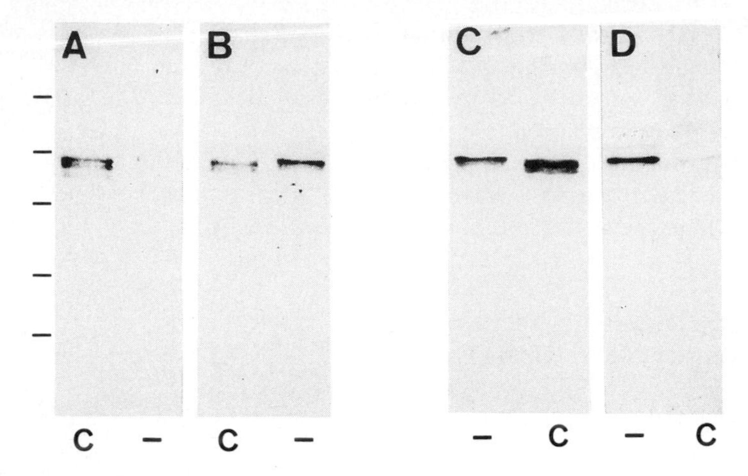

Fig. 7. Western blots of the soluble fraction of 7-day-old control (–) and colchicine-treated (c) neuronal cultures. The blots were incubated with the monoclonal antibody tau-1 (**A**), the B19 antiserum to tau (**B** and **C**), and the monoclonal antibody AT8 (**D**). Bars on the left indicate the position of the following molecular mass markers (from *top* to *bottom*) 97.4 kDa (phosphorylase b), 58.1 kDa (catalase), 39.8 kDa (alcohol dehydrogenase), 29.0 kDa (carbonic anhydrase), 20.1 kDa (trypsin inhibitor)

labeled by the AT8 antibody in colchicine-treated cultures than in control cultures (Fig. 7D).

Injection of purified tau in RAT-1 fibroblasts (Drubin and Kirschner 1986) or transfection of recombinant tau in 3T3 cells (Lo et al. 1993) has been reported to increase the resistance of microtubules (Drubin and Kirschner 1986) or the cell shape (Lo et al. 1993) to the effect of colchicine. We similarly found tau species associated to the colchicine-resistant microtubules; these tau species migrate at the same level as in control cultures and are similarly not immunoreactive with the AT8 antibody. This indicates that cold-stable and colchicine-resistant microtubules both contain tau species unphosphorylated on Ser^{202} and that microtubules containing this phosphorylated tau are cold- and colchicine-sensitive.

In conclusion, our data strongly suggest that AT8-immunoreactive phosphorylated tau (i.e., tau phosphorylated on Ser^{202}) is not a marker of late developmental steps like neurite stabilization and synapse formation, but is rather expressed during the period of intense neuritic outgrowth. The observation that AT8-immunoreactive phosphorylated tau is not associated with stable microtubules, and previous results indicating that phosphorylated tau is less efficient than unphosphorylated tau in promoting tubulin polymerisation, together suggest that a high level of this phosphorylated tau during the period of intense neuronal development would provide developing cells with a less stable and less rigid microtubule network, more adapted to these developing cells which need a high degree of plasticity. The subsequent disappearance of AT8-immunoreactive phosphorylated tau and the expression of adult-type tau isoforms might act in concert to provide neurones with a more stable and rigid microtubule network during their maturation in the second and third postnatal weeks.

The observation that AT8-immunoreactive phosphorylated tau is transiently present during postnatal development might be important for understanding some aspects of the pathophysiology of Alzheimer's disease: AT8-immunoreactive phosphorylated tau is not present in the normal adult brain but reappears in Alzheimer's disease, associated with neurofibrillary lesions. In addition to phosphorylated Ser^{202}, other amino acids phosphorylated in PHF-tau were observed to be phosphorylated in foetal tau (Brion et al. 1993; Watanabe et al. 1993). These results suggest that the developing brain may be a useful experimental system in which to study the mechanisms which control tau phosphorylation.

Acknowledgements. This study was supported by grants from the Belgian FRSM (3.4504.91), the "Fonds de Recherche Divry," the "Fondation M.T. de Lava," Alzheimer Belgique and the European Neuroscience Program. The authors thank Dr. Vandevoorde (Innogenetics) for providing the AT8 antibody.

References

Bass PW, Pienkowski TP, Kosik KS (1991) Processes induced by tau expression in Sf9 cells have an axon-like microtubule organization. J Cell Biol 115: 1333–1344

Benowitz LI, Apostolides PJ, Perrone-Bizzozero N, Finklestein SP, Zwiers H (1988) Anatomical distribution of the growth-associated protein GAP-43/B50 in the adult rat brain. J Neurosci 8: 339–352

Biernat J, Mandelkow E-M, Schröter C, Lichtenberg-Kraag B, Steiner B, Berling B, Meyer H, Mercken M, Vandermeeren A, Goedert M, Mandelkow E (1992) The switch of tau protein to an Alzheimer-like state includes the phosphorylation of two serine-proline motifs upstream of the microtubule binding region. EMBO J 11: 1593–1597

Binder LI, Frankfurter A, Rebhun I (1985) The distribution of tau in the mammalian central nervous system. J Cell Biol 101: 1371–1378

Bloom GS, Luca FC, Valee RB (1985) Microtubule-associated protein 1B: identification of a major component of the neuronal cytoskeleton. Proc Natl Acad Sci USA 82: 5404–5408

Brady ST, Tytell M, Lasek RJ (1984) Axonal tubulin and axonal microtubules: biochemical evidence for stability. J Cell Biol 99: 1716–1724

Bramblett GT, Goedert M, Jakes R, Merrick SE, Trojanowski JQ, Lee VM-Y (1993). Abnormal tau phosphorylation at Ser^{396} in Alzheimer's disease recapitulates development and contributes to reduced microtubule binding. Neuron 10: 1089–1099

Brion JP, Passareiro H, Nunez J, Flament-Durand J (1985) Mise en évidence immunologique de la protéine tau au niveau des lésions de dégénérescence neurofibrillaire de la maladie d'Alzheimer. Arch Biol (Brux) 95: 229–235

Brion JP, Guilleminot J, Couchie D, Nunez J (1988) Both adult and juvenile tau microtubule-associated proteins are axon specific in the developing and adult rat cerebellum. Neuroscience 25: 139–146

Brion JP, Hanger DP, Bruce MT, Couck AM, Flament-Durand J, Anderton BH (1991a) Tau in Alzheimer neurofibrillary tangles: N- and C-terminal regions are differentially associated with paired helical filaments and the location of a putative abnormal phosphorylation site. Biochem J 273: 127–133

Brion JP, Hanger DP, Couck AM, Anderton BH (1991b) A68 proteins in Alzheimer's disease are composed of several tau isoforms in a phosphorylated state which affects their electrophoretic mobilities. Biochem J 279: 831–836

Brion JP, Smith C, Couck AM, Gallo JM, Anderton BH (1993) Developmental changes in tau phosphorylation: fetal-type tau is transiently phosphorylated in a manner similar to paired helical filament-tau characteristic of Alzheimer's disease. J Neurochem 61: 2071–2080

Caceres A, Kosik KS (1990) Inhibition of neurite polarity by tau antisense oligonucleotides in primary cerebellar neurons. Nature 343: 461–463

Calvert R, Anderton BH (1985) A microtubule-associated protein (MAP1) which is expressed at elevated levels during development of the rat cerebellum. EMBO J 4: 1171–1176

Cleveland DW, Hwo SY, Kirschner MW (1977) Purification of tau, a microtubule-associated protein that induces assembly of microtubules from purified tubulin. J Mol Biol 116: 207–225

Couchie D, Nunez J (1985) Immunological characterization of microtubule-associated proteins specific for the immature brain. FEBS Lett 188: 331–335

Couchie D, Faivre-Bauman A, Puymirat J, Guilleminot J, Tixier-Vidal A, Nunez J (1986) Expression of microtubule-associated proteins during the early stages of neurite extension by brain neurons cultured in a defined medium. J Neurochem 47: 1255–1261

Couchie D, Legay F, Guilleminot J, Lebargy F, Brion J-P, Nunez J (1990) Expression of Tau protein and Tau mRNA in the cerebellum during axonal outgrowth. Exp Brain Res 82: 589–596

Crandall JE, Jacobson M, Kosik KS (1986) Ontogenesis of microtubule-associated protein 2 (MAP2) in embryonic mouse cortex. Dev Brain Res 28: 127–133

Delacourte A, Defossez A (1986) Alzheimer's disease: tau proteins, the promoting factors of microtubule assembly, are major components of paired helical filaments. J Neurol Sci 76: 173–186

Dotti CG, Banker GA, Binder LI (1987) The expression and distribution of the microtubule-associated proteins tau and microtubule-associated protein 2 in hippocampal neurons in the rat in situ and in cell culture. Neuroscience 23: 121–130

Drewes G, Lichtenberg-Kraag B, Döring F, Mandelkow E-M, Biernat J, Goris J, Dorée M, Mandelkow E (1992) Mitogen activated protein (MAP) kinase transforms tau protein into an Alzheimer-like state. EMBO J 11: 2131–2138

Drubin DG, Kirschner MW (1986) Tau protein function in living cells. J Cell Biol 103: 2739–2746

Ferreira A, Busciglio J, Caceres A (1987) An immunocytochemical analysis of the ontogeny of the microtubule-associated proteins MAP-2 and tau in the nervous system of the rat. Dev Brain Res 34: 9–31

Ferreira A, Busciglio J, Caceres A (1989) Microtubule formation and neurite growth in cerebellar macroneurons which develop in vitro: evidence for the involvement of the microtubule-associated proteins, MAP-1a, HMW-MAP2 and Tau. Dev Brain Res 49: 215–228

Francon J, Lennon AM, Fellous A, Mareck A, Pierre M, Nunez J (1982) Heterogeneity of microtubule-associated proteins and brain development. Eur J Biochem 129: 465–471

Gallo J-M, Hanger DP, Twist EC, Kosik KS, Anderton BH (1992) Expression and phosphorylation of a three-repeat isoform of tau in transfected non-neuronal cells. Biochem J 286: 399–404

Goedert M, Jakes R (1990) Expression of separate isoforms of human tau protein: Correlation with the tau pattern in brain and effects on tubulin polymerization. EMBO J 9: 4225–4230

Goedert M, Wischik CM, Crowther RA, Walker JE, Klug A (1988) Cloning and sequencing of the cDNA encoding a core protein of the paired helical filament of Alzheimer disease:

identification as the microtubule-associated protein tau. Proc Natl Acad Sci USA 85: 4051–4055

Goedert M, Spillantini MG, Jakes R, Rutherford D, Crowther RA (1989) Multiple isoforms of human microtubule-associated protein tau: sequences and localization in neurofibrillary tangles of Alzheimer's disease. Neuron 3: 519–526

Goedert M, Jakes R, Crowther RA, Six J, Lübke U, Vandermeeren M, Cras P, Trojanowski JQ, Lee VM-Y (1993) The abnormal phosphorylation of tau protein at Ser-202 in Alzheimer disease recapitulates phosphorylation during development. Proc Natl Acad Sci USA 90: 5066–5070

Goslin K, Schreyer DJ, Skene JHP, Banker G (1988) Development of neuronal polarity: GAP-43 distinguishes axonal from dendritic growth cones. Nature 336: 672–674

Grundke-Iqbal I, Iqbal K, Quinlan M, Tung YC, Zaidi MS, Wisniewski HM (1986) Microtubule-associated protein tau: a component of Alzheimer paired helical filaments. J Biol Chem 261: 6084–6089

Jacobson RD, Virag I, Skene JHP (1986) A protein associated with axon growth, GAP-43, is widely distributed and developmentally regulated in rat CNS. J Neurosci 6: 1843–1855

Kanai Y, Takemura R, Oshima T, Mori H, Ihara Y, Yanagisawa M, Masaki T, Kirokawa N (1989) Expression of multiple tau isoforms and microtubule bundle formation in fibroblasts transfected with a single tau cDNA. J Cell Biol 109: 1173–1184

Kanemaru K, Takio K, Miura R, Titani K, Ihara Y (1992) Fetal-type phosphorylation of the τ in paired helical filaments. J Neurochem 58: 1667–1675

Kenessey A, Yen S-HC (1993) The extent of phosphorylation of fetal tau is comparable to that of PHF-tau from Alzheimer paired helical filaments. Brain Res 629: 40–46

Knaus P, Betz H, Rehm H (1986) Expression of synaptophysin during postnatal development of the mouse brain. J Neurochem 47: 1302–1304

Kobayashi S, Ishiguro K, Omori A, Takamatsu M, Arioka M, Imahori K, Uchida T (1993) A cdc2-related kinase PSSALRE/cdk5 is homologous with the 30 kDa subunit of tau protein kinase II, a proline-directed protein kinase associated with microtubule. FEBS Lett 335: 171–175

Kosik KS, Finch EA (1987) MAP2 and tau segregate into dendritic and axonal domains after the elaboration of morphologically distinct neurites: an immunocytochemical study of cultures rat cerebrum. J Neurosci 7: 3142–3153

Kosik KS, Joachim CL, Selkoe DJ (1986) The microtubule-associated protein, tau, is a major antigenic component of paired helical filaments in Alzheimer's disease. Proc Natl Acad Sci USA 83: 4044–4048

Kosik KS, Orecchio LD, Bakalis S, Neve RL (1989) Developmentally regulated expression of specific tau sequences. Neuron 2: 1389–1397

Ledesma MD, Correas I, Avila J, Díaz-Nido J (1992) Implication of brain cdc2 and MAP2 kinases in the phosphorylation of tau protein in Alzheimer's disease. FEBS Lett 308: 218–224

Lee VMY, Balin BJ, Otvos L, Trojanowski JQ (1991) A68 proteins are major subunits of Alzheimer disease paired helical filaments and derivatized forms of normal tau. Science 251: 675–678

Lindwall G, Cole RD (1984) Phosphorylation affects the ability of tau protein to promote microtubule assembly. J Biol Chem 259: 5301–5306

Litman P, Barg J, Rindzoonski L, Ginzburg I (1993) Subcellular localization of tau mRNA in differentiating neuronal cell culture: Implications for neuronal polarity. Neuron 10: 627–638

Liu W-K, Moore WT, Williams RT, Hall FL, Yen S-H (1993) Application of synthetic phospho- and unphospho- peptides to identify phosphorylation sites in a subregion of the tau molecule, which is modified in Alzheimer's disease. J Neurosci Res 34: 371–376

Lo MMS, Fieles AW, Norris TE, Dargis PG, Caputo CB, Scott CW, Lee VM-Y, Goedert M (1993) Human tau isoforms confer distinct morphological and functional properties to stably transfected fibroblasts. Mol Brain Res 20: 209–220

Mangin G, Couchie D, Charrière-Bertrand C, Nunez J (1989) Timing of expression of τ and its encoding mRNAs in the developing cerebral neocortex and cerebellum of the mouse. J Neurochem 53: 45–50

Mattson MP (1992) Effects of microtubule stabilization and destabilization on tau immunoreactivity in cultured hippocampal neurons. Brain Res 582: 107–118

Matus A (1991) Microtubule-associated proteins and neuronal morphogenesis. J Cell Sci 100 Suppl. 15: 61–67

Mercken M, Vandermeeren M, Lübke U, Six J, Boons J, Van De Voorde A, Martin JJ, Gheuens J (1992) Monoclonal antibodies with selective specificity for Alzheimer tau are directed against phosphatase-sensitive epitopes. Acta Neuropathol (Berl) 84: 265–272

Migheli A, Butler M, Brown K, Shelanski ML (1988) Light and electron microscope localization of the microtubule-associated tau protein in rat brain. J Neurosci 8: 1846–1851

Miller MW (1988) Development of projection and local circuit neurons in neocortex. In: Peters A, Jones EG (eds) Cerebral cortex. Volume 7. Development and maturation of cerebral cortex. Plenum Press, New York and London, pp 133–175

Nukina N, Ihara Y (1986) One of the antigenic determinants of paired helical filaments is related to tau protein. J Biochem (Tokyo) 99: 1541–1544

Oestreicher AB, Gispen WH (1986) Comparison of the immunocytochemical distribution of the phosphoprotein B-50 in the cerebellum and hippocampus of immature and adult rat brain. Brain Res 375: 267–279

Papasozomenos SC, Binder LI (1987) Phosphorylation determines two distinct species of tau in the central nervous system. Cell Motil Cytoskel 8: 210–226

Riederer B, Matus A (1985) Differential expression of distinct microtubule-associated proteins during brain development. Proc Natl Acad Sci USA 82: 6006–6009

Riederer BM, Innocenti GM (1991) Differential distribution of tau proteins in developing cat cerebral cortex and corpus callosum. Eur J Neurosci 3: 1134–1145

Riederer B, Cohen R, Matus A (1986) MAP5: a novel brain microtubule-associated protein under strong developmental regulation. J Neurocytol 15: 763–775

Schoenfeld TA, McKerracher L, Obar R, Vallee RB (1989) MAP1A and MAP1B are structurally related microtubule associated proteins with distinct developmental patterns in the CNS. J Neurosci 9: 1712–1730

Scott CW, Klika AB, Lo MMS, Norris TE, Caputo CB (1992) Tau protein induces bundling of microtubules in vitro: Comparison of different tau isoforms and a tau protein fragment. J Neurosci Res 33: 19–29

Scott CW, Vulliet PR, Caputo CB (1993) Phosphorylation of tau by proline-directed protein kinase ($p34^{cdc2}/p58^{cyclin\ A}$) decreases tau-induced microtubule assembly and antibody SMI33 reactivity. Brain Res 611: 237–242

Sygowski LA, Fieles AW, Lo MMS, Scott CW, Caputo CB (1993) Phosphorylation of tau protein in tau-transfected 3T3 cells. Mol Brain Res 20: 221–228

Takemura R, Kanai Y, Hirokawa N (1991) In situ localization of tau mRNA in developing rat brain. Neuroscience 44: 393–407

Tucker RP (1990) The roles of microtubule-associated proteins in brain morphogenesis: A review. Brain Res Rev 15: 101–120

Tucker RP, Garner CC, Matus A (1989) In situ localization of microtubule-associated protein mRNA in the developing and adult rat brain. Neuron 2: 1245–1256

Vanisberg MA, Maloteaux JM, Octave JN, Laduron PM (1991) Rapid agonist-induced decrease of neurotensin receptors from the cell surface in rat cultured neurons. Biochem Pharmacol 42: 2265–2274

Watanabe A, Hasegawa M, Suzuki M, Takio K, Morishima-Kawashima M, Titani K, Arai T, Kosik KS, Ihara Y (1993) In vivo phosphorylation sites in fetal and adult rat tau. J Biol Chem 268: 25712–25717

Wischik CM, Novak M, Thogersen HC, Edwards PC, Runswick MJ, Jakes R, Walker JE, Milstein C, Roth M, Klug A (1988) Isolation of a fragment of tau derived from the core of the paired helical filament of Alzheimer disease. Proc Natl Acad Sci USA 85: 4506–4510

Wood JG, Mirra SS, Pollock NJ, Binder LI (1986) Neurofibrillary tangles of Alzheimer disease share antigenic determinants with the axonal microtubule-associated protein tau. Proc Natl Acad Sci USA 83: 4040–4043

Yamada KM, Spooner BS, Wessels MK (1970) Axon growth: role of microfilaments and microtubules. Proc Natl Acad Sci USA 66: 1206–1212

Modifications of Phosphorylated Tau Immunoreactivity Linked to Excitotoxicity in Neuronal Cultures

*J. Hugon**, *P. Sindou*, *M. Lesort*, *P. Couratier*, *F. Esclaire*, and *C. Yardin*

Introduction

Glutamate is one of the major excitatory neurotransmitters in the human brain (Fonnum 1984) but is also a potent neurotoxin producing in vitro and in vivo neuronal degradation and death (Meldrum and Garthwaite 1991). Three types of post-synaptic receptors are described: N-Methyl D aspartate, AMPA/Kainate, and metabotropic according to their principal pharmacological agonists. A large variety of molecules can activate these three post-synaptic receptors (Seeburg 1993). Glutamate has been implicated in the pathophysiology of many neurodegenerative disorders both acute, such as stroke and hypoglycemia (Simon et al. 1984; Wieloch 1985), and chronic, such as amyotrophic lateral sclerosis (Couratier et al. 1993), Parkinson's disease (Turski et al. 1991), Huntington's disease (Young et al. 1988), AIDS dementia complex (Lipton et al. 1991) and Alzheimer's disease (Koh et al. 1990; Kowall and Beal 1991; Mattson et al. 1992). Glutamate is also able to produce an intracellular signal transduction into neurons leading to protein phosphorylations (Scholz and Palfrey 1991). Tau is a microtubule-associated protein which favours microtubule polymerisation and stabilisation (Kosik 1993). One of the major neuropathological hallmarks of Alzheimer's disease is neurofibrillary tangles (NFT) associated with neuronal degeneration. NFT are composed of paired helical neurofilaments (PHF). Tau is one of the principal constituents of PHF but PHF tau is abnormally phosphorylated (Brion et al. 1991; Goedert 1993; Lee and Trojanowski 1992). The goal of the following experiments was to detect changes in tau immunoreactivity observed in neuronal cultures after glutamate exposure. We used a new monoclonal antibody, AT8 (provided by Innogenetics), raised against an abnormally phophorylated tau site at serine 202. Immunocytochemistry with confocal laser microscopy and immunoblot studies were carried out in these experiments.

* Unité de Neurobiologie Cellulaire, Laboratoire d'Histologie, Faculté de Médecine, 2 rue du Docteur Marcland, 87025 Limoges Cédex, France

K.S. Kosik et al. (Eds.)
Alzheimer's Disease: Lessons from Cell Biology

Materials and Methods

Neuronal Cell Cultures

The cortices of 18-day-old embryonic rats were dissected free of meninges and minced with scissors in PBS glucose without Ca^{2+} and Mg^{2+}. Then the cortices were mechanically dissociated using a Pasteur pipette. The resulting cell suspension was centrifuged (300 × g for 10 minutes) and the supernatant was removed. The resulting pellet was resuspended in MEM Earle's salts. Pretreatment of culture dishes was carried out with poly-L-lysine (PM: 30,000–70,000) for one hour at 37°C under humidified 95% air – 5% CO_2. This procedure and the washout of the dishes were performed with PBS glucose without Ca^{2+} and Mg^{2+}. The cortical cells were placed in MEM Earle's salts containing insulin (5 μg/ml), progesterone (2×10^{-8} mol), transferrin (100 μg/ml), selenium (3×10^{-8} mmol) and putrescine (100×10^{-6} mol) and 5% fetal calf serum. Cells were transferred into 35 mm dishes at a concentration of 1.25×10^6 cells/ml.

Poisoning Procedure

Neuronal cell cultures were poisoned with glutamate at concentrations of 200 μM or 500 μM for 30 seconds to 4 minutes or with 500 μM NMDA under the same conditions. Control cultures were not poisoned.

Confocal Laser Microscopic Study

Cell cultures were fixed with paraformaldehyde (4%) one and two minutes after the end of glutamate exposure. Cultures not treated with glutamate served as controls. An immunocytochemical study was carried out with AT8 antibody (Innogenetics). Indirect immunofluorescence was evaluated with a confocal laser microscope (Tracor Nooran Instruments). A quantitative image analysis was performed (Image 1) to assess AT8 immunoreactivity in the cytoplasm of cortical neurons. More than 300 neurons were analysed each time in a series of three different experiments.

After glutamate exposure and cell fixation, and prior to AT8 immunocytochemical study, some culture dishes were treated with alkaline phosphatase (Boerhinger) for 18 hours at 37°C (alkaline phosphatase was diluted 1:10 in Tris buffer; pH: 8).

Immunoblot Analysis

Neuronal cells were homogenized in sample buffer (0.5 M Tris, pH 6.8; 1.0% SDS; 1.0 mM Dithiothreitol) and heat treated for three minutes. Protein concentration was carried out on diluted sodium dodecyl sulfate

(SDS) samples using a bicinchoninic (BCA) protein assay reagent (Pierce). Proteins (50 μg) were separated on 10% SDS-polyacrylamide gel electrophoresis (SDS-page). Resolved proteins were then electrophoretically transferred to nitrocellulose membranes and tau AT8 immunodetection was performed immediately thereafter. AT8 antibody used at a dilution of 1/500 was incubated for three hours with nitrocellulose bands. Western blots were revealed with a second antibody against murine immunoglobulins conjugated with peroxydase (DAKO) prior to use of the PAP complex (DAKO). Western blots were stained with diamino benzidine (5 mg/10 ml). A semi-quantitative evaluation of tau (AT8) on immunoblots was realized using a scanning densitometer (BOP.BIOCOM Lecphor). The procedure was repeated twice.

Prior to glutamate exposure and biochemical analysis, cycloheximide was added to the culture medium in some dishes for one hour at the 10^{-4} M concentration.

All results were statistically evaluated by analysis of variance (ANOVA).

Results

AT8 Tau Modifications Using Laser Confocal Immunocytochemistry (Fig. 1)

Results are expressed as mean percentages of controls using arbitrary units obtained with the confocal laser microscope. Glutamate produces an increase in AT8 tau immunoreactivity both one and two minutes after the end of a five-minute glutamate exposure (500 μM); the percentage was 225.8% of controls one minute after the poisoning Pretreatment of exposed cultures with alcaline phophatase prior to immunocytochemical procedures dramatically reduced AT8 tau immunolabelling to the control level. Cultures exposed to 500 μM NMDA (in a Mg^{++}-free, glycine-supplemented medium) revealed a similar increase in AT8 tau immunoreactivity. The maximum effect was observed after a 15-minute NMDA exposure (250% of controls). Combined exposure using NMDA (500 μM) plus the NMDA antagonist MK801 (20 μM) dramatically decreased the AT8 tau reactivity to the control level (100%).

AT8 Tau Modifications Using Immunoblot Analysis (Fig. 2)

AT8 tau immunodetection was carried out from one to four minutes after the end of a five-minute glutamate exposure (200 μM) in neuronal cultures. Similar to what we observed with immunocytochemical studies, AT8 immunoblot profiles revealed a marked augmentation in glutamate-exposed cultures. AT8 tau appears on immunoblots as a single band with a molecular weight near 55 Kd. The increase in tau immunostaining was observed in all

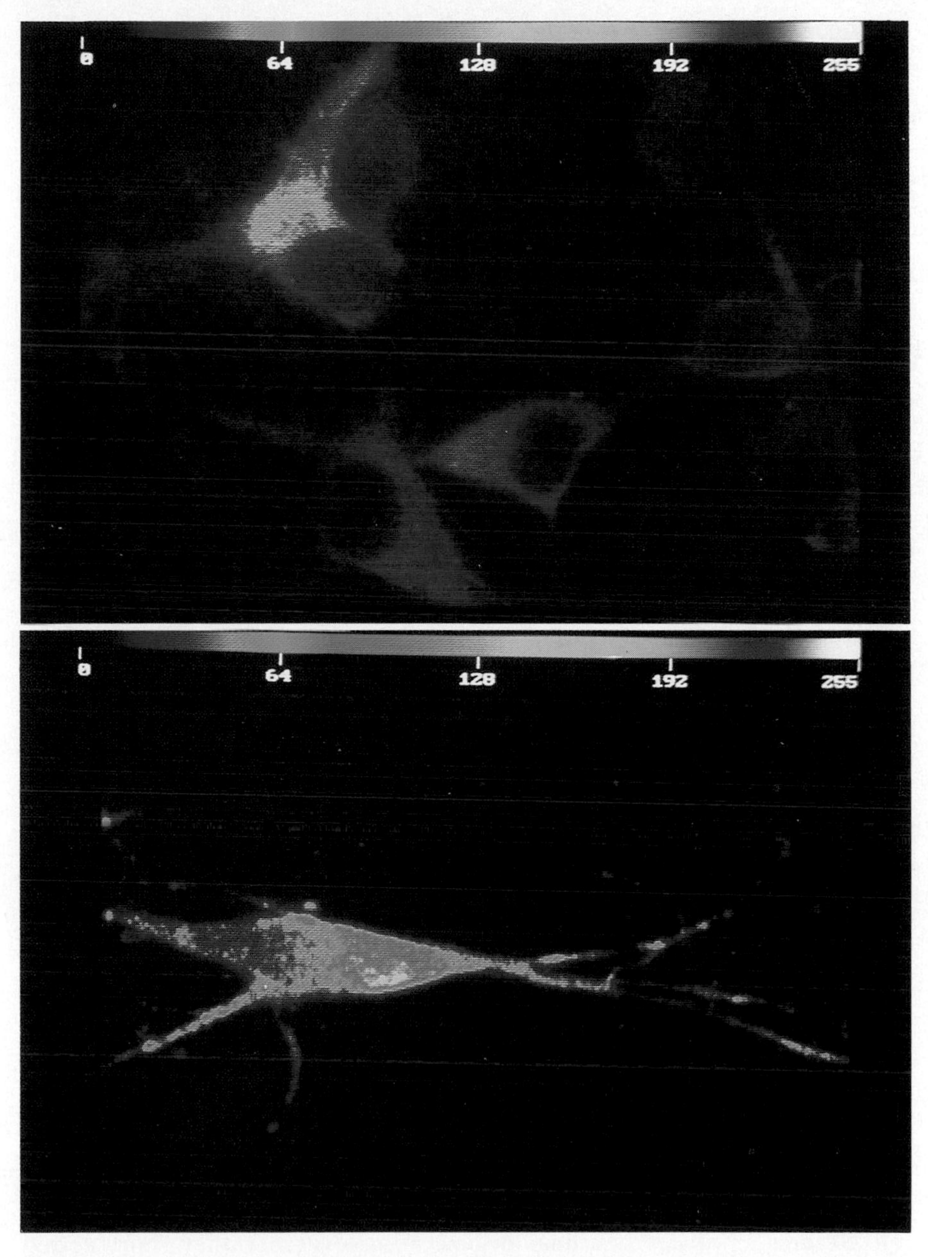

Fig. 1. Laser confocal microscopy: primary cortical cultures. *Upper view*, control cultures not exposed to glutamate. *Lower view*, cultures exposed to 500 μM glutamate for five minutes and fixed two minutes after the end of glutamate exposure. An increased immunostaining is observed. Immunofluorescence microscopy was performed with the anti tau AT8 antibody (provided by Innogenetics Belgium) which is raised against an abnormally phosphorylated site at serine 202

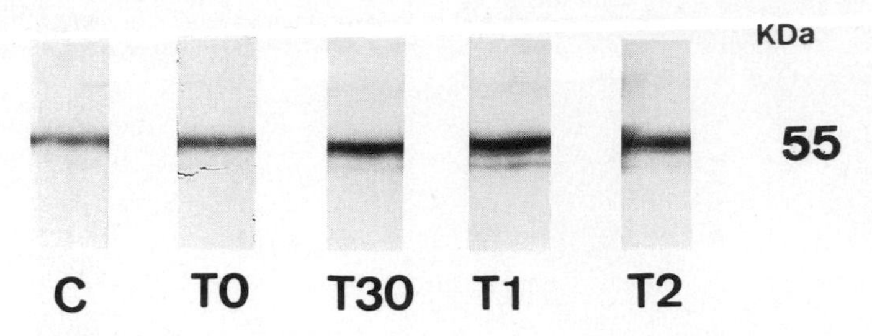

Fig. 2. Immunoblot analysis of primary cortical cultures exposed to 500 μM glutamate for five minutes. C, control cultures. T0, T30, T1, T2, cultures exposed to 500 μM glutamate and processed immediately (T0) or 30 seconds (T30), one minute (T1), or two minutes (T2) after the end of glutamate exposure. An increased immunolabelling is noted at T30, T1, and T2

glutamate-exposed cultures, but the maximum density was detected one minute after the end of glutamate exposure (355% of controls). The intensity of tau immunolabelling decreased slightly two and four minutes after glutamate exposure. Pretreatment of glutamate-exposed cultures by cycloheximide, a protein synthesis inhibitor, did not modify AT8 tau immunoblot profiles compared to glutamate-exposed cultures without cycloheximide. In a similar way neuronal cultures treated with 500 μM of NMDA displayed a striking AT8 tau immunolabelling, reaching a peak after 10 minutes of NMDA exposure.

Discussion

Our study demonstrates that glutamate toxicity produces in neuronal cultures a rapid and transient modification of tau immunoreactivity using AT8, a monoclonal antibody which is raised against an abnormally phosphorylated Tau protein (serine 202) present in Alzheimer's PHF but also present in fetal Tau protein.

Glutamate Toxicity

The concentration of glutamate (200 and 500 μM) used to poison neuronal cell cultures produces neuronal degeneration over a rather short period of time (Choi 1987). We have previously shown that glutamate at these concentrations in the same culture system model is responsible for a dose-dependent decrease in neuronal survival, with less than 10% of neurons surviving 12 hours after a brief exposure to 500 μM glutamate (Sindou et al. 1992).

Tau Modification

In previous reports several authors have shown that glutamate exposure in neuronal cultures produces immunoreactive modifications of tau protein using ALZ50. Specifically Mattson (1990) showed that an increase in tau immunostaining is observed after brief glutamate exposure. Similarly, glutamate produces an augmented immunostaining using tau_2 monoclonal antibody. These modifications were linked to the concentration of glutamate applied to neuronal cultures. A combined biochemical study (Sautiere et al. 1992) displayed that tau modifications consist in a shift of immunodetections from the lowest to the highest molecular weight tau isoforms associated with an acidification of tau protein.

Tau Phosphorylation

A large number of studies have demonstrated that PHF tau is composed of abnormally phophorylated tau proteins. An abnormal triplet of Tau proteins with increased molecular weights (55 Kd, 64 Kd, 69 Kd) is detected in the brain of Alzheimer's patients (Delacourte and Defossez 1986). A recent report has shown that the extent of phosphorylation of fetal tau is comparable to that of PHF tau from Alzheimer PHF (Goedert et al. 1993; Kenessey and Yen 1993). Antibodies used in the study of Goedert et al. (1993) include AT8, which is raised against a peptide comprising an abnormally phosphorylated epitope at serine 202, as desribed in our report. We can therefore postulate that, in neuronal cultures where a small proportion of fetal Tau is still present, glutamate toxicity is able to reproduce the abnormal biochemical pathway leading to the phosphorylation of tau protein (at serine 202) observed in Alzheimer's PHF. Previous work has demonstrated that glutamate can produce neuronal phosphorylation of proteins (Scholz and Palfrey 1991).

Neuronal Degeneration and Tau Modification

A rise in intracellular calcium plays an important role in neuronal degeneration produced by glutamate or NMDA toxicity (Choi 1987). NMDA receptors are permeable to Ca^{2+} when receptor ion-channels are opened. Calcium, which is a key factor in the excitotoxic cascade, can activate a large number of deleterous enzymes including kinases, phosphatases, proteases, etc. We do not know in our experimental model which kinase is activated by intracellular calcium influx to produce tau phosphorylation. It is plausible to postulate that modifications of enzyme activities linked to tau phosphorylation and produced by glutamate (increased kinase activities or decreased phosphatase activities) are directly or indirectly triggered by the rise in intracetllular calcium. Further work will have to decipher which enzymes are implicated in tau phosphorylation produced by glutamate. This could represent a useful pharmacological target to act on abnormal phosphorylation of tau protein.

Acknowledgements. We thank M.L. Autef and I. Teissandier for typing the manuscript. F. Forestier for technical help and Innogenetics for providing AT8 antibody. This work was supported by La Fondation pour le Recherche Médicale et Le Conseil Régional du Limousin.

References

Brion JP, Hanger DP, Couck AM, Anderton BH (1991) A68 proteins in Alzheimer's disease are composed of several tau isoforms in a phosphorylated state which affects their electrophoretic mobilities. Biochem J 279: 831–836

Choi DW (1987) Ionic dependence of glutamate neurotoxicity. J Neurosci 7: 369–379

Couratier P, Hugon J, Sindou P, Vallat JM, Dumas M (1993) Cell culture evidence for neuronal degeneration in amyotrophic lateral sclerosis being linked to glutamate AMPA/Kainate receptors. Lancet 341: 265–268

Delacourte A, Defossez A (1986) Alzheimer's disease: tau proteins, the promoting factors of microtubule assembly, are major components of paired helical filaments. J Neurol Sci 76: 173–186

Fonnum F (1984) Glutamate: a neurotransmitter in mammalian brain. J Neurochem 42: 1–11

Goedert M (1993) Tau protein and the neurofibrillary pathology of Alzheimer's disease. Trends Neurosci 16: 460–465

Goedert M, Jakes R, Crowther RA, Six J, Lobke U, Vandermeeren I, Cras P, Troganows JQ, Lee VM-Y (1993) The abnormal phosphorylation of tau protein at Ser-202 in Alzheimer disease recapitulates phosphorylation during development. Proc Natl Acad Soc USA 90: 5066–5070

Kenessey A, Yen S-H C (1993) The extent of phosphorylation of fetal tau is comparable to that of PHF-tau from Alzheimer paired helical filaments. Brain Res 629: 40–46

Koh JY, Yang LL, Cotman CW (1990) β-amyloid protein increases the vulnerability of cultured cortical neurons to excitotoxic damage. Brain Res 533: 315–320

Kosik KS (1993) The molecular and cellular biology of Tau. Brain Pathol 3: 39–43

Kowall NW, Beal MF (1991) Glutamate, glutaminase and taurine immunoreactive neurons develop neurofibrillary tangles in Alzheimer's disease. Ann Neurol 29: 162–167

Lee V M-Y, Trojanowski JQ (1992) The disordered neuronal cytoskeleton in Alzheimer's disease. Curr Opin Neurobiol 2: 653–656

Lipton SA, Sucher NJ, Kaiser PK, Dreyer EB (1991) Synergistic effects of HIV coat protein and NMDA receptor-mediated neurotoxicity. Neuron 7: 111–118

Mattson MP (1990) Antigenic changes similar to those seen in neurofibrillary tangles are elicited by glutamate and Ca^{2+} influx in cultured hippocampal neurons. Neuron 2: 105–117

Mattson MP, Cheng B, Davis D, Bryant K, Lieberbury I, Rydel A (1992) β-amyloid peptides destabilize calcium homeostasis and render human cortical neurons vulnerable to exocitotoxicity. J Neurosci 12: 376–389

Meldrum B, Garthwaite J (1991) Excitatory amino acid neurotoxicity and neurodegenerative disease. TIPS Spec Rept 54–62

Sautiere PE, Sindou P, Couratier P, Hugon J, Wattez A, Delacourte A (1992) Tau antigenic changes induced by glutamate in rat primary culture model: a biochemical approach. Neurosci Lett 140: 206–210

Scholz WK, Palfrey HC (1991) Glutamate-stimulated protein phosphorylation in cultured hippocampal pyramidal neurons. J Neuronsci 11: 2422–2432

Seeburg PH (1993) The molecular biology of mammalian glutamate receptor channels. TINS 16: 359–365

Simon RP, Swan JH, Griffiths T, Meldrum BS (1984) Blockade of N-methyl-D-aspartate receptors may protect against ischemic damage in the brain. Science 2267: 850–852

Sindou P, Couratier P, Barthe D, Hugon J (1992) A dose-dependent increase of tau immunostaining is produced by glutamate toxicity in primary neuronal cultures. Brain Res 572: 242–246

Turski L, Bressler K, Rettig K-J, Loeschmann PA, Wachtel H (1991) Protection of substantia nigra from MPP+ neurotoxicity by N-methyl-D-aspartate antagonists. Nature 349: 414–419

Wieloch T (1985) Hypoglycemia-induced neuronal damage prevented by an N-methyl-D-aspartate antagonist. Science 230: 681–683

Young AB, Greenamyre JT, Hollingworth Z, Albin R, d'Amato C, Shoulson I, Penney JQ (1988) NMDA receptor losses in putamen from patients with Huntington's disease. Science 241: 981–983

Neurofilaments and Motor Neuron Disease

*D.W. Cleveland** and *Z. Xu*

Summary

Motor neuron disease is clinically characterized by progressive muscle wasting leading to total muscle paralysis. In the most prominent disease, amyotrophic lateral sclerosis, or ALS, a long history of pathological study has firmly established that the primary lesion site is in spinal and cortical motor neurons. Although widespread loss of these neurons is seen late in the pathogenic progression, the earliest detectable abnormalities are aberrant accumulations of neurofilaments in cell bodies and proximal axons. To test whether accumulation of neurofilaments directly contributes to the pathogenic process, transgenic mice that produce high levels of neurofilaments in motor neurons have been generated. These transgenic mice show most of the hallmarks observed in motor neuron disease, including swollen perikarya with eccentrically localized nuclei, proximal axonal swellings, axonal degeneration and severe skeletal muscle atrophy. These data indicate that excessive accumulation of neurofilaments can trigger selective motor neuron failure. Since such accumulation is seen both in sporadic and familial ALS, it is almost certain to be a key intermediate in the pathway of pathogenesis leading to neuronal loss.

Introduction

Neurofilaments assembled from three polypeptide subunits, NF-L (68 kd), NF-M (95 kd) and NF-H (115 kd), are 10 nm diameter filaments that are the most abundant structures in most neurons and are particularly prominent in large myelinated axons, such as those elaborated by spinal motor neurons. Mounting evidence has strengthened the view that neurofilaments play a critical role in the development and maintainence of axonal caliber (Friede and Samorajski 1970; Hoffman et al. 1987; Cleveland et al. 1991; Yamasaki et al. 1992; Eyer and Peterson 1994), a crucial determinant for conduction

* Departments of Biological Chemistry and Neuroscience, 725 North Wolfe Street, The Johns Hopkins University School of Medicine, Baltimore, MD 21205, USA

K.S. Kosik et al. (Eds.)
Alzheimer's Disease: Lessons from Cell Biology

velocity of axons and perhaps also a trigger for myelination (Arbuthnott et al. 1980; Voyvodic 1989).

In addition to a function in supporting the growth and maintenance of axonal caliber in normal neurons, several lines of evidence have supported a role of neurofilaments in the pathogenesis of several types of neurodegenerative diseases. This is certainly so for the most frequent motor neuron disease, amyotrophic lateral sclerosis or ALS (Banker 1986; Carpenter 1968; Inoue and Hirano 1979; Hirano et al. 1984a,b; Mulder 1984, 1986), a late-onset disease usually initiating in the fourth or fifth decade and characterized by progressive loss of motor neurons, which in turn leads to wasting of skeletal muscle, paralysis and ultimately death (for review, see Mulder 1984, 1986; Gomez 1986). The majority of cases are sporadic, although in ~10% of incidences the disease is inherited as an autosomal dominant. In 15–20% of these familial cases (known as FALS), the disease is known to arise from point mutations in the enzyme superoxide dismutase 1 (SOD1). The primary cause(s) of the remaining cases (>97% of all cases) has not been elucidated and in no case is the mechanism(s) of selective, age-dependent motor neuron failure known.

A long history of pathological examination dating from the middle of the last century has firmly established that the primary lesion in motor neuron disease lies predominantly in the cortical and spinal motor neurons. Various degrees of motor neuron loss in either upper (Betz cell) or lower (α-) motor neurons are seen as the major pathological hallmark (Chou 1992), although in those diseases which progress rapidly (e.g., infantile muscular atrophy and adult diseases which progress relatively rapidly), significant numbers of surviving motor neurons are observed in post mortem examination. Most of these remaining motor neurons display various abnormal morphologies, including swollen cell bodies and dispersal of the rough endoplasmic reticulum (often called Nissl substance). Further, numerous large axonal swellings that sometimes reach the size of a cell body are commonly found (Carpenter 1968; Chou 1992). These swollen structures stain strongly with silver (Fig. 1), suggesting that they are rich in neurofilaments (Carpenter 1968; Hughes and Jerrome 1971; Chou and Fakadej 1971; Inoue and Hirano 1979; Hirano et al. 1984). The use of immunocytochemistry and electron microscopy has unequivocally established that the swollen perikarya and axons contain abundant swirls of neurofilaments (Carpenter 1968; Hughes and Jerrome 1970; Chou and Fakadej 1971; Hirano et al. 1984a,b) and have lead Hirano (1991) to conclude "several apparently early changes [in ALS] are truly due to the accumulation of neurofilaments in anterior horn cells [the α- motor neurons]."

Motor neuron disease with symptoms resembling infantile spinal muscular atrophy has also been described in a number of animal species. Remarkable neurofilament accumulation in the anterior horn α-motor neurons has been found in all such cases (Delahunta and Shively 1974; Vandevelde et al. 1976; Higgins et al. 1977; Shields and Vandevelde 1978;

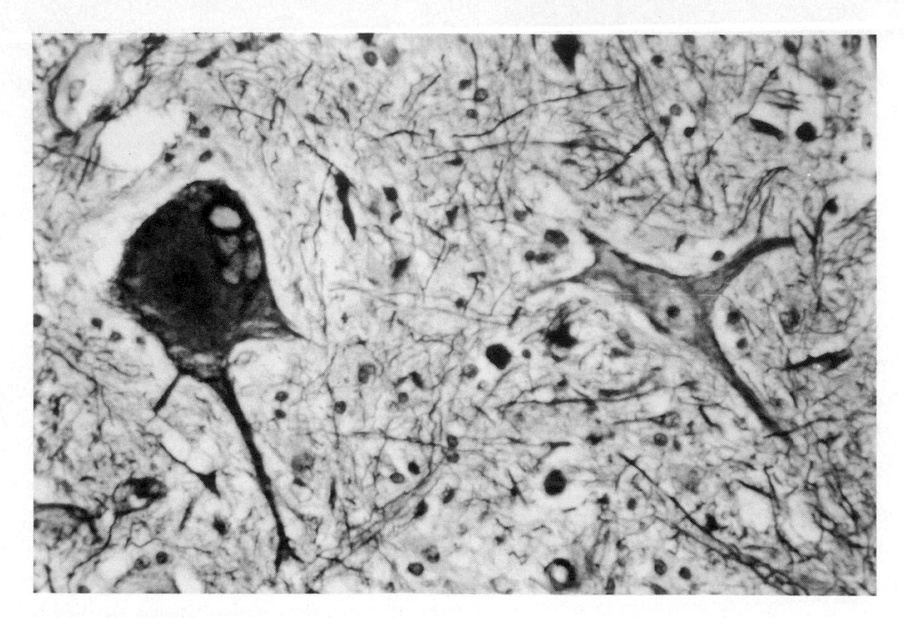

Fig. 1. Swollen neurons in the anterior horn of the spinal cord from human ALS. Swollen neurons with neurofilament accumulation from a case of familial ALS stained by Bielschowsky's silver impregnation. Scale bar, 50 μm. Reproduced with permission from Hirano et al. (1984)

Cork et al. 1982; Higgins et al. 1983). These studies have demonstrated that aberrant neurofilament accumulation is a common intermediate in both early- and late-onset motor neuron disease.

Despite these widely observed examples of neurofilament misaccumulation in motor neurons in humans and animals with motor neuron disease, a central unsolved question has been whether the aberrant accumulation is merely a harmless byproduct of the pathogenic process or a central element in the pathogenic pathway that leads to neuronal dysfunction and ultimately cell death. To distinguish between these two possibilities, we have used genetic methods to test whether forcing neurons to increase expression of neurofilaments is sufficient to yield motor neuron disease. As we show, the clear outcome is that excessive accumulation of neurofilaments in perikarya and proximal axons results in neuronal dysfunction, including increased axonal degeneration (Xu et al. 1993; Cote et al. 1993). These collective abnormalites in motor neurons in turn trigger severe skeletal muscle atrophy. These results demonstrate that misaccumulation of neurofilaments can cause motor neuron disease and strongly support the view that neurofilamentous aggregates are an integral part of the pathogenic pathway in motor neuron degeneration.

Results

Accumulation of High Levels of NF-L Result in Morphologic Characteristics of Motor Neuron Disease

To examine the consequence of forcing increased expression of wild-type murine NF-L, several lines of transgenic mice that carried additional copies of the authentic mouse NF-L gene were produced. To obtain high levels of NF-L expression, the putative NF-L gene promoter was substituted with the strong viral promoter from murine sarcoma virus. Immunoblotting revealed that two lines of mice accumulated amounts up to twice the normal level of wild type murine NF-L in peripheral nerves (Fig. 2). Immunohistochemistry of spinal cord and sciatic nerves revealed that accumulated NF-L, both endogenous and transgene encoded, was present only in the neurons, with none detectable in the surrounding Schwann cells or oligodendrocytes (see also Monteiro et al. 1990). Although no overt phenotypic change was observed in any of these transgenic mice (Monteiro et al. 1990), when two independent, highly expressing mouse lines were mated, doubly transgenic mice were obtained that accumulated four times the normal level of NF-L (Fig. 2). During the first 21 days postnatally, all of these double transgenic animals were significantly smaller in size than their littermates (around 1/3–2/3 of the normal weight) and all displayed progressive reduction in kinetic activity, culminating in eventual death during the third postnatal week. With careful feeding, two of these doubly transgenic animals survived

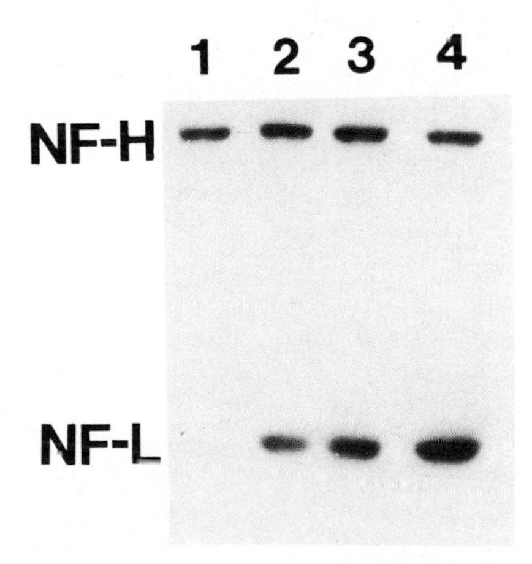

Fig. 2. Accumulation of excess NF-L in sciatic nerves of singly and doubly transgenic mice determined by immunoblot. Total proteins extracted from the sciatic nerve of (lane 1) wild-type, (lane 2) line MSV-NF-L103, (lane 3) line MSV-NF-L58 and (lane 4) a doubly transgenic mouse were taken from 20-day-old animals, separated by SDS-PAGE and immunoblotted with a mixture of anti-NF-L and phosphorylation independent NF-H monoclonal antibodies

past three weeks of age. Genomic DNA blotting revealed that eight of eight animals showing this slow growth phenotype were doubly transgenic mice, whereas all other littermates were wild type or singly transgenic.

To determine the consequence of increased NF-L accumulation, four of the doubly transgenic animals were sacrificed by perfusion at day 21 or 22 and their tissues were examined morphologically. In all of these animals, striking changes were observed in the anterior horn motor neurons of the spinal cord. Compared with non-transgenic littermates, almost all of the motor neurons (at all levels of the spinal cord) displayed enlarged, ballooned perikarya with depleted rough endoplasmic reticulum (often referred to as chromatolysis) and eccentrically positioned nuclei (Fig. 3). These were strikingly similar to what has been seen in many reported cases of motor neuron disease (e.g., Fig. 1).

At higher magnification, massive accumulation of filaments in all motor neuron compartments (cytoplasm, dendrites and axons) was confirmed (not shown, but see Xu et al. 1993). Accompanying these filaments were two obvious axonal abnormalities. First, as in ALS, within the anterior horn of a two-month-old doubly transgenic mouse numerous axonal swellings were present (Fig. 4a). These swellings were filled with neurofilaments (Fig. 4b). Second, in the ventral roots containing the corresponding motor axons from doubly transgenic animals of various ages, degenerating axons were present (Fig. 4c). While the proportion of degenerating axons in four ventral roots of doubly transgenic animals was less than 0.2%, this was a markedly

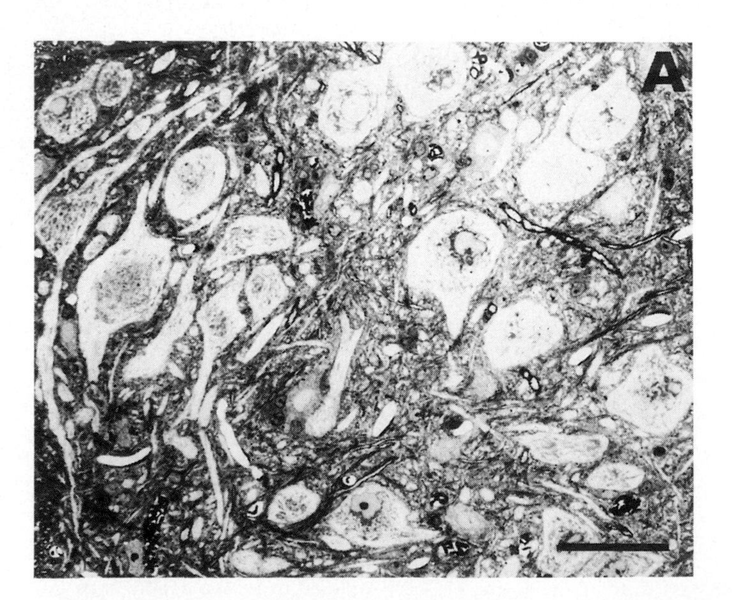

Fig. 3. Morphology of spinal anterior horn motor neurons expressing high levels of NF-L. The anterior horn of a spinal cord from a 21-day-old MSV-NF-L58/MSV-NF-L103 doubly transgenic animal. One micron section stained with toliudine blue. Bar, 60 μm

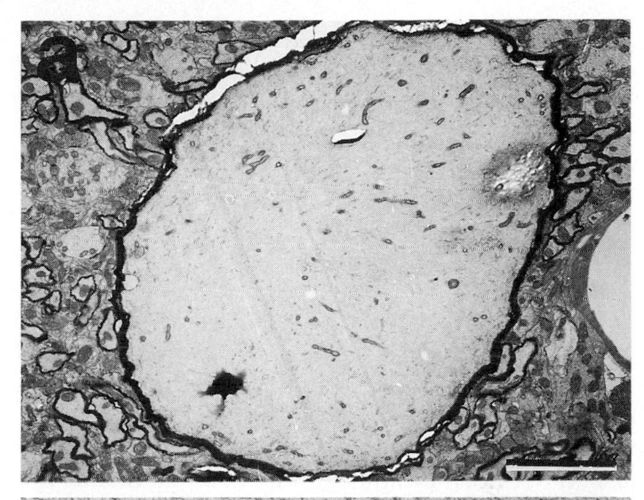

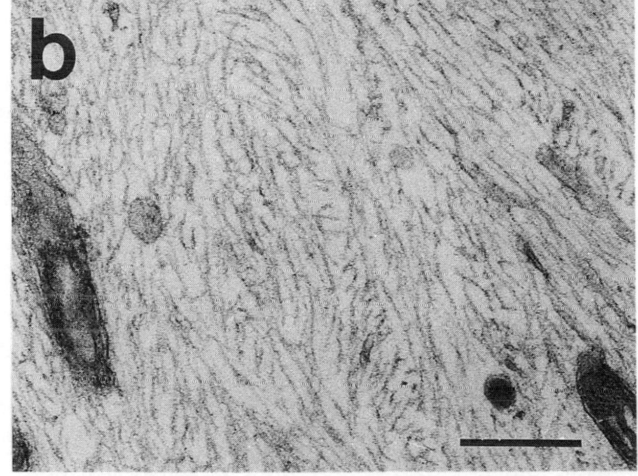

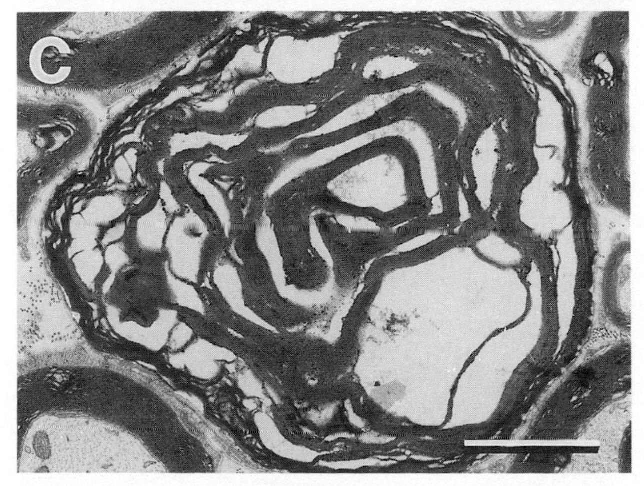

Fig. 4. Axonal abnormalities in the doubly transgenic NF-L mice. **A** Light microscopic view of an axonal swelling in the proximal axon of an anterior horn motor neuron from a doubly transgenic animal. Bar, 4 μm. **B** Higher magnification view of an area inside the axonal swelling. Bar, 0.4 μm. **C** Degenerated axon from a ventral root of a doubly transgenic NF-L mouse. Bar, 2 μm

increased frequency compared to control littermates, where no degenerating axons were found.

The two doubly NF-L transgenic mice that survived past three weeks of age gradually recovered and, by two months of age, were 4/5 of the weight of littermate controls. The non-progressive course of transgene-mediated pathology in these animals was initially a surprise. However, quantitation of protein blots of sciatic nerve extracts from both doubly and singly NF-L transgenic mice revealed that, although NF-L accumulation initially rises significantly above littermate controls, after three weeks of age it gradually falls back to about the same as in wild type animals. Concomitant with this decrease in excessive NF-L, morphological examination of one doubly transgenic animal revealed restoration of a nearly wild type appearence at nine months of age. Similarly, neurofilament density in axons also declined with age. Since the only difference between the transgene and the wild type gene lies in their promoters, the most reasonable explanation for the decline in transgene encoded NF-L is an age-dependent reduction in activity of the MSV promoter used to drive transgene transcription. In any event, loss of both phenotypic and morphological abnormalities in neurons coincident with the age-dependent reduction in NF-L content, combined with the absence of abnormalities in either singly transgenic mouse line, strongly supports the view that only those increases in neurofilament economy above a threshold level result in obvious neurological abnormalities.

Abnormal Morphological Changes in Motor Neurons Are Accompanied by Severe Muscle Atrophy

The low body weight phenotype of the doubly transgenic mice was accompanied by a progressive loss in kinetic activity of the doubly transgenic animals. Postmortem examination of 21-day-old animals revealed widespread skeletal muscle atrophy. For example, Figure 5A,B displays cross sections of the anterior tibial muscle from a doubly transgenic and an age-matched control animal. Individual muscle fibers in the transgenic sample are <20% of the cross sectional area of the wild-type, a phenotype consistent with denervation-induced muscle atrophy. However, as noted previously for the singly transgenic lines (Monteiro et al. 1990), transgene encoded NF-L is not expressed exclusively in neurons but also accumulates in skeletal muscles. To distinguish whether atrophy was a consequence of nerve dysfunction or a direct effect of NF-L accumulation in muscle, we evaluated the level of NF-L in muscles from animals of different ages. In contrast to the decline of transgene expression in neurons as the animals age beyond 21 days, NF-L accumulation in muscle continues to increase up to two months of age and remains higher than the level seen in a two and one-half-week-old animal for at least 11 months thereafter. Since the muscle was most severely atrophic between two and three weeks of age but recovered to

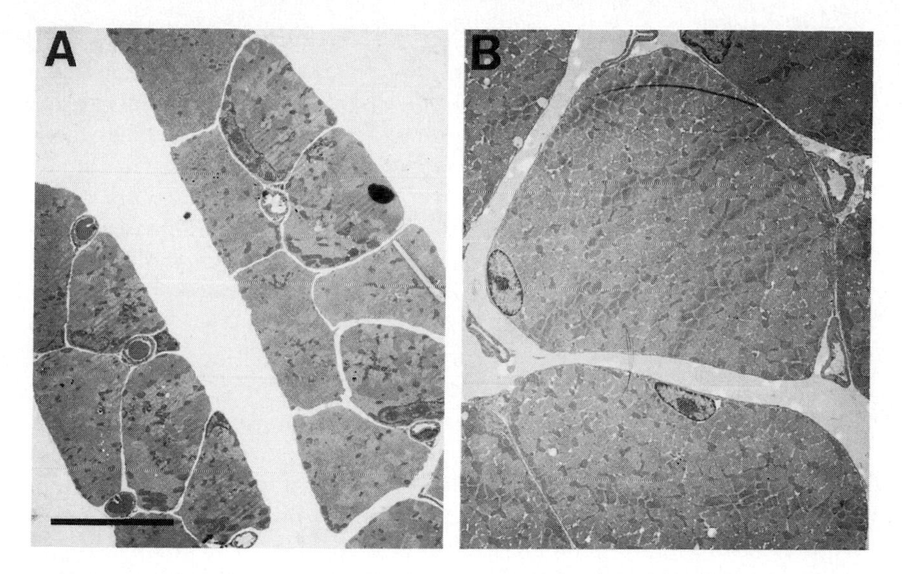

Fig. 5. Severe muscle atrophy in NF-L doubly transgenic mice. Electron micrographs of cross sections of muscle fibers from anterior tibial muscle of a 21-day-old (**A**) doubly transgenic mouse or a non-transgenic littermate (**B**). Bar, 10 μm. Reproduced with permission from Xu et al. (1993)

nearly normal size by two months despite the increasing burden of transgene encoded NF-L, we conclude that there is no correlation between the level of NF-L accumulation in muscle and muscle atrophy. In contrast, the muscle atrophy correlates well with the peak accumulation of NF-L in neurons. These data strongly support the view that the predominant cause of muscle atrophy is the dysfunction of motor neurons resulting from the excessive accumulation of neurofilaments.

Discussion

Neurofilamentous accumulations in perikarya, dendrites and axons occur in a variety of degenerative, toxic, and heritable diseases. Particularly striking examples have been reported in different types of motor neuron diseases, including familial ALS (Hughes and Jerrome 1970; Takahashi et al. 1972; Hirano et al. 1984b), sporadic ALS (Carpenter 1968; Schochet et al. 1969; Hirano et al. 1984a), and infantile spinal muscular atrophy (Byers and Banker 1961; Chou and Fakadej 1971; Wiley et al. 1987), as well as in spontaneous motor neuron disease in various animal species including dog (Delahunta and Shively 1974; Cork et al. 1982), zebra (Higgins et al. 1977), rabbit (Shields and Vandevelde 1978), cat (Vandevelde et al. 1976) and pig (Higgins et al. 1983). The present results, in conjunction with similar findings using transgenic technology of force accumulation of human NF-H in mouse motor neurons (Cote et al. 1993), provide an unambiguous demonstration that primary alterations in neurofilament economy can 1) lead to structural changes of the type seen in these neurodegenerative disorders and 2) ultimately lead to axonal breakdown and loss.

The morphological effects of overproducing neurofilaments in motor neurons bear most striking resemblance to those observed in rapidly progressing disease such as the early stages of ALS (Inoue and Hirano 1979; Hirano et al. 1984a,b) and infantile spinal muscular atrophy (Byers and Banker 1961; Chou and Fakadej 1971; Wiley et al. 1987). This finding raises the probability that marked neurofilament accumulation in perikarya and proximal axons is an early pathological change that precedes the widespread neuronal loss. Consistent with this is the observation that in virtually all the reported cases of spinal muscular atrophy from various animal species, a large number of swollen neurons with high neurofilament accumulation is a prominent feature. Further, in dogs with rapidly progressing spinal muscular atrophy, severe neurofilamentous accumulations are accompanied by only a minor motor neuron loss. Only in the cases in which the disease progresses more slowly is more prominent motor neuron loss observed (Cork et al. 1982). In this context, in many human examples a relatively low frequency of swollen perikarya and a higher proportion of degenerating axons may simply reflect the slow progression of disease which allows compromised neurons to initiate subsequent degeneration. From a slightly different perspective, the remarkable extent of neurofilament accumulation in both naturally occurring motor neuron disease and in the transgenic mice further indicates a great degree of tolerance of the neuron for the substantial increases in total "neurofilament burden," suggesting that filament-induced degeneration is a slow process. This is consistent with the gradual progression of many of the disorders with neurofilament accumulation.

If it is true, as our data imply, that neurofilament accumulation may be an active participant in the pathogenesis of motor neuron disease, then what mechanisms can lead to the accumulation of neurofilaments in the perikarya and proximal axons? The various possibilities (diagrammed in Fig. 6) include increased synthesis, decreased degradation, and defective transport of neurofilaments. Slowed degradation seems unlikely in the pathogenesis of motor neuron disease because increased neurofilament stability would be expected to lead to more extreme distal accumulations of neurofilaments, which has not been observed. While a concomitant increase in neurofilament synthesis is a possibility, the most plausible mechanism (and one which appears to be consistent with the neuropathological findings) is an alteration in slow axonal transport of neurofilaments. This alteration could be derived from defects either in the machinery that moves the filaments or in the filaments themselves. Further experimental support for this view emerges from various toxin-induced neuropathies. Intoxication with aluminum or β,β'-iminodiproionitrile (IDPN) selectively disrupts transport of neuro-

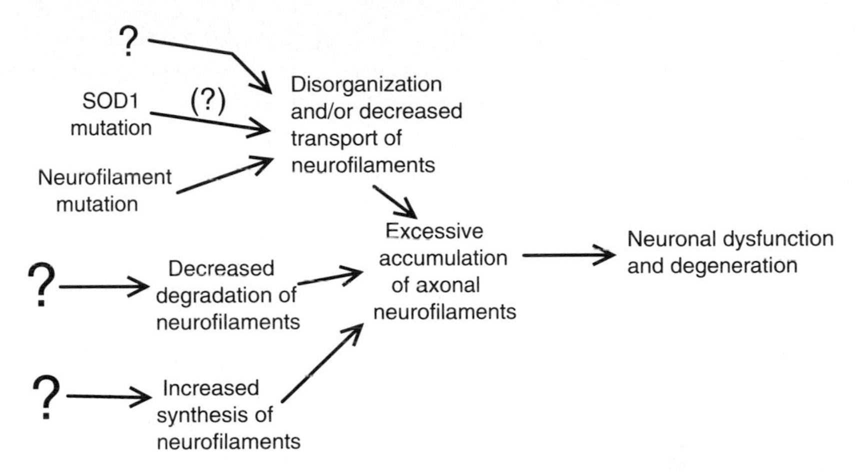

Fig. 6. Schematic model for involvement of neurofilaments in motor neuron disease. Adapted from Xu et al. (1993)

filaments (Griffin et al. 1978; Bizzi et al. 1984; Troncoso et al. 1985), leading to prominent axonal swelling. Like overexpression of NF-L in mice, these neurotoxins cause neuropathy without actually killing most motor neurons.

If neurofilamentous accumulations are important intermediates in the neurodegenerative process, how can this be reconciled with the recent discovery that 15–20% of the familial cases of ALS (≈1.5–2% of total ALS cases) result from mutations in the enzyme superoxide dismutase (SOD1; Rosen et al. 1993; Deng et al. 1993; Bob Brown, personal communication)? How too can we explain the puzzle that SOD1 mutations lead to the selective death of motor neurons even though SOD1 is thought to be expressed in most (possibly all) cells? Since SOD1 acts to limit oxidative damage by converting oxygen radicals into peroxide (Fridovich 1986), if the sites of primary damage are proteins, we suggest that it is reasonable that the most affected proteins will be those that have long turnover times, since proteins with short half lives are quickly replaced. In this context, neurofilaments are slowly transported proteins with transit times from synthesis to arrival near the nerve terminus of as long as one to two years (calculated for a meter-long axon and a transport rate of 1–2 mm/day; see Oblinger and Lasek 1985). It is conceivable, therefore, that damaged proteins of this class of slowly transported proteins gradually accumulate to poison the transport machinery. Such a mechanism can explain the particular vulnerability of motor neurons in motor neuron disease: whatever the primary cause, if damage to neurofilaments or their transport ultimately results in neuronal degeneration, then cells that normally have the highest neurofilament burden will be the ones first and most severely affected. Consistent with this prediction, both in our mice and in human motor neuron diseases, the neurons that are most severely affected by accumulation of neurofilaments are among the largest neurons with the longest axons that normally contain abundant amounts of neurofilaments. Indeed, recent work has demonstrated that the prominent neurofilamentous accumulations in two familial ALS kindreds are the pathologic outcome of mutation in SOD1 (T. Siddique, A. Clark and G. Rouleau, personal communications). [In fact, perhaps the best studied familial ALS kindred (the one studied by Hirano et al. (1984b) and demonstrated in Fig. 1) has disease caused by an SOD1 mutation (T. Siddique, personal communication).]

Our evidence has established one point of pathological significance for human motor neuron disease: primary changes in the cytoskeleton, and specifically in neurofilaments, are sufficient to produce most of the pathological changes encountered in neurodegenerative diseases such as ALS. Combined with the finding that mutations in other genes (e.g., SOD1) lead to the same neurofilamentous abnormalities that typify sporadic ALS, our data further promote the suggestion (diagrammed in Fig. 6) of a common pathogenetic sequence that includes neurofilament accumulation as a central pathological intermediary leading to subsequent axonal swelling and degeneration. Indeed, cytoskeletal abnormalities may increase the susceptibility of the neuron to other insults, so that multiple factors could culminate in production of disease.

Acknowledgements. We thank Dr. A. Hirano (Montifiore Medical Center, New York) for providing the micrograph of neurons from patients with familial ALS. We thank Drs. Teppu Siddique, Arthur Clark and Guy Rouleau for providing results prior to publication. We also thank Drs. John Griffin, Don Price, Michael Lee and our colleagues at Johns Hopkins for their many contributions throughout the course of this work. This work has been supported by grants from the NIH to D.W.C. Z-S.X. has been supported by a postdoctoral fellowship from the Muscular Dystrophy Association.

References

Arbuthnott ER, Boyd IA, Kalu KU (1980) Ultrastructural dimensions of myelinated peripheral nerve fibres in the cat and their relation to conduction velocity. J Physiol 308: 125–157

Banker BQ (1986) The pathology of the motor neuron disorders. In: Engel AG, Banker BQ (eds) Myology. McGraw-Hill, New York, pp 2031–2066

Bizzi A, Crane RC, Autilio-Cambetti L, Gambetti P (1984) Aluminum effect on slow axonal transport: a novel impairment of neurofilament transport. J Neurosci 4: 722–731

Byers RK, Banker BQ (1961) Infantile spinal muscular atrophy. Arch Neurol 5: 140–164

Carpenter S (1968) Proximal axonal enlargement in motor neuron disease. Neurology 18: 841–851

Chou SM (1992) Pathology-light microscopy of amyotrophic lateral sclerosis. In: Smith RA (ed) Handbook of amyotrophic lateral sclerosis. Marcel Dekker, New York, pp 133–181

Chou SM, Fakadej AV (1971) Ultrastructure of chromatolytic motor neurons and anterior spinal roots in a case of Werdnig-Hoffmann Disease. J Neuropathol Exp Neurol 30: 368–379

Cleveland DW, Monteiro MJ, Wong PC, Gill SR, Gearhart JD, Hoffman PN (1991) Involvement of neurofilaments in the radial growth of axons. J Cell Sci 15: 85–95

Cork LC, Griffin JW, Choy C, Padula CA, Price DL (1982) Pathology of motor neurons in accelerated hereditary canine spinal muscular atrophy. Lab Invest 46: 89–99

Cork LC, Troncoso JC, Klavano GG, Johnson ES, Sternberger LA, Sternberger NH, Price DL (1988) Neurofilamentous abnormalitics in motor neurons in spontaneously occurring animal disorders. J Neuropathol Exp Neurol 47: 420–431

Cote F, Collard J-F, Julien J-P (1993) Progressive neuronopathy in transgenic mice expressing the human neurofilament heavy gene: a mouse model of amyotrophic lateral sclerosis. Cell 73: 35–46

Delahunta A, Shively JN (1974) Neurofibrillary accumulation in a puppy. Cornell Vet 65: 240–247

Deng H-X, Hentati A, Tainer JA, Iqbal Z, Cayabyab A, Hung W-Y, Getzoff ED, Hu P, Herzfeldt B, Roos RP, Warner C, Deng G, Soriano E, Smyth C, Parge HE, Ahmed A, Roses AD, Hallewell RA, Pericak-Vance MA, Siddique T (1993) Amyotrophic lateral sclerosis and structural defects in Cu,Zn superoxide dismutase. Science 261: 1047–1051

Eyer J, Peterson A (1994) Neurofilament-deficient axons and perikaryal aggregates in viable transgenic mice expressing a neurofilament beta-galactosidase fusion protein. Neuron 12: 389–405

Fridovich I (1986) Superoxide dismutase. Adv Enzym 58: 61–97

Friede RL, Samorajski T (1970) Axon caliber related to neurofilaments and microtubules in sciatic nerve fibers of rats and mice. Anat Rec 167: 379–388

Gomez MR (1986) Motor neuron diseases in children. In: Engel AG, Banker BQ (eds) Myology. McGraw-Hill, New York, pp 1993–2012

Griffin JW, Hoffman PN, Clark AW, Carroll PT, Price DL (1978) Slow axonal transport of neurofilament proteins: impairment by β,β'-iminodipropionitrile administration. Science 202: 633–636

Higgins RJ, Vandevelde M, Hoff EJ, Jagar JE, Cork LC, Selbermann MS (1977) Neurofibrillary accumulation in the zebra (Equus burchelli). Acta Neuropathol 37: 1–5

Higgins RJ, Rings DM, Fenner WR, Stevenson S (1983) Spontaneous lower motor neuron disease with neurofibrillary accumulation in young pigs. Acta Neuropathol 59: 288–294

Hirano A (1991) Cytopathology of amyotrophic lateral sclerosis. Adv Neurol 56: 91–102

Hirano A, Donnenfeld H, Shoichi S, Nakano I (1984a) Fine structural observations of neurofilamentous changes in amyotrophic lateral sclerosis. J Neuropathol Exp Neurol 43: 461–470

Hirano A, Nakano I, Kurland LT, Mulder DW, Holley PW, Saccomanno G (1984b) Fine structural study of neurofibrillary changes in a family with amyotrophic lateral sclerosis. J Neuropathol Exp Neurol 43: 471–480

Hoffman PN, Cleveland DW, Griffin JW, Landes PW, Cowan NJ, Price DL (1987) Neurofilament gene expression: a major determinant of axonal caliber. Proc Natl Acad Sci USA 84: 3272–3476

Hughes JT, Jerrome D (1970) Ultrastructure of anterior horn motor neurons in the Hirano-Kurland-Sayre type of combined neurological system degeneration. J Neurol Sci 13: 389–399

Inoue K, Hirano A (1979) Early pathological changes of amyotrophic lateral sclerosis-autopsy findings of a case of 10 months' duration. Neurol Med (Tokyo) 11: 448–455

Monteiro MJ, Hoffman PN, Gearhart JD, Cleveland DW (1990) Expression of NF-L in both neuronal and nonneuronal cells of transgenic mice: increased neurofilament density in axons without affecting caliber. J Cell Biol 111: 1543–1557

Mulder DW (1984) Motor neuron disease. In: Dyck PJ, Thomas PK, Lambert EH, Bunge R (eds) Peripheral neuropathy. Saunders, Philadelphia, pp. 1525–1536

Mulder DW (1986) Motor neuron disease in adults. In: Engel AG, Banker BQ (eds) Myology. McGraw-Hill, New York, pp 2013–2029

Oblinger MM, Lasek RJ (1985) Selective regulation of two axonal cytoskeletal networks in dorsal root ganglion cells. In: O'Lague P (ed) Neurobiology: Molecular biological approaches to understanding neuronal function and development. The Shering Corp.: UCLA Symposium on Molecular and Cellular Biology. Vol. 24. New York, Liss, 1–91

Rosen DR, Siddique T, Patterson D, Figlewicz DA, Sapp P, Hentati A, Donaldson D, Goto J, O'Regan JP, Deng H-X, Rahmani Z, Krizus A, McKenna-Yasek D, Cayabyad A, Gaston SM, Berger R, Tanzi R, Halperin JJ, Herzfeld B, Van den Bergh R, Hung W-Y, Bird T, Deng G, Mulder DW, Smyth C, Laing NG, Soriano E, Pericak-Vance M, Haines J, Rouleau GA, Gusella JS, Horvitz HR, Brown Jr RH (1993) Mutations in Cu/Zn superoxide dismutase gene are associated with familial amyotrophic lateral sclerosis. Nature 362: 59–62

Schochet Jr SS, Hardman JM, Ladewig PP, Earle KM (1969) Intraneuronal conglomerates in sporadic motor neuron disease. Arch Neurol 20: 548–553

Shields RP, Vandevelde M (1978) Spontaneous lower motor neuron disease in rabbits (Oryctolagus cuniculus). Acta Neuropathol 44: 85–90

Takahashi K, Nakamura H, Okada E (1972) Heraditary amyotrophic lateral sclerosis. Arch Neurol 27: 292–299

Troncoso JC, Hoffman PN, Griffin JW, Hess-Kozlow KM, Price DL (1985) Aluminum intoxication: a disorder of neurofilament transport in motor neurons. Brain Res 342: 172–175

Vandevelde M, Greene CE, Hoff EJ (1976) Lower motor neuron disease with accumulation of neurofilaments in a cat. Vet Pathol 13: 428–435

Voyvodic JT (1989) Target size regulates caliber and myelination of sympathetic axons. Nature 342: 430–432

Wiley CA, Love S, Skoglund RR, Lampert PW (1987) Infantile neurodegenerative disease with neuronal accumulation of phosphorylated neurofilaments. Acta Neuropathol 72: 369–376

Xu Z-S, Cork LC, Griffin JW, Cleveland DW (1993) Increased expression of neurofilament subunit NF-L produces morphological alterations that resemble the pathology of human motor neuron disease. Cell 73: 23–33

Yamasaki H, Bennett GS, Itakura C, Mizutani M (1992) Defective expression of neurofilament protein subunits in hereditary hypotrophic axonopathy of quail. Lab Invest 66: 734–743

Lurcher, Cell Death and the Cell Cycle

N. Heintz, L. Feng, J. Gubbay, S. Cheng,*
J. Zuo, P.L. De Jager, and *D.J. Norman*

Summary

The mouse neurologic mutant *Lurcher* carries a semidominant genetic lesion that results in severe neurologic dysfunction (Philips 1960). Classical studies have established that the *Lurcher* mutation results in cell autonomous death of cerebellar Purkinje cells beginning in the second postnatal week (Caddy and Biscoe 1975). We have examined the expression of terminal markers for Purkinje cell differentiation, including the Kv3.3b potassium channel (Goldman-Wohl et al. 1994), and demonstrated that they are expressed prior to cell death in *Lurcher* animals. Detailed genetic studies have allowed identification of *Lc*/*Lc* homozygotes prior to their death in the first postnatal day, and histologic studies of these animals indicate that *Lurcher* homozygotes are missing large neurons in several hindbrain nuclei. These studies establish that the *Lurcher* gene causes dose-dependent cell death of specific neuronal populations following their differentiation in the cerebellum and hindbrain.

In considering possible causes for cell loss in *Lc* mice, I speculated that gross perturbations in signal transduction within terminally differentiated neurons can cause reactivation of programmed cell death (Heintz 1993). This process is formally analogous to the activation of oncogenes in stem cells, which perturb signal transduction sufficiently to alter cell growth control, eventually leading to transformation. We suggest that this type of activation, or failure in cellular homeostasis, in differentiated neurons can only result in programmed cell death because this is the only efferent pathway available to cells that have permanently exited the cell cycle. As an initial indication that this idea may pertain in the case of the *Lurcher* disease, we examined the mode of death of Purkinje cells in mutant animals. We found that the *Lurcher* gene causes activation of programmed cell death in Purkinje cells, and we suggest that this may be typical of many neurologic diseases. Cloning of the *Lurcher* gene product will allow identification of the molecules and mechanisms that participate in this process, providing a first test of this general idea.

---

* Howard Hughes Medical Institute, The Rockefeller University, 1230 York Avenue, New York, NY 10021, USA

K.S. Kosik et al. (Eds.)
Alzheimer's Disease: Lessons from Cell Biology

Introduction

It is clear from studies of neurodegeneration in mammals and insects that a very large number of genes can be mutated to result in neurologic diseases. These disorders share the common property that the basis of the disease is death of a relatively specific population of differentiated neurons in the adult brain. For example, there are now known to be at least 30 different mouse mutations that can result in inherited neurologic disease in mice. Many of these mouse neurologic disorders result in the death of specific neuronal populations: for example, there are at least eight different genetic loci that cause postnatal death of cerebellar Purkinje cells. For the past several years, my laboratory has been engaged in experiments to identify the primary genetic lesion in several of these diseases and to understand the molecular mechanisms that result in neuronal death in response to these lesions. It is hoped that a precise definition of the mechanisms resulting in neuronal death in neurologic mutant mice will directly inform efforts to understand neurodegeneration in human disease. In this essay, I will discuss our progress toward achieving these goals, and review a rather speculative hypothesis regarding a possible common pathway for neuronal cell death in a subset of these diseases.

Lurcher Mutant Mice

Classical studies have established that the mouse *Lurcher* (*Lc*) mutation causes a semidominant neurodegenerative disease that results in ataxic and uncoordinated behavior typical of mammalian cerebellar dysfunction (Phillips 1960). *Lc*/+ animals are viable and show cell autonomous death of Purkinje neurons beginning during the second postnatal week (Wetts and Herrup 1982a,b). As a secondary consequence of Purkinje cell loss, cerebellar granule neurons and neurons in the inferior olive also die. *Lc*/*Lc* mice are not viable, dying shortly after birth. Thus, the phenotypic consequences of the *Lc* mutation are dose dependent, resulting in either postnatal cerebellar disease or lethality.

To identify the primary mutation that causes the *Lurcher* phenotype, my laboratory has engaged in a positional cloning strategy. Thus, a large intersubspecific backcross has been employed to precisely position the *Lc* mutation on the mouse genetic map (Norman et al. 1991). Three very closely linked molecular markers have been identified that flank the *Lc* gene by approximately 0.5 cM, corresponding to a physical distance of approximately 1.6 megabases of DNA. Our preliminary physical mapping experiments indicate that these flanking markers are linked on a single genomic DNA fragment of 1.1 megabases, which is within the range expected from our genetic analysis. Using the three closely linked molecular markers referred to above, we have isolated more than a dozen large yeast artificial chromosome (YAC) clones containing mouse genomic DNA surrounding the *Lc* locus, and we have constructed a YAC contig that spans approximately 1.5 megabases. Although these clones could contain the *Lc* gene, we have not yet correlated the genetic map with the YAC contig to establish whether this is the case. Thus, our positional cloning efforts are rapidly approaching identification of the *Lc* gene.

In considering the possible functions of the *Lurcher* gene product, we have addressed several issues regarding the biology of the *Lurcher* disorder. The first question we have addressed is whether the *Lurcher* gene causes the death of terminally differentiated Purkinje cells, or whether *Lurcher* is a developmental mutation that prevents differentiation of Purkinje cells, resulting consequently in the death of these Purkinje cells because they have not properly formed. This is an important consideration, because the timing of Purkinje cell death in *Lc*/+ animals corresponds to their last phase of differentiation: dendritic growth and synaptogenesis. To address this issue, we have assayed the expression of a marker for Purkinje cell terminal differentiation that we have recently characterized as a gene that is expressed only upon terminal differentiation and that continues to be expressed in the adult. This gene, the Kv3.3b K^+ channel, encodes a *Shaw* type potassium channel whose expression in cerebellum is restricted to Purkinje cells and their post synaptic partners in the cerebellar deep nuclei (Goldman-Wohl et al. 1994). Using *in situ* hybridization, we have shown that the Kv3.3b mRNA is present in cerebellar Purkinje cells in *Lc* mutant mice prior to their death as a result of the mutation. We believe these results strongly argue that the *Lurcher* gene causes Purkinje cell death only after terminal differentiation of these cells, and that *Lc* is not a developmental mutation. Previous studies using genetic crosses between *staggerer* (*sg*) and *Lurcher* mutant mice demonstrate that *Lc* Purkinje cells must differentiate in order to die as a consequence of the mutation (Messer et al. 1991). In *sg*/*sg* *Lc*/+ animals, Purkinje cell differentiation cannot occur because of the *sg* mutation, and the *Lc* gene cannot kill these undifferentiated Purkinje cells. Thus, the *Lurcher* mutation causes true neurodegenerative disease in mice.

A second point that we have addressed to increase our understanding of the *Lc* phenotype is the cause of death in *Lc*/*Lc* homozygotes. The opportunity to perform these studies was dependent on our genetic information, since there are no obvious gross phenotypic differences between newborn *Lc*/*Lc* and *Lc*/+ pups prior to birth. However, using the molecular markers we have identified during our positional cloning efforts, we have been able to distinguish the genotypes of these pups at the *Lc* locus and definitively identify *Lc*/*Lc* animals at birth. Using this information, we have closely examined the homozygous *Lc* pups during embryogenesis and in the perinatal period and have discovered an obvious and devastating defect in several hindbrain nuclei. Large neurons in these nuclei are completely missing in *Lc*/*Lc* animals. For example, if one examines the fifth motor nucleus for the trigeminal ganglion, it is immediately apparent that the large neurons typical of this nucleus are not present in *Lc*/*Lc* animals. In addition, it appears as if the smaller interneurons in the hindbrain are significantly decreased in number. Although we have not yet definitively demonstrated that these cells die as they terminally differentiate in the hindbrain, our results are consistent with the idea that two doses of the *Lc* mutation cause cell death of large neurons in the hindbrain as they terminally differentiate, and that the death of these cells may result in the secondary cell death of hinbrain interneurons. The model that arises from this work is that the *Lc* mutation causes dose-dependent and neuron-specific cell death as specific large neurons of the hindbrain and cerebellum terminally differentiate and become physiologically active.

The Thanatogene Hypothesis

Our ideas concerning programmed cell death as a common effector pathway for cell loss in a variety of neurodegenerative diseases (Heintz 1993) are based on a relatively simple set of facts concerning the properties of differentiated neurons and of the neurodegenerative disorders that have been described:

1) The most fundamental biological fact to be considered is that *differentiated neurons do not and cannot divide*. There is not a single case in the clinical literature in which a brain tumor is thought to have arisen from a terminally differentiated neuron. Given the fact that there are over 10^{11} neurons in the mammalian brain, it is certain that genetic disturbances similar to those resulting in transformation and neoplasia must occur in this long-lived cell population.

2) A second relevant fact is that *there are a large number of genes (thanatogenes) that can be mutated to cause neurodegeneration*. As mentioned above, the number of genetically well-characterized loci that can lead to death of relatively specific neuronal cell populations in mammals is at least 50. I think that it is very unlikely that these genes all act as loss of function mutations in survival pathways identified in studies of neurotrophic factors.

3) The third important fact for this discussion is that *expression of known cellular oncogenes in terminally differentiated neurons causes cell death, rather than transformation*. Thus, transgenic mice expressing SV40 T-antigen (a very well-characterized oncogene) in dividing cell populations rapidly develop tumors, whereas transgenic mice expressing this gene in terminally differentiated neurons develop neurodegenerative disease in a dose- and cell type-specific manner (Feddersen et al. 1992).

4) A fourth important fact for our discussion is that *programmed cell death is a fundamental and shared pathway that participates in development of all regions of the nervous system*. Molecular genetic dissection of this

pathway in C. *elegans* has identified several genes that are critical for this pathway (Chalfie 1984; Ellis et al. 1991).

Consideration of these facts and of the pathway leading to transformation and neoplastic growth has led me to propose that some, perhaps many, *neurodegenerative disorders could arise by reactivation of the programmed cell death pathway as a consequence of genetic disturbances similar to oncogene activation in stem cell populations*. As in the case of transformation, the specificity of neurodegeneration in these diseases might result either from biochemical activation of a signal transduction molecule that is cell specific or from misregulation of a common gene product in a given neuronal cell population. For example, transformation of B cells can result from rearrangements that juxtapose the c-myc gene with the immunoglobulin transcriptional enhancer. In this case, one sees neoplastic transformation specifically in B cells because the c-myc gene is specifically transcriptionally activated only in that cell population, not because the biochemical role of c-myc in B cells is different than it is in all other cells. Given that oncogene expression in postmitotic neurons has been demonstrated to cause dose-dependent neurodegeneration in transgenic mice (Feddersen et al. 1992), one can easily imagine that this type of event might explain some cases of spontaneous neurodegenerative disease. This is simply one example of a variety of genetic mechanisms that have been discovered in studies of neoplastic transformation that could potentially be relevant to activation of programmed cell death in neurons.

The Significance of the Thanatogene Hypothesis

If this type of reasoning is correct, there are several consequences that are biologically interesting and relevant to neurodegenerative disease. The first of these is that programmed cell death may have been selected in evolution as the only efferent pathway that can be utilized in mature neurons in response to gross metabolic disturbances because this pathway leads to clonal elimination rather than clonal expansion of the affected cell. This would seem to be rather beneficial to the organism because it would protect it from the statistical certainty that neoplastic transformation will occur in this very long-lived cell population in a large fraction of individuals within the population. Secondly, it suggests that the number of thanatogenes will be quite large, and they will be biochemically recognizable as signal transduction molecules. Finally, whether these ideas concerning the molecular mechanisms that lead to neurodegeneration are correct or incorrect, *they have served to focus our attention on reactivation of programmed cell death as a common pathway for destruction of neurons in a wide variety of neurodegenerative diseases*. If programmed cell death can be identified as the mechanism for cell death in several disorders, then it is possible that definition of the cell death pathway at the molecular level can lead to palliative therapies that are generally useful even in cases where the primary defect that elicits the pathway is not understood.

Programmed Cell Death in *Lurcher* Mice

As an initial indication that the *Lurcher* thanatogene causes loss of Purkinje cells in the cerebellum due to reactivation of programmed cell death, we have performed an in-depth study of dying Purkinje cells in postnatal *Lc*/+ animals. Our first such studies examined the morphology of these cells as they died to determine whether they showed the characteristic morphologic features of cells undergoing apoptosis. The term apoptosis (a term commonly used interchangeably with programmed cell death) was coined by Kerr et al. in 1972 to described the morphological changes associated with many types of cells undergoing programmed cell death (for reviews, see Schwartzman and Cidlowski 1993; Ucker 1991; Vaux 1993; Williams et al. 1992; Williams and Smith 1993; Wyllie et al. 1980; Wyllie 1981, 1987) and is distinct from necrosis. Cells undergoing apoptotic death require metabolic energy and gene expression (reviewed in Fesus et al. 1991) and actively kill themselves through a unique series of morphologically identifiable stages. Nuclear chromatin becomes condensed and nuclei may fragment although most intracellular organelles (except in some cases the endoplasmic reticulum) remain intact. Blebbing of nuclear and cellular membranes is also a common event in apoptotic death. Fragments of cells dying in this manner are phagocytosed by macrophages and neighboring cells so no leakage of intracellular macromolecules occurs and the inflammatory response typical of necrotic death is not seen.

Examination of Purkinje cells in *Lc*/+ mutant mice during the second postnatal week in both the light and electron microscopes reveals that *Lc*/+ Purkinje cells show the morphological features of apoptotic cell death. Thus, we have observed that *Lc*/+ Purkinje cells display blebbing of both nuclear and cellular membranes prior to death and that Purkinje cell nuclei contain large clumps of condensed chromatin that first appear lining the nuclear membrane and adhered to the nucleolus and then proceed to fill the nucleus. In late stages of the cell death program, *Lurcher* Purkinje cell nuclei are entirely electron-dense due to the condensation of chromatin, and in some cases fragmentation of these condensed nuclei can be observed. Finally, we have observed that Purkinje cell axons in *Lc*/+ animals also show morphology characteristic of neurons dying of programmed cell death, including axonal swellings and terminal bulbs reminiscent of peripheral neurons undergoing programmed cell death when deprived of nerve growth factor.

Another characteristic of programmed cell death is fragmentation of nuclear DNA during the process of cell death. This is typically assayed using gel electrophoresis to visualize DNA laddering as cells die in tissue culture.

However, in cases where cell death is affecting only a small subpopulation of cells in otherwise intact tissue, an alternative in situ approach must be employed. In this case, incorporation of dNTPs into fragmented DNA in tissue slices is assayed by incubation of those slices with DNA polymerase in vitro (Wood et al. 1993). Cell nuclei containing damaged DNA are then visualized using antibodies to a derivatized dNTP (e.g., biotin, digoxygenin) that is incorporated during the polymerase reaction. To assay the state of DNA in *Lc*/+ Purkinje cells prior to their death, we have compared the results of this assay with control slices prepared from wild type animals. Although we can detect programmed cell death of granule cell precursors in the external germinal layer of both wild type and *Lc*/+ animals, Purkinje cell nuclei are labeled only in the *Lc*/+ animals. These results demonstrate DNA degradation in Purkinje cells as a consequence of the *Lc* mutation, further supporting the idea that these cells are dying by programmed cell death.

Although both the morphologic and DNA fragmentation evidence we have obtained in examining dying *Lc*/+ Purkinje cells is consistent with reactivation of programmed cell death in this mutation, we sought a third, independent line of evidence to support these studies. Although no molecular markers that characterize programmed cell death in neurons are yet available, the expression of the SGP2 (clusterin, TRPM2; Jenne and Tschoop 1992) gene has been correlated with programmed cell death during prostate regression and T-cell negative selection (reviewed in May and Finch 1992). Since it seems reasonable to assume that at least some aspects of the cell death program are shared between different dying cell types, we asked whether the SGP2 gene is expressed in *Lc*/+ Purkinje cells during their death. Using in situ hybridization, it is clear that SGP2 mRNA is present in dying *Lc*/+ Purkinje cells, although it is not expressed at any stage in the differentiation of +/+ Purkinje cells. This observation strongly supports our conclusion that the *Lc* gene reactivates programmed cell death in Purkinje cells following their terminal differentiation.

A Model for *Lurcher* Gene Action

Our studies of programmed cell death in response to the *Lurcher* mutation demonstrate that this gene can cause reactivation of programmed cell death in Purkinje cells. Characterization of the phenotype of *Lc*/*Lc* animals suggests that the *Lc* gene can also kill large neurons in several hindbrain nuclei. Thus, we propose that the *Lurcher* neurologic disorder results from dose-dependent reactivation of programmed cell death in specific neuronal cell populations. Given the considerations discussed above, we speculate that the *Lc* gene may sufficiently perturb the balance of signal transduction in these cell types and that upon terminal differentiation and synaptogenesis, this failure in cellular homeostasis exceeds a threshold that results in activation of apoptosis. Although we do not know the mechanism by which this reactivation occurs, we suggest that molecules involved in integrating transduction signals in stem cells to control cell growth may also be critical for integrating this information in postmitotic neurons. Thus, these "cell cycle" regulatory molecules may play an important role in terminally differentiated neurons by providing a mechanism for their clonal elimination under conditions that might be expected to cause clonal expansion in dividing cell populations.

References

Caddy KWT, Biscoe TF (1975) Preliminary observations on the cerebellum in the mutant mouse *Lurcher*. Brain Res 91: 276–280

Chalfie M (1984) Neuronal development in *Caenorhabditis elegans*. Trends Neurosci 7: 197–202

Ellis RE, Yuan J, HR Horvitz (1991) Mechanisms and functions of cell death. Ann Rev Cell Biol 7: 663–698

Feddersen RM, Ehlenfeldt R, Yunis WS, Clark HB, Orr HT (1992) Disrupted cerebellar cortical development and progressive degeneration of Purkinje cells in SV40 T antigen transgenic mice. Neuron 9: 955–966

Fesus L, Davies PPJA, Piacenti M (1991) Apoptosis: molecular mechanisms in programmed cell death. Eur J Cell Biol 56: 170–177

Goldman-Wohl DS, Chan E, Baird D, Heintz N (1994) Kv3.3b: A novel *Shaw* type potassium channel expressed in terminally differentiated cerebellar Purkinje cells and deep cerebellar nuclei. J Neurosci 14: 511–522

Heintz H (1993) Cell death and the cell cycle: a relationship between transformation and neurodegeneration? Trends Biochem Sci 18: 157–159

Jenne DE, Tschhopp J (1992) Clusterin: the intriguing guises of a widely expressed glycoprotein. Trends Biochem Sci 17: 154–159

Kerr JFR, Wyllie AH, Currie AR (1972) Apoptosis: a basic biological phenomenon with wide-ranging implications in tissue kinetics. Br J Cancer 26: 239–257

May PC, Finch CE (1992) Sulfated Glycoprotein 2: new relationships of this multifunctional protein to neurodegeneration. Trends Neurosci 15: 391–395

Messer A, Eisenberg B, Plummer J (1991) The *Lurcher* cerebellar mutant phenotype is not expressed on *staggerer* mutant background. J Neurosci 11(8): 2295–2302

Norman DJ, Fletcher C, Heintz N (1991) Genetic mapping of the *Lurcher* locus on mouse chromosome 6 using an intersubspecific backcross. Genomics 9: 147–153

Phillips RJS (1960) "Lurcher," a new gene in linkage group XI of the house mouse. J Genet 57: 35–42

Schwartzman RA, Cidlowski JA (1993) Molecular regulation of apoptosis: genetic controls on cell death. Cell 74: 777–779

Ucker DS (1991) Death by suicide: one way to go in mammalian cellular development? New Biol 3: 103–109

Vaux DL (1993) Toward an understanding of the molecular mechanisms of physiological cell death. Proc Natl Acad Sci USA 90: 786–789

Wetts R, Herrup, K (1982a) Interaction of granule, Purkinje and inferior olivary neurons in *Lurcher* chimeric mice. I. Qualitative studies. J Embryol Exp Morphol 68: 87–98

Wetts R, Herrup, K (1982b) Interaction of granule, Purkinje and inferior olivary neurons in *Lurcher* chimeric mice. II. Granule cell death. Brain Res 250: 358–362

Williams GT, Smith CA (1993) Molecular regulation of apoptosis: genetic controls on cell death. Cell 74: 777–779

Williams GT, Smith CA, McCarthy NJ, Grimes EA (1992) Apoptosis: final control point in cell biology. Trends Cell Biol 2: 263–267

Wood KA, Dipasquale B, Youle RJ (1993) In situ labeling of granule cells for apoptosis-associated DNA fragmentation reveals different mechanisms of cell loss in developing cerebellum. Neuron 11: 621–632

Wyllie AH (1981) Cell death: a new classification separating apoptosis from necrosis. In: Bowen ID, Lockshin RA (eds) Cell death in biology and pathology. London, Chapman & Hall, pp 9–34

Wyllie AH (1987) Cell death. Int Rev Cytol 17: 755–785

Wyllie AH, Kerr JFR, Currie AR (1980) Cell death: the significance of apoptosis. Int Rev Cytol 68: 251–306

Neuronal Death, Proinflammatory Cytokines and Amyloid Precursor Protein: Studies on *Staggerer* Mutant Mice

B. Brugg, Y. Lemaigre-Dubreuil, G. Huber, B. Kopmels, N. Delhaye-Bouchaud, E.E. Wollman,* and *J. Mariani*

Summary

In Alzheimer's, disease (AD), a combination of genetic predisposition and environmental factors may contribute to changes in β-amyloid precursor protein (APP) expression, β-amyloid deposition and neuronal loss. Head injury and acute infection that triggers inflammatory processes are known to be risk factors. In the present in vivo study we show that peripheral endotoxin injection induces a phasic increase of IL-1β and IL-6 mRNA in mouse cerebellum, followed within 24 hours by an increase in the APP-KPI/APP-695 ratio. In the cerebellum of the *staggerer* mouse mutant, where a severe deficit of Purkinje and granule cells occurs, elevated basal levels of IL-1β and IL-6 mRNA and an increase in the APP-KPI/APP-695 ratio compared to wild type mice was observed. LPS stimulation further accentuated these differences for cytokines and APP isoforms. Our in vivo studies suggest that interaction loops between cytokines and APP could play an important role in the regulation of APP expression in degenerating Alzheimer's disease brain tissue.

Introduction

The 4-kilodalton amyloid β peptide (A4; Glenner and Wong 1984; Masters et al. 1985), which forms amyloid deposits in AD and is potentially neurotoxic (Yankner et al. 1990; Frautschy et al. 1990; Koh et al. 1990), is derived from a larger glycoprotein referred to as the β-amyloid precursor protein (APP; Bahmanyer et al. 1987; Kang et al. 1987). In contrast to the initially cloned 695 residues isoform (APP_{695}), the two major alternative transcripts (APP_{751} and APP_{770}) contain an exon with strong homology to the Kunitz family of serine protease inhibitors (KPI) (Kitaguchi et al. 1988; Ponte et al. 1988; Tanzi et al. 1988). The metabolic processing of APP molecules has been extensively studied; in addition to the processing by a protease (α-

* Université Pierre et Marie Curie, Institut des Neurosciences, (URA CNRS 1488) Lab. de Neurobiologie du Développement, 75005 Paris, France

K.S. Kosik et al. (Eds.)
Alzheimer's Disease: Lessons from Cell Biology

secretase) that cuts the APP in the A4 region (Esch et al. 1990), an alternate processing exists that leads to potentially amyloidogenic peptides. These peptides are produced and released by most cells during normal metabolism (Haass et al. 1992; Weidemann et al. 1989; Shoji et al. 1992) and are present in soluble forms in biological fluids (Seubert et al. 1992). However, in AD A4 peptide is produced and released in abnormally high amounts (Cai et al. 1993; Citron et al. 1992). One important factor that could lead to this excess seems to be a change in the ratio of the different isoforms, with an increase in the KPI-bearing ones (Quon et al. 1991; Johnson et al. 1990). Several etiological factors have been shown to be involved in alteration of APP expression and metabolism and pathogenesis of AD. They are schematically illustrated in Figure 1. Most attention has been given recently to genetic factors. Mutations in the APP gene chromosome 21 (Chartier-Harlin et al. 1991; Goate et al. 1991) and of a still unidentified gene on chromosome 14 have been shown to correlate with early-onset familial AD and some late-onset cases (Chartier-Harlin et al. 1991; Goate et al. 1991; Schellenberg et al. 1992). Mutations in the APP gene may lead to alternative processing of the APP and an elevated release of the A4 (Cai et al. 1993; Lewis et al. 1988). In late-onset AD cases, Apolipoprotein E4 gene dosage (on chromosome 19) appears to be a major risk factor (Corder et al. 1993); it may contribute in combination with other etiological factors (Selkoe 1993) to age-associated changes in APP isoforms, A4 deposition and neuronal loss. Preliminary studies show that anti-inflammatory drugs slow down AD, thereby supporting the idea that inflammatory processes (head injury, acute infections) could trigger changes in APP expression and processing (Abraham et al. 1990; Allsop et al. 1990; Mortimer et al. 1991; Roberts et al. 1991; Schnabel 1993; Vandenabeele and Fiers 1991). Elevated amounts of Interleukin-1β (IL-1β) and Interleukin-6 (IL-6), two pro-inflammatory cytokines, have been described in neurodegenerative diseases such as AD and Down's syndrome in which astroglial proliferation occurs (Bauer et al.

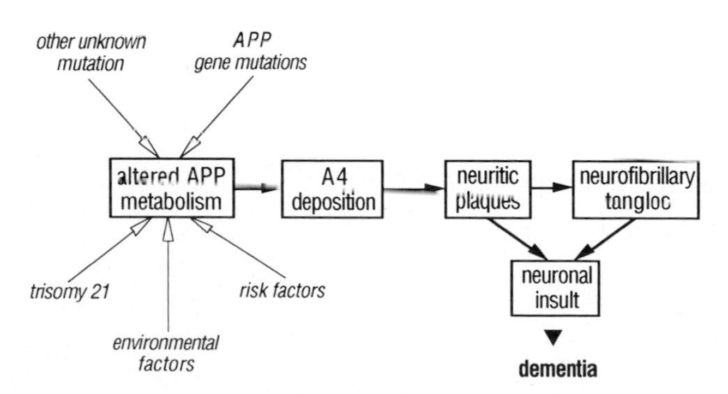

Fig. 1. Suggested pathogenic pathways for the progression of Alzheimer's disease. Modified from Hardy and Allsop (1991)

1991; Griffin et al. 1989). Cell culture studies suggest that IL-1β could be responsible for an altered β-APP gene regulation in neurons, astrocytes and endothelial cells (Buxbaum et al. 1992; Forloni et al. 1992; Goldgaber et al. 1989; Gray and Patel 1993). It has not yet been demonstrated in vivo that the amount of these cytokines and APP regulation are linked and play a crucial role in neurodegenerative processes.

In the present study wild type and *staggerer* mutant mice (Fig. 2) were used as a model system to investigate the relationship between inflammatory processes, neuronal death and APP in the brain. The homozygous *staggerer* (*sg*/*sg*) animal shows a severe deficiency in the number of Purkinje cells during the first postnatal weeks, and the surviving cells (about 25%) are also intrinsically deficient: the cell bodies are small and ectopic and the dendritic arbor is reduced in size and branching complexity. Due to the absence of many of their Purkinje cells targets, almost two thirds (60%) of the inferior olivary neurons and all the cerebellar granule neurons die in *staggerer* mice during the first postnatal month, with this degeneration being associated with a marked astrogliosis (Shojaeian et al. 1985; Zanjani et al. 1990; Herrup 1983; Herrup and Mullen 1979a,b).

Materials and Methods

Animals

Staggerer (*sg*/*sg*) mutant mice were obtained by intercrossing heterozygous (+/*sg*) animals and were then identified based on clinical symptoms or on macroscopical examination of the brain when young animals were needed for cell culture. The *staggerer* mutation was bred on a C57BL/6J background. Controls were age- and strain-matched mice (+/+).

Experimental Procedures

Western Blots

A monoclonal antibody (anti-APP) was used to detect soluble and membrane-bound full-length forms of β-APP_{695} and $APP_{751/770}$. An affinity purified polyclonal antibody (anti KPI) made against synthetic peptide (Löffler and Huber 1992) of the KPI domain was used to specifically detect β-$APP_{751/770}$ in soluble and membrane fractions.

In each experiment, repeated three times, five *staggerer* (postnatal day 30) and five control mice were injected intraperitoneally with LPS (0.5 μg/g body weight, carrier: RPMI) or RPMI alone. Twenty-four hours later the animals were sacrificed, and the pooled brain regions were processed as described elsewhere (Löffler and Huber 1992) to obtain Western blots from

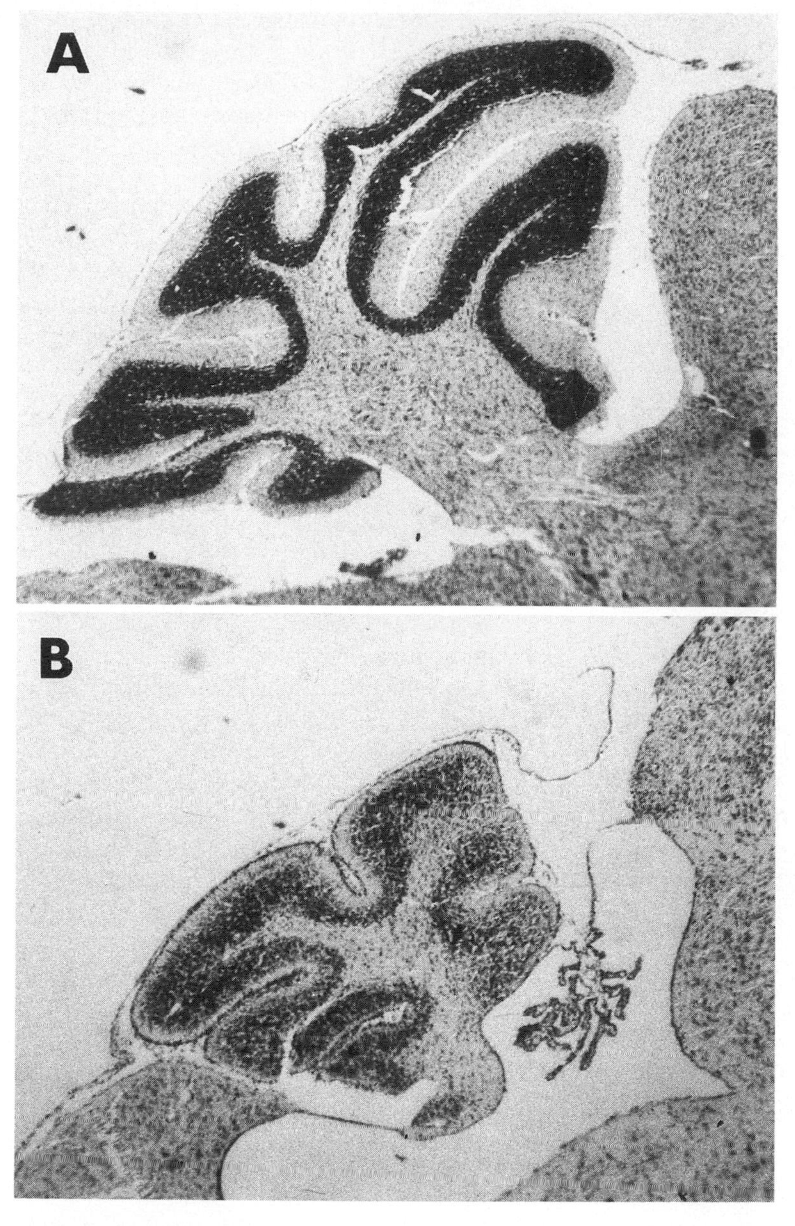

Fig. 2. Representative parasagittal cresyl violet stained section through the cerebellum of wild type (**A**) and *staggerer* mutant mice (**B**)

soluble and membrane fractions. Proteins on nitrocellulose membranes were stained with Ponceau Red for visualization, then destained, and non-specific binding sites were blocked using 50 g/L skim milk powder in PBS for 30 minutes. Membranes were then incubated overnight with a polyclonal affinity-purified rabbit anti-KPI (0.1 μg/ml; Löffler and Huber 1992) or anti-APP (0.8 μg/ml, Boehringer). Biotinylated second antibody (Boehringer) and horseradish peroxidase conjugated streptavidin (Boehringer) were used to visualize immunoreactive protein bands. Immunoreactive bands were compared densitometrically using a laser scanner. Measurements were performed in the linear range, which was determined using dilution curves of brain extracts.

RT-PCR

Total RNA was isolated from different brain regions by the RNA^B method (Bioprobe, Paris, France). Each RNA sample was isolated from five pooled brain regions from *staggerer* mutant and control mice (postnatal day 30). For all quantitative comparisons RNA quality and concentration were measured by spectrometry OD$^{280/260\,nm}$ and 5 μg samples of total RNA were separated on 1% agarose-formamid gels. Total RNA (10 μg) was denatured for 3 minutes at 65°C and immediately shock-cooled on ice. RNA was reverse transcribed by the Riboclone® cDNA synthesis system (Promega, USA). Reverse transcriptase, 50 U (H.C., Promega), was added to the 20 μl reaction cocktail containing Oligo $(dT)_{15}$ primers. The reaction was allowed to proceed at 42°C for 90 minutes and then terminated at 70°C for 5 minutes. PCR amplification of reverse transcribed RNA was carried out using mouse amplimer sets IL-1β, IL-6, and control primers actin or G3PDH (Clontech, Palo Alto, CA, USA) and in the presence of P^{32} ATP. For quantitative PCR we used mouse IL-1β and IL-6 PCR MIMICs competitor DNA (Clontech; Siebert and Larrick 1992). The PCR products were analyzed on 1.5% agarose BET gels, photographed and scanned. The BET stained bands, containing the P^{32} labeled PCR products, were cut out of the agarose gel and measured in a beta-counter. The resulting APP isoforms, IL-1β and IL-6 mRNA data, were normalized using the corresponding competitor DNA and G3PDH values.

Microglial Cell Cultures

The cerebella obtained from 5- to 6-day-old animals of either (+/+) or (sg/sg) were trypsinized and transferred in a BME medium complemented with 10% fetal calf serum, glutamine (1 mM), penicillin (100 u/ml) and streptomycin (10 μg/ml). After centrifugation the cells were cultivated in the same medium. Three to four weeks later microglial cells were purified from astrocytes by differential adhesion. Homogeneity of the microglial cultures was verified by immunocytochemistry with anti Mac and anti GFAP antibodies. The cultures were stimulated with 5 μg/ml of LPS for four hours and processed for Northern blots.

Northern Blots

Total RNA was extracted from cultured cells using the guanidium/cesium chloride method (Chirgwing et al. 1979).

Northern blots were performed with 10 mg of total microglial RNA. After electrophoretic migration on a 1.2% formaldehyde/agarose gel, the RNA was transferred onto a nylon membrane as described by Maniatis et al. (1987). The transferred RNA was probed with a 0.5-kbp PstI-HindIII fragment from the murine cDNA IL-1β and with a 0.6-kbp PstI-TaqIb actin fragment. The cDNA was then labeled with [^{32}P]dCTP using a multiprime kit (Amersham). Prehybridization and hybridization were carried out at 65°C in 5xDenhart's/6xSSC (1xSSC = 0.15 M NaCl, 0.015 M CH_3 COONa)/0.1 SDS and 10 mg salmon sperm. Washes were also performed at 65°C in 2xSSC/0.1 SDS and 0.2xSSC/0.1 SDS for 30 minutes each. To ensure that an equal amount of RNA was present in every sample, controls were systematically performed on each blot by hybridization with an actin probe. Autoradiograms were quantified with an image analyzer by measuring the mean optical density of each specific hybridization spot after delimitation of its surface area. Normalization with respect to the actin signal was obtained by calculating the ratio between the measured optical density of actin and IL-1β.

Results

Analysis of APP Isoforms

The different β-APP isoforms in the cerebral cortex and the cerebellum of wild type and *staggerer* animals were characterized by Western blot analysis of both soluble and membrane-bound proteins before and after i.p. administration of lipopolysaccharide (LPS), which mimicked an acute infection.

Anti-APP antibody revealed a band of 110–116 kDa, representing the released form of APP_{695}, and another band of 123–125 kDa; the latter was selectively detected by the anti APP antibody directed against KPI domain and represented the released form of $APP_{751/770}$ (KPI-APP; Löffler and Huber 1992).

In wild type mice, APP_{695} was the predominant form in brain-soluble proteins and was more abundant in the cerebral cortex than in the cerebellum. Twenty-four hours after LPS stimulation, no change could be detected in the cerebral cortex, whereas in the cerebellum a decrease of APP_{695} and an increase of KPI-APP occurred simultaneously. By contrast, in membrane-bound proteins, the KPI-APP isoforms were predominant. In the cerebral cortex, APP expression did not change after 24 hours of LPS stimulation. In the cerebellum, APP_{695} decreased but no changes for the membrane-bound KPI-APP were noticed. These results suggest that overproduction of KPI-APP induced by LPS is accumulated in the soluble fraction.

In the *staggerer* cerebral cortex, for both soluble and membrane-bound proteins, the different APP isoforms levels were comparable to those of wild type mice and did not change after LPS stimulation. In *staggerer* cerebellum, basal levels of the different soluble APP isoforms were very similar to those described above in the case of LPS-stimulated wild type mice, i.e., an increase of KPI-APP and a decrease of APP_{695}, and the levels were further increased after 24 hours of LPS stimulation. The shift was also present in the membrane fraction and seems to be a specific feature of the *staggerer* cerebellum, but did not change after a 24 hours of LPS stimulation.

Densitometrical analysis revealed a KPI/695 APP (membrane-bound + soluble proteins) ratio of ≈ 1 in the cortex and cerebellum of wild type mice. Figure 3 illustrates the measurements of a typical experiment from five pooled brains. After 24 hours of LPS stimulation the ratio increased to ≈ 2 in the cerebellum but not in the cerebral cortex. In the cerebellum of the *staggerer* mutant the basal values were ≈ 2.5 and increased to ≈ 4 after 24 hours of LPS treatment. Interestingly, this LPS effect was only observed after a time period of 24 hours; after 8 hours the APP protein isoform levels had not changed. The densitometrical studies also showed that the total amount of APP protein was the same in both *staggerer* and wild type brains after 24 hours of LPS administration (data not shown).

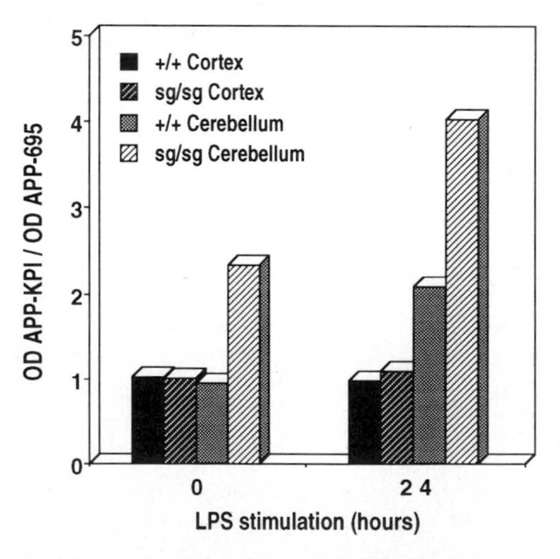

Fig. 3. Total APP isoform ratios ($APP_{751/770}/APP_{695}$) in cerebral cortex and cerebellum of wild type and *staggerer* mutant mice before (0 hours) and after (24 hours) peripheral LPS administration

Analysis of Interleukins RNAs

The same brain regions were also analyzed for their IL-1β and IL-6 mRNA expression by quantitative RT-PCR (Fig. 4a,b,c). The basal IL-1β mRNA levels in cerebral cortex were in the range of 2×10^{-20} mole/μg total RNA and did not differ in wild type and *staggerer* animals (Fig. 4a; lanes 1, 2); no IL-6 mRNA could be detected (Fig. 4b; lanes 1, 2). Basal IL-1β mRNA levels in cerebellum were five-fold higher in *staggerer* than in wild type mice (Fig. 4a; lanes 7, 8). IL-6 mRNA could only be detected in *staggerer* cerebellum (2×10^{-22} mole/μg total RNA) but not in wild type mice (Fig. 4b; lanes 7, 8).

In agreement with other studies (Fontana et al. 1984), an LPS mediated IL-1/IL-6 mRNA induction was observed in wild type mouse brain already after two hours (data not shown). This IL-1β mRNA increase was present eight hours after LPS injection (cerebral cortex: Fig. 4a; lane 3. cerebellum: Fig. 4a; lane 9) and was still detectable, although attenuated, after 24 hours of LPS injection in the cerebral cortex (Fig. 4a; lane 5) and in the cerebellum (Fig. 4a; lane 11). IL-6 mRNA was also induced in the cerebral cortex (Fig. 4b; lane 3) and in the cerebellum eight hours after LPS injection (Fig. 4b;

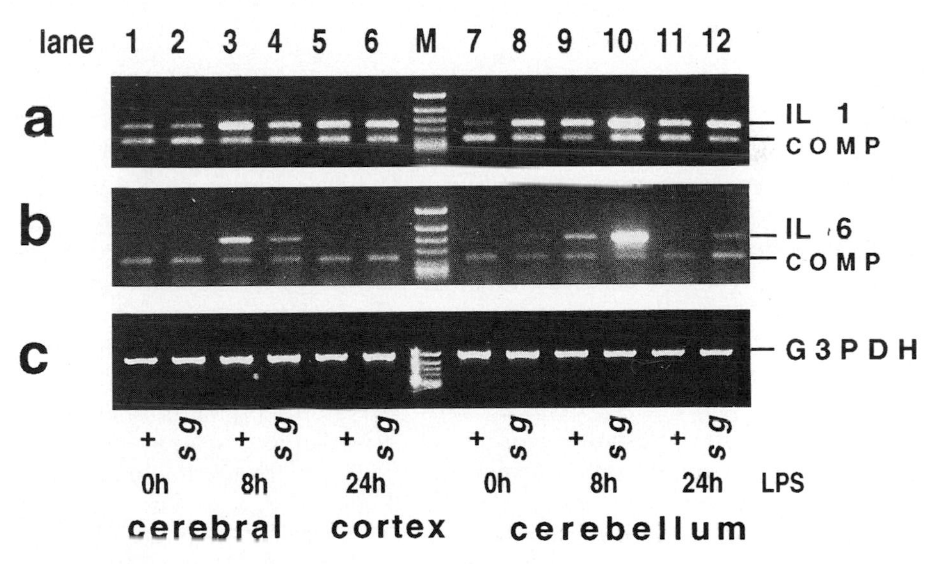

Fig. 4. RT-PCR detection of IL-1β and IL-6 mRNAs level in cerebral cortex and cerebellum of wild type ($^+$) and *staggerer* mutant mice (^{sg}) before LPS (lanes: 1, 2, 7, 8), after 8 hours of LPS (lanes: 3, 4, 9, 10) and after 24 hours of LPS stimulation (lanes; 5, 6, 11, 12). **a** IL-1 oligonucleotide primers amplified a 563 bp fragment corresponding to IL-1 mRNA. IL-1 competitor (comp) c-DNA (10^{-21} mol) amplified with the same primers resulted in a 420 bp fragment. **b** IL-6 oligonucleotide primers amplified a 638 bp fragment corresponding to IL-6 mRNA. The amplified IL-6 competitor c-DNA (10^{-22} mol) gave a 450 bp fragment. **c** G3PDH control amplification (housekeeping gene) gave a PCR fragment size of 983 bp; M, Mol. weight DNA markers ΦX 174 digested by Hinc II

lane 9), but disappeared in both brain regions after 24 hours (Fig. 4b; lanes 5, 11).

In *staggerer* cerebellum, eight hours after LPS injection, IL-1β mRNA levels were higher (2.5-fold; Fig. 4a, lane 10) than in wild type animals. In the same time period, a 10-fold induction was observed for IL-6 mRNA (Fig. 4b; lane 10), which dropped to basal levels after 24 hours (Fig. 4b; lane 12). By contrast, the response in the cerebral cortex for both cytokines was less in *staggerer* than in wild type mice (Fig. 4a,b; lanes 3, 4), possibly because of a desensitization of cytokine-producing cells by elevated levels of circulating IL-1/IL-6 in the cerebral fluid of *staggerer* mice (Kopmels et al. 1990).

Interleukin-1β in Microglial Cell Culture

In Northern blots, microglial cell cultures purified up to 95% homogeneity (Fig. 5A) did not express any IL 1β signal in the absence of LPS stimulation. After stimulation by 5 μg LPS the expression of IL 1β mRNA was induced and was two or three times higher in (sg/sg) cultures than in (+/+) cultures (Fig. 5B).

Discussion

This study showed for the first time that peripheral endotoxin (LPS) injection induced an increase in IL-1/IL-6 mRNA and a shift in APP isoforms ratio towards the KPI-APP isoforms in wild type mouse cerebellum. This result suggests that proinflammatory cytokines can regulate in vivo cerebellar APP expression. In the cerebral cortex, IL-1/IL-6 mRNA induction was not associated with any changes in APP expression. In the cerebellum, the APP shift towards the KPI-containing forms appeared after a relatively short time delay (24 hours) why it is unlikely due to a change in cellular composition, such as for instance IL-1β induced astrocytes proliferation (Giulian and Lachman 1985). It corresponds more to a LPS-mediated activation of certain cell types. In situ hybridization reveals that APP_{695} is the predominant form expressed in neurons, whereas both KPI-APP and APP_{695} are absent in glia cells in normal brain (Bahnmanyer et al. 1987; Lewis et al. 1988; Bendotti et al. 1988) but could be detected after brain lesions (Solà et al. 1993). In vitro, IL-1 induces KPI-APP isoforms in cortical neurons, brain-derived endothelial cells and astrocytes (Buxbaum et al. 1992; Forloni et al. 1992). These results are in agreement with the finding that the transcriptional factor AP-1, present in inducible genes, is also found in the APP gene (Vandenabeele and Fiers 1991; Goldgaber et al. 1989; Heinrich et al. 1990).

In the *staggerer* cerebellum where neuronal degeneration occurs, basal and LPS induced levels of IL-1β and IL-6 mRNA were higher than in

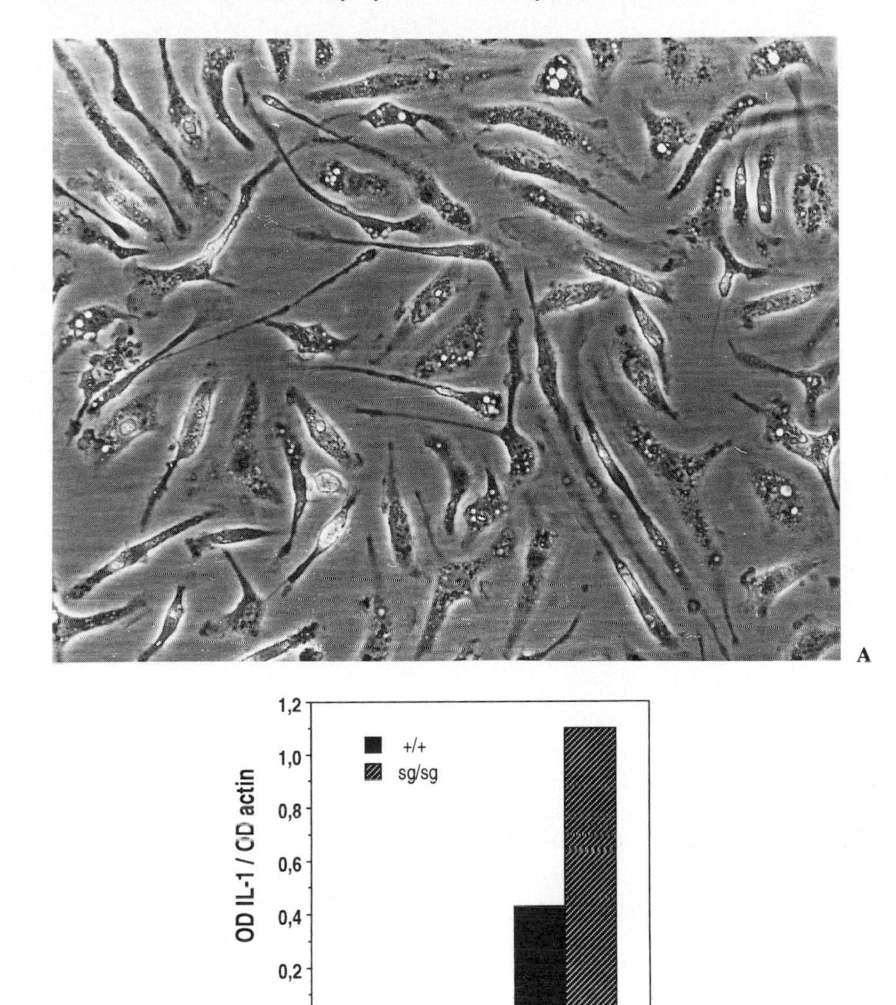

Fig. 5. A Cerebellar microglia cell culture form *staggerer* mutant mice. **B** Northern blots from cerebellar microglia cell culture (+/+ and sg/sg) without (0) and with 5 μg/ml LPS stimulation

controls. This could result from a change in the number of cytokine-producing cells, such as various types of glial cells and also granular neurons (Bandtlow et al. 1990). However, the cytokine hyperexpression does not involve granule cells, which have disappeared in sg/sg cerebellum (Zanjani et al. 1990), or microglia cells, which decreased by 75% in the *staggerer* cerebellum, i.e., to the same degree as Purkinje cells (Pajak et al., in preparation.). By contrast, pronounced astrogliosis of this cerebellum could play a role, especially for the elevated amount of IL-6 mRNA. Another possibility could be a change in the activation state of cytokine-producing cells, similar to the one we have already described in peripheral macrophages of *staggerer* and several other cerebellar mutant mice (Kopmels et al. 1990, 1992; Wollman et al. 1992). The hyperexpression of IL-1β mRNA that we observed in *staggerer* microglial cultures strongly supports this hypothesis.

The shift towards the KPI-APP isoforms observed in the unstimulated *staggerer* cerebellum is similar to that induced by LPS in wild type cerebellum. A similar shift is also described in AD brain regions, where neuronal loss is observed and thus be involved in an aberrant proteolytic process leading to β-amyloid deposition and neuronal death (Quon et al. 1991; Johnson et al. 1990; Nordstedt et al. 1991; Higgins et al. 1990). In both cases this shift in APP isoforms could be related to a change in cellular distribution and/or activation of certain cell types during the degeneration process.

A feature of the *staggerer* cerebellum was the increase in membrane KPI-APP, which was absent in wild type animals after LPS injection. This can be explained by the chronic elevated IL-1/IL-6 levels in the *staggerer* mouse, inducing an overproduction of KPI-APP membrane-bound forms in neurons and astrocytes. In contrast, LPS stimulation induced only a transient increase of IL-1/IL-6 mRNA levels and the resulting increase of the membrane-bound KPI-APP was immediately processed by the secretase, producing secreted non-amyloidogenic APP forms (Esch et al. 1990). A sustained increase of cytokine levels, for example with an infection or after head injury, could lead to an overproduction of secreted KPI-APP which could inhibit the secretase and lead to an accumulation of membrane-bound KPI-APP forms, as seen in the cerebellum of *staggerer* mice and in AD brain. It has been shown that the *staggerer* gene acts intrinsically in the Purkinje cells, with the loss of olivary and granule neurons being extrinsic

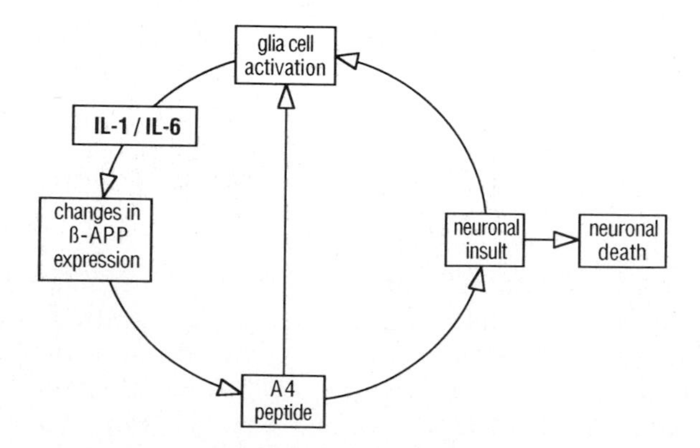

Fig. 6. Suggested interaction loops between proinflammatory cytokines (Il-1 and IL-6), APP, and neuronal insult in pathological situations such as *staggerer* mouse cerebellum or Alzheimer's disease

(Herrup and Mullen 1979a). An intrinsic effect of the gene does not exclude the role of external signals in the cascade of events leading to neuronal death, as shown recently for the *weaver* gene (Goa et al. 1992). Work is now in progress to elucidate the loops (Fig. 6) between cytokines, APP isoforms and the mutated *staggerer* gene that could ultimately elucidate the neurodegeneration observed. These "lethal loops" between inflammation and neuronal death could aggravate the lesions and be of general significance in other neurodegenerative diseases such as AD.

Acknowledgements. The authors want to thank G. Ferguson for improvement of the English. This work has been supported by grants from AFM, Fondation de France, INSERM and FRM. B.B. was a recipient of DRED and Swiss National Foundation fellowships.

References

Abraham CR, Shirihama T, Potter H (1990) The protease inhibitor a 1-antichymotrypsin is associated solely with amyloid deposits containing the β-protein and is localized in specific cells of both normal and diseased brain. Neurobiol Aging 11: 525–527

Allsop D, Haga S, Bruton CJ, Ishini T, Roberts GW (1990) Neurofibrillary tangles in some cases of dementia pugilistica share antigens with amyloid beta-protein of Alzheimer's disease. Am J Pathol 136: 255–260

Bahmanyer S, Higgins GA, Goldgaber D, Lewis DA, Morison JW, Wilson MC Shaker SK, Gajdusek JDC (1987) Localization of amyloid protein mRNA in brains from patients with Alzheimer's disease. Science 237: 77–80

Bandtlow CE, Meyer M, Lindholm D, Spranger M, Heumann R, Thoenen H (1990) Regional and cellular codistribution of interleukin 1 beta and nerve growth factor synthesis. J Cell Biol 111: 1701–1711

Bauer J, Strauss S, Schreiter-Gasser U, Ganter U, Schegel P, Witt I, Volk B, Berger M (1991) Interleukin-6 and alpha-2-macroglobulin indicate an acute phase state in Alzheimer's disease cortices. FEBS Lett 258: 111–114

Bendotti C, Forloni GL, Morgan R, O'hara BF, Oster-Granite ML, Reeves RH, Gearhart JD, Coyle JT (1988) Neuroanatomical localization and quantification of amyloid precursor protein mRNA by in situ hybridization in the brain of normal, aneuploid and lesioned mice. Proc Natl Acad Sci USA 85: 3628–3632

Buxbaum JD, Oishi M, Chen HI, Pinkas-Kramarski R, Jaffe EA, Gandy SE, Greengard P (1992) Cholinergic agonist and interleukin 1 regulate processing and secretion of the Alzheimer β/A4 amyloid protein precursor. Proc Natl Acad Sci USA 89: 10075–10078

Cai X-D, Golde TE, Younkin SG (1993) Release of excess amyloid β protein from a mutant amyloid β protein precursor. Science 259: 514–516

Chartier-Harlin M-C, Crawford F, Houlden H, Warren A, Hughes D, Fidani L, Goate A, Rossor M, Roques P, Hardy J, Mullan M (1991) Early-onset Alzheimer's disease caused by mutations at codon 717 of the β-amyloid precursor protein gene. Nature 353: 844–846

Chirgwing JM, Przybyla AE, Macdonald RJ, Rutter, WJ (1979) Isolation of biologically active ribonucleic acid from sources enriched in ribonuclease. Biochemistry 18: 5294–5299

Citron M, Oltersdorf T, Haass C, McConlogue L, Hung AY, Seubert P, Vigo-Pelfrey C, Lieberburg I, Selkoe DJ (1992) Mutation of the β-amyloid precursor protein in familial Alzheimer's disease increases β-protein production. Nature 360: 672–674

Corder EH, Saunders AM, Strittmatter WJ, Schmechel DE, Gaskell PC, Small GW, Roses AD, Haines JL, Pericak-Vance MA (1993) Gene dose of Apolipoprotein E type 4 allele and the risk of Alzheimer's disease in late onset families. Science 261: 921–923

Esch FS, Keim PS, Beattie EC, Blacher RW, Culwell AR, Oltersdorf T, McClure D, Ward PJ (1990) Cleavage of amyloid β peptide during constitutive processing of its precursor. Science 248: 1122–1124

Fontana A, Weber E, Dayer JM (1984) Synthesis of Interleukin-1/endogenous pyrogen in the brain of endotoxin-treated mice: a step in fever induction? J Immunol 133: 1696–1698

Forloni G, Demicheli D, Giorgi S, Bendotti C, Angeletti N (1992) Expression of amyloid precursor protein mRNAs in endothelial, neuronal and glial cells: modulation by interleukin-1. Mol Brain Res 16; 128–134

Frautschy SA, Baird A, Cole GM (1991) Effects of injected Alzheimer β-cores in rat brain. Proc Natl Acad Sci USA 88: 8362–8366

Giulian D, Lachman LB (1985) Interleukin-1 stimulation of astroglial proliferation after brain injury. Science 228: 497–500

Glenner GG, Wong CW (1984) Alzheimer's disease: initial report of the purification and characterization of a novel cerebrovascular amyloid protein. Biochem Biophys Res Commun 120: 885–890

Goa W-Q, Liu X-L, Hatten ME (1992) The weaver gene encodes a nonautonomus signal for CNS neuronal differentiation. Cell 68: 841–854

Goate A, Chartier-Harlin M-C, Mullan M, Brown J, Crawford F, Fidani L, Giuffra L, Haynes A, Irving N, James L, Mant R, Newton P, Rocke K, Roques P, Talbot C, Pericak-Vance M, Roses A, Williamson R, Rossor M, Owen M, Hardy J (1991) Segregation of a missence mutation in the amyloid precursor gene with familial Alzheimer's disease. Nature 349: 704–706

Goldgaber D, Harris HW, Hla T, Maciag T, Donnely RJ, Jacobson JS, Vitek MD, Gajdusek DC (1989) Interleukin-1 regulation of synthesis of amyloid protein precursor in human endothelial cells. Proc Natl Acad Sci USA 86: 7606–7610

Gray CW, Patel AJ (1993) Regulation of β-amyloid precursor protein isoform mRNAs by transforming growth factor-β1 and interleukin-1β in astrocytes. Mol Brain Res 19: 251–256

Griffin WST, Stanley LC, Ling C, White L, MacLeod V, Perrot LJ, White CL, Arora C (1989) Brain interleukin 1 and S-100 immunoreactivity are elevated in Down syndrome and Alzheimer's disease. Proc Natl Acad Sci USA 86: 7611–7615

Haass C, Schlossmacher MG, Hung AY, Vigo-Pelfrey C, Mellon A, Ostaszewski BL, Lieberburg I, Koo EH, Schenk D, Telplow DB, Selkoe DJ (1992) Amyloid β-peptide is produced by cultured cells during normal metabolism. Nature 359: 322–325

Hardy J, Allsop D (1991) Amyloid deposition as the central event in the aetiology of Alzheimer's disease. Trends Pharmacol Sci 12: 383–388

Heinrich PC, Castell JV, Andus T (1990) Interleukin-6 and the acute phase response. Biochem J 265: 621–636

Herrup K (1983) Role of *staggerer* gene in determining cell number in cerebellar cortex. I Granule cell death is an indirect consequence of *staggerer* gene action. Dev Brain Res 11: 267–274

Herrup K, Mullen R (1979a) *Staggerer* chimeras: intrinsic nature of Purkinje cell defect and implication for normal cerebellar development. Brain Res 178: 443–457

Herrup K, Mullen R (1979b) Regional variation and absence of large neurons in the cerebellum of the "*staggerer*" mouse. Brain Res 172: 1–12

Higgins GA, Oyler GA, Neve RL, Chen KS, Gage FH (1990) Altered levels of amyloid protein precursor transcripts in the basal forebrain of behaviorally impaired aged rats. Proc Natl Acad Sci USA 87: 3032–3037

Johnson SA, McNeill T, Cordell B, Finch CE (1990) Relation of neuronal APP-751/APP-695 mRNA ratio and neuritic plaque density in Alzheimer's disease. Science 248: 854–857

Kang J, Lemaire HG, Unterbeck A, Salbaum JM, Masters CL, Greeschik K-H, Multhaupt G, Beyreuther, K, Muller-Hill B (1987) The precursor of Alzheimer's disease amyloid A4 protein resembles a cell-surface receptor. Nature 325: 733–736

Kitaguchi NYT, Shiojiri S, Ito H (1988) Novel precursor of Alzheimer's disease amyloid protein shows protease inhibitory activity. Nature 331: 530–532

Koh J-Y, Yang LL, Cotman CW (1990) β-Amyloid protein increase the vulnerability of cultured cortical neurons to excitotoxic damage. Brain Res 533: 315–320

Kopmels B, Wollman EE, Guastavino JM, Delhaye-Bouchaud N, Fradelizi D, Mariani J (1990) Interleukin-1 hyperproduction by in vitro activated peripheral macrophages. J Neurochem 55: 1980–1985

Kopmels B, Mariani J, Delhaye-Bouchaud N, Audibert F, Fradelizi D, Wollman EE (1992) Evidence for an hyperexcitability state of *staggerer* mutant mice macrophages. J Neurochem 58: 192–199

Lewis DA, Higgins GA, Young WG, Goldgaber D, Gajdusek, DC, Wilson C, Morrison JH (1988) Distribution of precursor amyloid-β-protein messenger RNA in human cerebral cortex: relationship to neurofibrillary tangles and neurite plaques. Proc Natl Acad Sci USA 85: 1691–1695

Löffler J, Huber G (1992) β-Amyloid precursor protein isoforms in various rat brain regions and during brain development. J Neurochem 59: 1316–1324

Maniatis T, Fritch EF, Sambrook J (1987) Molecular cloning A laboratory manual. Cold Spring Harbor Laboratory

Master CL, Simms G, Weinman NA, Multhaup G, McDonald, BL, Beyreuther, K (1985) Amyloid plaque core protein in Alzheimer's disease and Down syndrome. Proc Natl Acad Sci USA 82: 4245–4249

Mortimer JA, van Dujin CM, Chandra V (1991) Head trauma as a risk factor for Alzheimer's disease: a collaborative re-analysis of case-control studies. Int J Epidemiol 20: S28

Nordstedt C, Gandy SE, Alafuzoff I, Caporaso GL, Iverfeldt K, Greeb JA, Winblad B, Greengard P (1991) Alzheimer β/A4 amyloid precursor protein in human brain: Aging-associated increases in holoprotein and in a proteolytic fragment. Proc Natl Acad Sci USA 88: 8910–8914

Ponte P, Gonzales-DeWhitt P, Schilling J, Miller J, Hsu D, Greenberg B, Davis K, Wallas W, Lieberburg I, Fuller F, Cordell B (1988) A new A4 amyloid mRNA contains a domain homologous to serine proteinase inhibitors. Nature 331: 525–527

Quon D, Wang Y, Catalano R, Scardina JM, Murakami K, Cordell B (1991) Formation of β-amyloid protein deposits in brains of transgenic mice. Nature 352: 239–241

Roberts GW, Gentleman SM, Lynch A, Graham DI (1991) β-amyloid protein deposition in brain after head trauma. Lancet 338: 1422–1423

Schellenberg GD, Bird TD, Wijsman EM, Orr HT, Anderson, L, Nemens E, White JA Bonycastle L, Weber JL, Alonso ME, Potter H, Heston, LL, Martin GM (1992) Genetic linkage evidence for a familial Alzheimer's disease locus on chromosome 14. Science 258: 668–671

Schnabel J (1993) New Alzheimer's therapy suggested. Science 260: 1719–1720

Selkoe DJ (1993) Physiological production of the β-amyloid protein and the mechanism of Alzheimer's disease. Trends Neurosci 16: 403–408

Seubert P, Vigo-Pelfrey C, Esch F, Lee M, Dovey H, Davies D, Sinna S, Schlossmacher M, Whaley J, Swindlehurst C, McCormack R, Wolfert R, Selkoe D, Lieberburg I, Schenk D (1992) Isolation and quantification of soluble Alzheimer's β-peptide from biological fluids. Nature 359: 325–227

Siebert, PD, Larrick JW (1992) PCR MIMICS – Competitive DNA fragments for use as internal standards in quantitative PCR. Bio Techniques 14: 244–249

Shojaeian H, Delhaye-Bouchaud N, Mariani J (1985) Decreased number of cells in the inferior olivary nucleus of the developing *staggerer* mouse. Dev Brain Res 21: 141–146

Shoji M, Golde TE, Ghiso J, Cheung TT, Estus S, Shaffer LM, Cai X-D, McKay DM, Tintner R, Frangione B, Younkin SG (1992) Production of Alzheimer amyloid β protein by normal proteolytic processing. Science 258: 126–129

Solà C, García-Ladona FJ, Mengod G, Probst A, Frey P, Palacios JM (1993) Increased levels of the Kunitz protease inhibitor-containing βAPP mRNA in rat brain following neurotoxic damage. Mol Brain Res 17: 41–52

Tanzi R, McClatchey A, Lamperti ED, Villa-Komaroff L, Gusella JF, Neve RL (1988) Protease inhibitor domain encoded by an amyloid protein precursor mRNA associated with Alzheimer's disease brain. Nature 331: 528–530

Vandenabeele P, Fiers W (1991) Is amyloidogenesis during Alzheimer's disease due to an IL-1/IL-6-mediated "acute phase response" in the brain. Immunol Today 12: 217–219

Weidemann A, König G, Bunke D, Fischer P, Salbaum JM, Masters CL, Beyreuther K (1989) Identification, biogenesis, and localization of precursors of Alzheimer's disease A4 amyloid protein. Cell 57: 115–126

Wollman EE, Kopmels B, Bakalian A, Delhaye-Bouchaud N, Fradelizi D, Mariani J (1992) Cytokine and neuronal degeneration. In: Dantzer R, Rothwell N (eds) Cytokine and neuronal degeneration Pergamon Press Oxford, pp 187–203

Yankner BA, Duffy LK, Kirchner DA (1990) Neurotrophic and neurotoxic effects of amyloid β protein: reversal by tachykinin neuropeptides. Science 250: 279–282

Zanjani H, Mariani J, Herrup K (1990) Cell loss in the inferior olive of *staggerer* mutation mouse is an indirect effect of the gene. J Neurogenet 6: 229–241

Amyloid β-Peptide Induces Necrotic Cell Death in PC12 Cells

*J.B. Davis**, *C. Behl, F.G. Klier*, and *D. Schubert*

Introduction

Alzheimer's disease is characterised by senile plaques, neurofibrillary tangles and selective neuronal cell loss (Selkoe 1991). The senile plaques contain insoluble deposits whose major peptide constituent is β-amyloid protein (AβP). It has been suggested that the AβP is toxic to surrounding neurones, causing nerve cell death and the appearance of dystrophic neurites around the plaque deposits. The direct toxicity of AβP for neurones has been addressed in vivo, by injection of plaque cores (Frautschy et al. 1991) or purified AβP (Kowall et al. 1991) into rat brain, and in vitro, by addition of AβP to primary neuronal cultures (Yankner et al. 1990) and clonal nerve cell lines (Behl et al. 1992). As yet, however, little is known of the mechanism of cell death. It has been suggested that cell death occurs as a result of an imbalance in calcium homeostasis (Mattson et al. 1992) or as a result of oxidative stress (Behl et al. 1992), although the two may be linked. The cytotoxicity of AβP 1–40 has been associated with a fragment including residues 25–35 (Yankner et al. 1990). This study aimed at determining the mechanism of this β25–35 induced cytotoxicity and distinguishing between either an apoptotic or necrotic pathway.

Apoptosis Vs. Necrosis

Cell death may occur via two broad pathways, apoptosis or necrosis. Apoptosis (Wyllie et al. 1980) is a natural function that occurs throughout vertebrate development to control lymphocytes and nerve cell populations. It is often referred to as programmed cell death, as it may be triggered by physiological stimuli that activate a pathway, which is defined by a sequence of morphological and biochemical changes. Chromatin marginates toward the nuclear membrane and condenses to form apoptotic bodies. The nuclear changes are followed by a reduction in cell volume and finally vacuolisation

* The Salk Institute for Biological Studies, 10010 N. Torrey Pines Rd, La Jolla, CA 92037, USA. Present address: Department of Molecular Neuropathology (H26/1-023), SmithKline Beecham, Coldharbour Rd, Harlow CM19 5AD, UK

K.S. Kosik et al. (Eds.)
Alzheimer's Disease: Lessons from Cell Biology

and cellular fragmentation. Most intracellular organelles remain intact prior to phagocytosis of the apoptosing cell. Apoptosis may also be followed by various biochemical markers including DNA fragmentation, which is the effect of internucleosomal endonuclease activity resulting in a ladder of DNA on agarose gels (Hockenbery et al. 1990).

In contrast, necrosis is the result of pathology or an irrevocable disturbance to cellular homeostasis (Kerr et al. 1972). Necrosing cells swell rather than shrink, suffer immediate damage to organelles and finish with lysis of the plasma membrane and nuclear degradation. There is no DNA laddering, although DNA degradation may lead to a smear on agarose gels.

An example of apoptosis by neuronal cells is the degeneration of nerve growth factor (NGF)-differentiated PC12 cells following removal of serum and NGF from their growth medium (Mesner et al. 1992). In our experiments, this inducible apoptosis in PC12 cells was used as a positive control for comparison with the effect of added AβP fragment β25–35. Our investigation of the morphological and biochemical changes in PC12 cells following serum or NGF withdrawal and the changes induced by β25–35 lead to the conclusion that PC12 cells die via necrosis following exposure to β25–35.

Materials & Methods

Cell Culture

PC12 cells grown in DMEM/10%FCS/5% horse serum, untreated cells and NGF-treated (100 ng/ml for 5 days) cells were dispersed by trituration and plated at 3,000 cells/well in microtitre plates or 1×10^4 cells/35 mm dish. Cells were exposed to β25–35 and viability assays were conducted as described elsewhere (Behl et al. 1992).

Viability Assays

Cell viability was assessed using the 3-[4,5-dimethylthiazol-2-yl]-2,5,-diphenyltetrazolium bromide (MTT) assay and lactate dehydrogenase (LDH) release assay, performed according to the manufacturer's instructions (Promega). Assays were in triplicate; figures showed results from a typical experiment ± SEM.

Flow Cytometry

The methods used for flow cytometry were previously described (Dive et al. 1992; Sun et al. 1992). PC12 cells, 2×10^5 cells/60 mm dish, were treated with serum and NGF-free medium or serum-containing medium plus 10 μM β25–35 for 24 hours. Hoechst 33342 at a final concentration of 1 μg/ml was added for the final 10 minutes. Cells were then harvested and washed in phenol red-free, hepes buffered DMEM supplemented with 2% FCS at 4°C, and were resuspended in the same medium containing 2 μg/ml propidium iodide. Viable cells were then analysed after electronic gating of propidium iodide-positive cells and debris. Enhanced Hoechst 33342 staining was followed with UV excitation and emission at 400–500 nm. Data are the mean fluorescent peak heights of samples presented as the percentage of the value for control cells +/− standard deviation.

Analysis of DNA Fragmentation

PC12 cytoplasmic DNA was isolated after experimental treatments by the method of Hockenberry et al. (1990). Cytoplasmic supernatants were phenol-chloroform extracted and nucleic acids were precipitated. Samples of nucleic acids were RNase treated, run on 1.5% agarose/TBE gels and visualised by ethidium bromide staining.

Electron Microscopy

Cells were plated onto poly-lysine coated plates (100 ng/ml poly-L-lysine, overnight) and differentiated with 100 ng/ml NGF for five days. Cells were then treated with a 1- or 24-hour exposure to 20 μM β25–35 or NGF withdrawal, fixed in 2.5% glutaraldehyde in 0.1 M cacodylate buffer for 30 minutes at room temperature and postfixed in 1% osmium tetroxide. Double fixation in saturated thiocarbohydrazide-osmium was used to increase contrast. Samples were dehydrated through ethanol and the dishes were embedded in Epon 812, cured in vacuo for four hours at 60°C and sectioned.

Results

Amyloid β-peptide Fragment β25–35 Is Toxic for PC12 Cells

We have previously demonstrated the toxicity of AβP fragments β1–40 and β25–35 for PC12 cells (Behl et al. 1992). The dose responses for β25–35 are shown in Figure 1, using both the MTT and LDH release assays. β25–35 caused a maximum 60–70% drop in MTT reduction. As the MTT assay is thought to monitor mitochondrial activity, viable cells with damaged electron transport systems may lead to an overestimation of cell death. Measurement of cell lysis by release of LDH is an alternative to the MTT assay. LDH release reached a maximum of 45% at 48 hours of exposure to β25–35. An incubation time of 48 hours was necessary to detect LDH release, whereas MTT conversion is reduced after only a total of six hours of exposure (Fig. 2). This time difference suggests that a cell's energy metabolism, or ability to provide reducing potential, is affected early in the cell's demise. These biochemical findings agree with the electron microscopy (EM) data below,

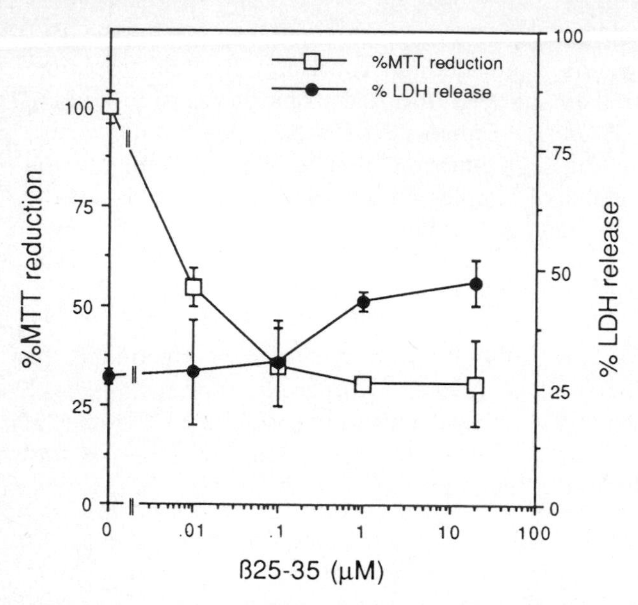

Fig. 1. β25–35 induced cell death in PC12 cells. PC12 cells were exposed to β25–35 for 48 hours and survival was assessed using the MTT assay and LDH release assay. Survival of control, untreated cells is shown as 100% in the MTT assay. LDH release was compared to the enzyme activity released upon total lysis of all cells

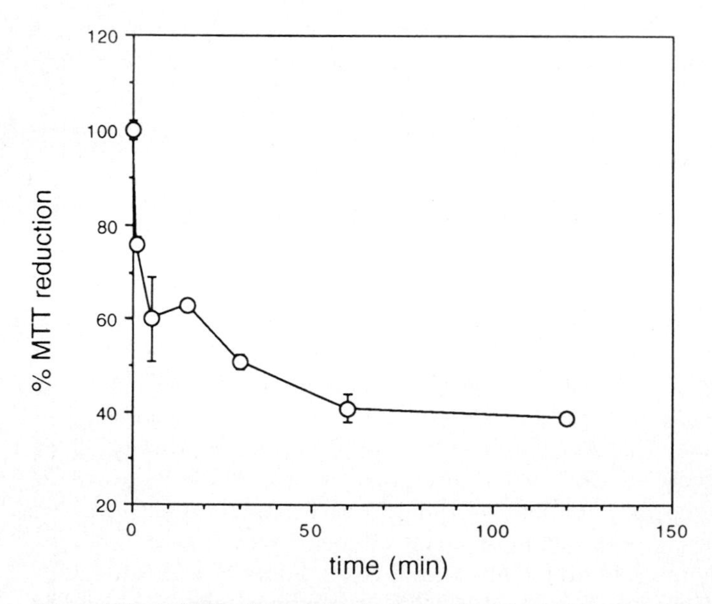

Fig. 2. Effect of β25–35 exposure time upon MTT reduction. PC12 cells were exposed to 5 μM β25–35 for the indicated time, washed once and incubated with MTT dye for four hours, after which the cells were solubilised and absorbance was measured

which demonstrated a rapid and early effect upon the cell's cytoplasmic organelles, including mitochondria.

Molecular Synthesis or Endonuclease Inhibitors and Inhibitors of Calcium Flux Fail to Block β25–35 Toxicity

A signature of apoptosis is its "programmed" nature. Hence, a number of inhibitory steps proposed to block steps in the apoptotic pathway have been successfully used to block cell death in apoptosing neurones. Inhibition of either protein or RNA synthesis rescues neurones upon withdrawal of neurotrophic factors (Scott and Davies 1990). The endonuclease inhibitor aurintricarboxylic acid (ATA) also rescues cells from apoptosis, inhibiting the formation of a DNA ladder after NGF withdrawal and glutamate excitotoxicity (Batistatou and Greene 1991; Samples and Dubinsky 1993). Similarly, blockade of calcium fluxes, either by removal of extracellular calcium or inhibition of release from intracellular stores, may block apoptosis (McConkey et al. 1989). In contrast, cycloheximide and actinomycin D were unable to rescue PC12 cells from β25–35 induced cell death (Fig. 3); neither were ATA, dantrolene or calcium withdrawal (data not shown). These data suggest that β25–35 toxicity in PC12 cells is not dependent upon *de novo*

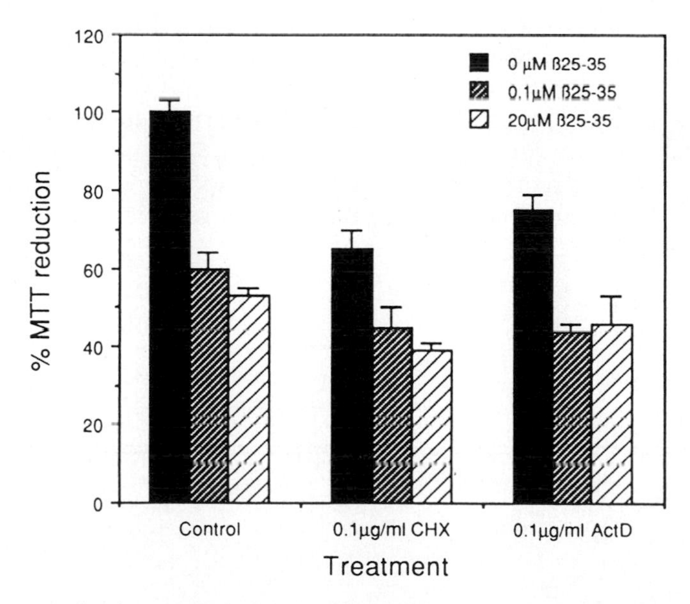

Fig. 3. Inhibition of macromolecular synthesis fails to block β25–35 toxicity. PC12 cells were exposed to β25–35 in the absence or presence of 0.1 μg/ml cycloheximide (CHX) or actinomycin D (ActD) for 24 hours and the effect upon MTT reduction was assayed after 24 hours. Actinomycin D and cycloheximide were highly toxic to the cells at higher concentrations

synthesis of proteins, or endonuclease activity or a calcium flux and, therefore, has none of these hallmarks of apoptosis.

Flow Cytometry Detected No Chromatin Condensation

Chromatin structure alters during apoptosis to allow increased Hoechst dye binding to DNA, which leads to a higher fluorescence. PC12 cells exposed to serum-free medium for 24 hours showed evidence of apoptosis, having a higher level of Hoechst fluorescence compared to control cells. In contrast, cells challenged with β25–35 for 24 hours showed no increased fluorescence (Table 1), suggesting that the cells did not undergo apoptosis but died via a necrotic pathway.

DNA Laddering Was Not Detected

An established biochemical marker of apoptosis, DNA fragmentation due to endonuclease activity, was assayed in PC12 cells exposed to either serum withdrawal or β25–35 for 12, 24 or 48 hours of exposure (Fig. 4). A DNA ladder typical of apoptosis was observed in DNA isolated from serum-deprived cells, whereas β25–35 treated cells displayed a DNA smear. No DNA degradation was detectable with less than 12 hours of exposure. The appearance of DNA degradation only after 12 hours of exposure compares with the EM observation of cytoplasmic damage after only one hour, and with the early drop in ability to reduce MTT (Fig. 2). These results suggest that necrotic events occur in the cytoplasm before any DNA degradation.

Electron Microscopy Demonstrated Early Cytoplasmic Changes

Ultrastructural changes in the morphology of NGF-treated PC12 cells following treatment with the β25–35 fragment were studied by transmission

Table 1. β25–35 treatment does not increase Hoechst 33342 binding[a]

	% change in Hoechst 33342 fluorescence	
	Experiment 1	Experiment 2
PC12 untreated	100 ± 9.3	100 ± 9.8
+β25–35 for 24 hours	106 ± 5.5	98 ± 10.1
Serum-free for 24 hours	172 ± 6.2	132 ± 13.4

[a] PC12 cells were treated with 10 μM β25–35, or serum-free medium, for 24 hours and then analysed by FACS after staining with Hoechst 33342 as described in the Methods. (Reprinted with permission from Behl et al. 1994a.)

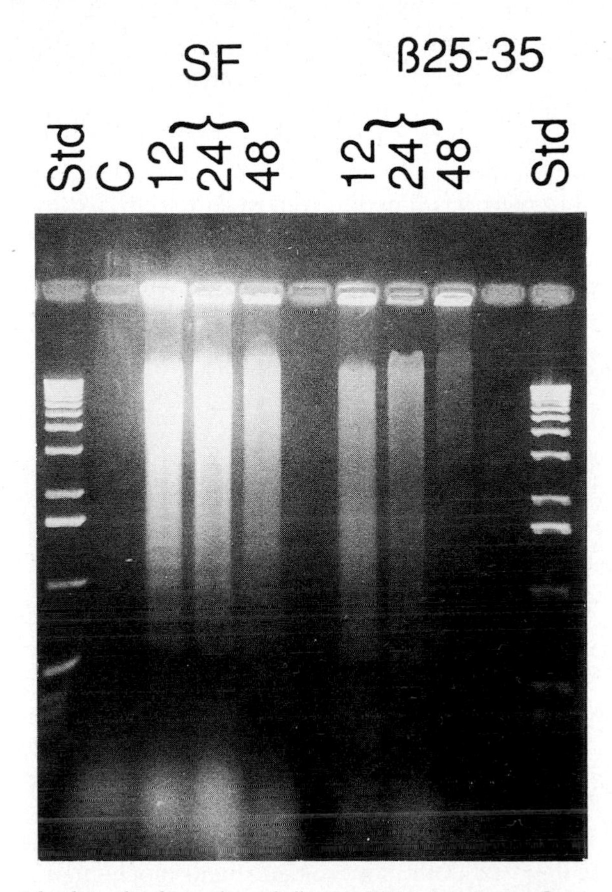

Fig. 4. β25–35 exposure does not lead to the formation of discreet DNA fragments. Cytoplasmic DNA was harvested from PC12 cells following treatment with 20 μM β25–35 or factor withdrawal (SF) for the indicated length of time. Thirty μg of total extracted nucleic acid was RNase digested and analysed on 1.5% agarose/TBE gels. DNA fragments were visualised by ethidium bromide staining. C, control; Std, DNA molecular weight standards. (Reprinted with permission from Behl et al. 1994a)

electron microscopy. NGF-differentiated PC12 cells were fixed after treatment with β25–35 or NGF withdrawal for one or 24 hours. Ultrastructural changes that occurred due to NGF withdrawal are depicted in Figure 5(I). The chromatin aggregation at the nuclear membrane after one hour and the formation of apoptotic bodies after 24 hours are characteristic of apoptosis. These contrasted with the chromatin of untreated and of β25–35 treated PC12 cells (Fig. 5(II)), which remained evenly dispersed and with little margination. PC12 cells treated with β25–35 had healthy nuclei but damaged cytoplasmic organelles (Fig. 5(II) B&C). After 24 hours of treatment the cytoplasm was highly vacuolated and membrane systems were severely disrupted. The most important observation, however, is that there was already significant cytoplasmic damage after only one hour of exposure to β25–35 (Fig. 5(II) B). The parallel damage to the PC12 neurites was equally striking. After only one hour of exposure to β25–35 the plasma membranes of the neurites were blebbed and disrupted. The filamentous structures within the neurites were also fragmented and in disarray. Hence, the β25–35 fragment was found to lead to a rapid breakdown of cytoplasmic membranes and organelles, including the structure of the extended neurites. Also, these changes occurred in the absence of nuclear changes. These observations match those that define necrosis, not apoptosis, and finalise our conclusion that the β25–35 fragment initiates a necrotic event in PC12 cells.

Discussion

Categorisation of a cell death process as either apoptotic or necrotic by comparison with the literature is fraught with difficulty due to inconsistencies between model systems. For these reasons a direct comparison was made between changes induced by β25–35 in PC12 cells and in the same cell type undergoing apoptosis due to serum and NGF withdrawal. In summary, none of the macromolecular synthesis inhibitors or calcium flux inhibitors that block apoptosis blocked β25–35 induced cell death. DNA laddering was not observed in β25–35 treated cells. Instead the DNA was found to be degraded in a random pattern not characteristic of endonuclease activity; neither was the endonuclease inhibitor ATA effective. Nuclear Hoechst 33342 dye binding was not increased in β25–35 treated cells. Most conclusively, the earliest ultrastructural changes observed were not nuclear but cytoplasmic. Having compared both morphological and biochemical parameters, the data clearly show that the β25–35 fragment is toxic to PC12 cells via necrotic event.

A recent publication by Loo et al. (1993) reports evidence of AβP causing apoptosis in neuronal primary cultures. The heterogeneity of most CNS primary cultures may explain the discrepancy between the two reports. It is possible that AβP may cause apoptosis in one population of cells indirectly, via the release of glutamate from cells dying from necrotic events. It is also possible that different cells are killed by the peptide fragments via separate pathways. The data presented here, however, are supported by similar findings in purified rat cortical cultures (Behl et al. 1994a) and by experiments demonstrating that *bcl*-2 is able to rescue PC12 cells from glutamate toxicity but not from AβP toxicity (Behl et al. 1993). Despite the distinction between necrosis and apoptosis, the mechanism of β25–35 cytotoxicity still remains unclear. The rescuing effect of vitamin E (Behl et al. 1992) suggests an oxidative pathway; it has also been suggested that peroxides are involved (Behl et al. 1994b), although their origin has yet to be determined.

Induction of a necrotic death process by β-peptide is consistent with the pathology of Alzheimer's disease, in which acute phase proteins are found

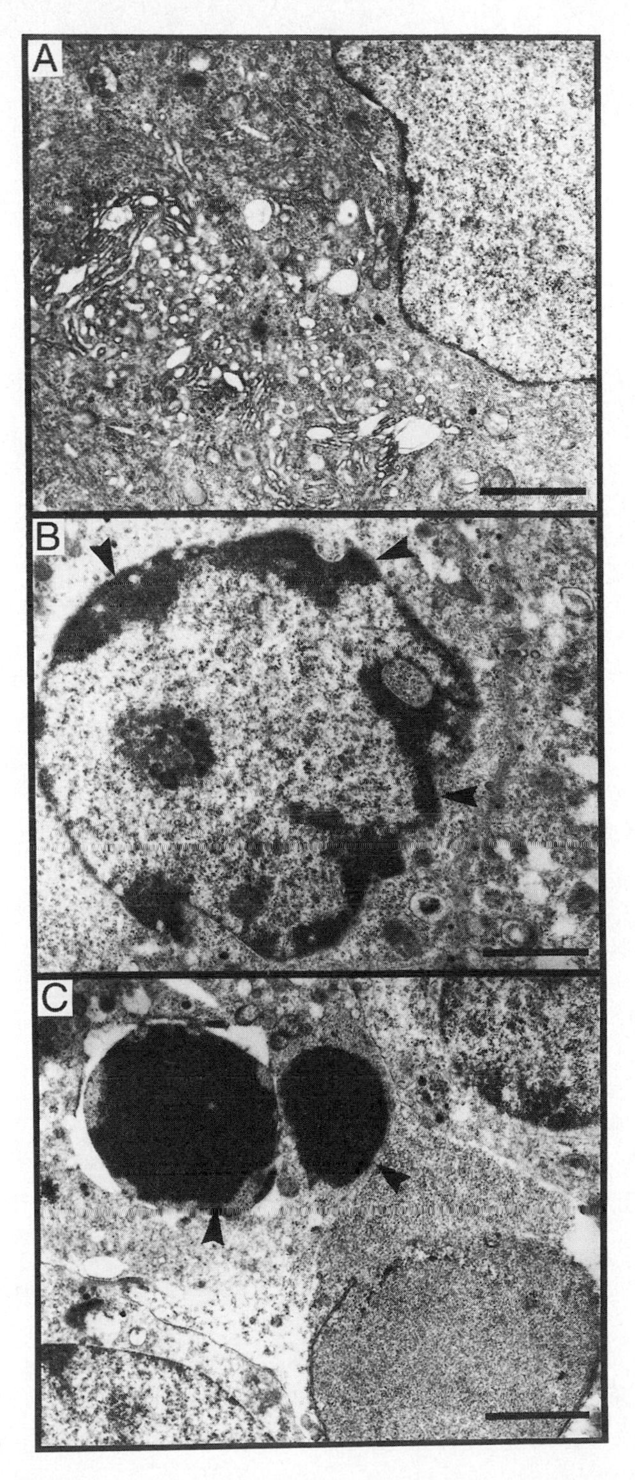

Fig. 5.(I)

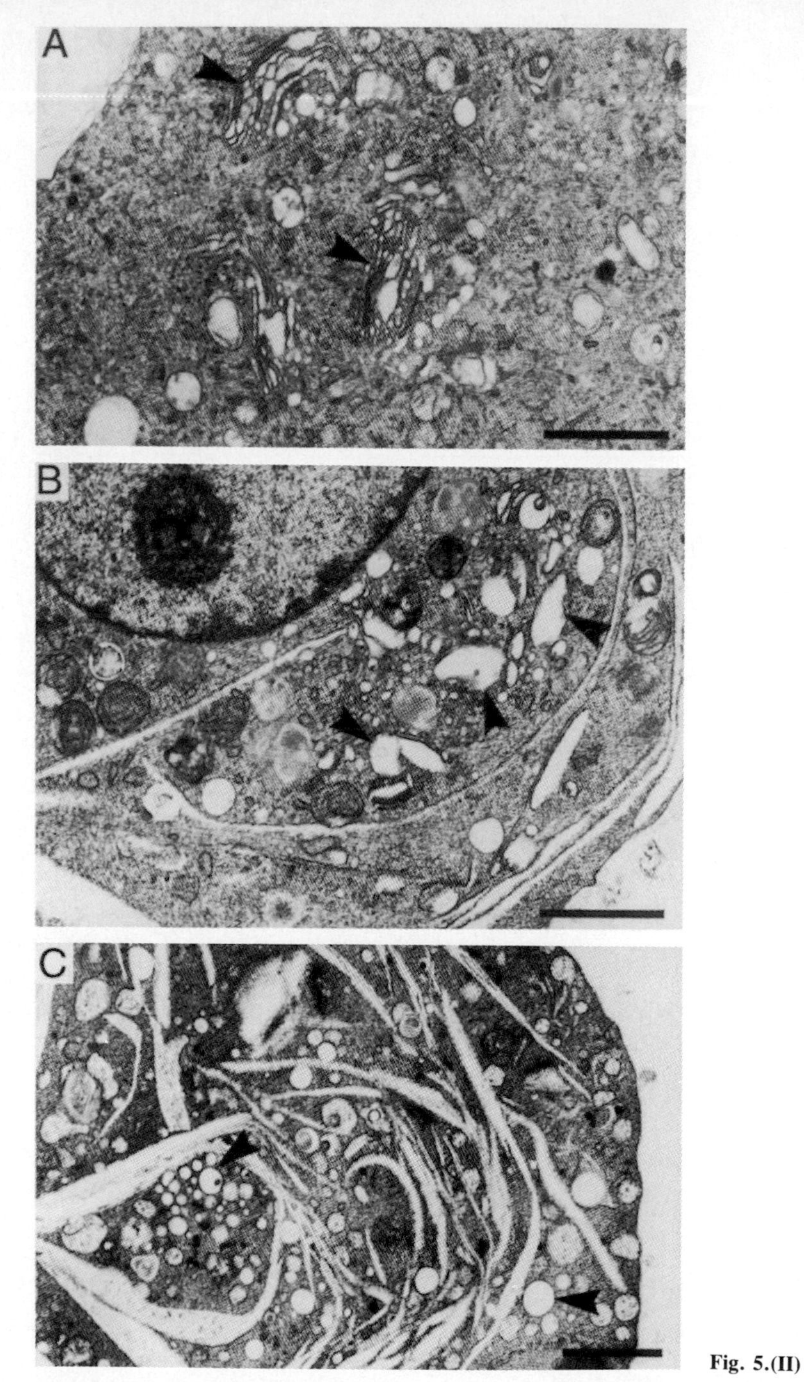

Fig. 5.(II)

Fig. 5. Electron microscopy of PC12 cells treated with β25–35. (I) PC12 cells were treated with NGF for 5 days, followed by 24 hours of NGF withdrawal (**B**, **C**) or without NGF withdrawal (control; **A**). Note chromatin aggregates (arrowheads in **B**) and apoptotic bodies in **C**. (II) In place of NGF withdrawal, cells were treated with 20 μM β25–35 (**A–C**). The effect upon the neurites is also shown (**D–F**). Samples were processed after one hour (**B**, **E**) and 24 hours (C,

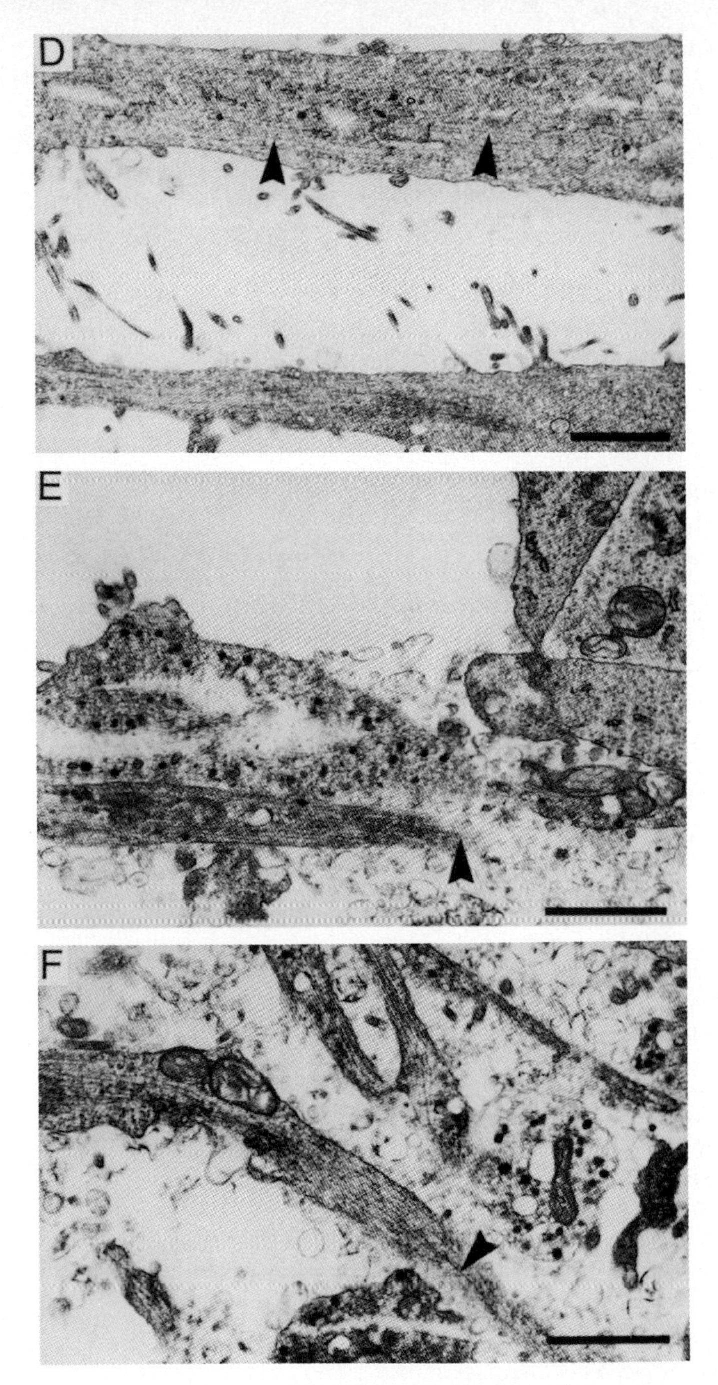

F). **A** and **D** represent control cells not exposed to β25–35. Note intact Golgi apparatus in **A**, well-defined microtubules in **D**, high vacuolization of the cytoplasm in **B** and **C** and the disintegration of the neurites in **E** and **F** as indicated by arrowheads. Bars represent 1 μM. (Reprinted with permission from Behl et al. 1994a)

in, and activated microglia around, the plaque. Furthermore, AβP is itself a chemoattractant for microglia (Davis et al. 1992). Taken together, these results suggest a pathology mediated by the neurotoxicity of overexpressed AβP and that the mechanism of toxicity is a necrotic one, possibly involving an oxidative mechanism.

Acknowledgements. The authors wish to thank Joe Trotter for excellent technical assistance and recognise the support from the Alzheimer's Association, the National Institutes of Health, the Von Briesen Trust and the Deutsche Forschungsgemeinschaft.

References

Batistatou A, Greene LA (1991) Aurintricarboxylic acid rescues PC12 cells and sympathetic neurons from cell death caused by nerve growth factor deprivation: correlation with suppression of endonuclease activity. J Cell Biol 115: 461–471

Behl C, Davis JB, Cole GM, Schubert D (1992) Vitamin E protects nerve cells from amyloid β protein toxicity. Biochem Biophys Res Comm 186: 944–950

Behl C, Hovey L, Krajewski S, Schubert D, Reed JC (1993) Bcl-2 prevents killing of neuronal cells by glutamate but not by amyloid beta protein. Biochem Biophys Res Comm 197: 949–956

Behl C, Davis JB, Klier FG, Schubert D (1994a) Amyloid β peptide induces necrosis rather than apoptosis. Brain Res 645: 253–264

Behl C, Davis JB, Leslie R, Schubert D (1994b) Hydrogen peroxide mediates amyloid β protein toxicity. Cell 77: 817–827

Davis JB, McMurray HF, Schubert D (1992) The amyloid beta-protein of Alzheimer's disease is chemotactic for mononuclear phagocytes. Biochem. Biophy Res Comm 189: 1096–1100

Dive C, Gregory CD, Phipps DJ, Evans DL, Milner AE, Wyllie AH (1992) Analysis and discrimination of necrosis and apoptosis (programmed cell death) by multiparameter flow cytometry. Biochim Biophys Acta 1133: 275–285

Frautschy SA, Baird A, Cole GM (1991) Effects of injected Alzheimer β-amyloid cores in rat brain. Proc Natl Acad Sci USA 88: 8362–8366

Hockenbery D, Nunez G, Milliman C, Schreiber RD, Korsmeyer SJ (1990) Bcl-2 is an inner mitochondrial membrane protein that blocks programmed cell death. Nature 348: 334–336

Kerr JFR, Wyllie AH, Currie AR (1972) Apoptosis: a basic biological phenomenon with wide-ranging implications in tissue kinetics. Br J Cancer 26: 239–257

Kowall NW, Beal MF, Busciglio J, Duffy LK, Yankner B (1991) An in vivo model for the neurodegenerative effects of β amyloid and protection by substance P. Proc Natl Acad Sci USA 88: 7247–7251

Loo DT, Agata C, Pike CJ, Whittemore ER, Wulencewicz AJ, Cotman CW (1993) Apoptosis is induced by β-amyloid in cultured nervous system neurons. Proc Natl Acad Sci USA 663: 234–249

Mattson MP, Cheng B, Davis D, Bryant K, Lieberburg I, Rydel R (1992) β-Amyloid peptides destabilize calcium homeostasis and render human cortical neurons vulnerable to excitotoxicity. J Neurosci 12: 376–389

McConkey DJ, Nicotera P, Hartzell P, Bellomo G, Wyllie AH, Orrenius S (1989) Glucocorticoids activate a suicide process in thymocytes through an elevation of cytosolic Ca2+ concentration. Arch Biochem Biophys 269: 365–370

Mesner PW, Winters TR, Green SH (1992) Nerve growth factor withdrawal-induced cell death in neuronal PC12 cells resembles that in sympathetic neurons. J Cell Biol 119: 1669–1680

Samples SD, Dubinsky JM (1993) Aurintricarboxylic acid protects hippocampal neurons from glutamate excitotoxicity in vitro. J Neurochem 61: 382–385
Scott SA, Davies AM (1990) Inhibition of protein synthesis prevents cell death in sensory and parasympathetic neurons deprived of neurotrophic factor in vitro. J Neurobiol 21: 630–638
Selkoe DJ (1991) The molecular pathology of Alzheimer's disease. Neuron 6: 487–498
Sun X-M, Snowden RT, Skilleter DN, Dinsdale D, Ormerod MG, Cohen GM (1992) A flow-cytometric method for the separation and quantitation of normal and apoptotic thymocytes. Anal Biochem 204: 351–356
Wyllie AH, Kerr JFR, Currie AR (1980) Cell death: the significance of apoptosis. Int RevCyto 68: 251–306
Yankner BA, Duffy LK, Kirschner DA (1990) Neurotrophic and neurotoxic effects of amyloid β protein: reversal by tachykinin neuropeptides. Science 25: 279–282

Linking Amyloid Precursor Protein Processing and Tau-Related Pathology in Alzheimer's Disease

K.S. Kosik, * *A. Ferreira, R. Knowles, N. Leclerc,*
and *S.M. Greenberg*

Summary

A key missing link in the pathogenesis of Alzheimer's disease is the connection between the complex processing of the amyloid precursor protein (APP) to form senile plaques and the polymerization of the tau protein into the paired helical filaments of neurofibrillary tangles. While both plaques and tangles are required for the clinical expression of Alzheimer's disease, understanding how the two component proteins of these structures are related is problematic. One possibility being explored by some investigators is that the Aβ fragment is directly toxic to neurons, and aggregates of this fragment can induce both a cytoskeletal reorganization and ultimately cell death. While these experiments are compelling when performed in culture, the presence of amyloid fibrils in the post-mortem brain often seems relatively benign. Brain tissue in the region of amyloid deposition, particularly diffuse amyloid, can be unremarkable, and neurofibrillary tangles often occur at sites anatomically distant from plaques.

We have found that a secreted fragment of APP can stimulate the mitogen-activated protein kinase (MAPK) in a ras-dependent manner and that stimulation along this pathway results in the enhanced phosphorylation of tau protein. Phosphorylation at several recently identified sites is one means by which tau function is mediated. Our recent observations suggest that tau effects its function not just by interacting with microtubules, but also with other elements of the cytoskeleton. One of these interactions that appears important in the initiation of processes is a direct or indirect interaction between tau and the actin system. In Sf9 cells expressing tau, the segregation of actin filments to one pole of the cell anticipates process formation. The induction of this type of organization suggests a role for tau in establishing polarity.

---

* Harvard Medical School and Center for Neurologic Diseases, Department of Medicine (Division of Neurology), Brigham and Women's Hospital, Boston, MA 02115, USA

K.S. Kosik et al. (Eds.)
Alzheimer's Disease: Lessons from Cell Biology

Introduction

It is almost a faith among Alzheimer researchers that, once the nature of the senile plaques and neurofibrillary tangles are fully unraveled, the disease will be solved. These two characteristic lesions have been known for nearly a century. However, only within the past six years have researchers identified the proteins in the senile plaques and neurofibrillary tangles. The principle gene product in the neurofibrillary tangle, an intracellular structure, is the microtubule-associated protein, tau. The principle gene product in the senile plaque, an extracellular structure, is β-amyloid (Aβ), which is a fragment of a much larger protein called the amyloid precursor protein (APP). A central problem in Alzheimer research is how the two signature lesions – the neurofibrillary tangles and the senile plaques – are related to each other. A rare, but exceedingly instructive mutation in the APP gene is one of the fundamental causes of the disease. This discovery has squarely pointed to the APP gene as a pathogenetic locus in the disease process. On the other hand, the vast majority of so-called sporadic cases, often with a more advanced age-at-onset, do not appear to have mutations in the APP gene; hence additional pathogenetic mechanisms must be sought.

Amyloid deposition is necessary but not sufficient for disease expression. Also necessary are the neurofibrillary tangles, found in neuronal cells bodies and in the dystrophic neurites, which consist of swollen and distorted neuronal processes. Both lesions contain highly insoluble intra-neuronal filaments (PHF) that self-assemble from tau protein, presumably due to the presence of excess phosphate on the molecule. Neither amyloid deposition in senile plaques nor the formation of tau-filaments alone is unique for Alzheimer's disease; their combinations is requisite for the diagnosis. Underscoring the necessity for both lesions is the frequent observation that Alzheimer type amyloid deposits without neurofibrillary tangles occur frequently in normal aging. The very high percentage of amyloid deposits in normal elderly individuals emphasizes the importance of learning how amyloid can, in some cases, lead to the neuritic dystrophy and tangles, a necessary concomittant of Alzheimer's disease.

Mitogen-Activated Protein Kinases (MAPKs)

A number of integrated pathways serve to transduce the signals that regulate cellular function. One theme among these regulatory pathways is that the delivery of a molecular signal to a cell surface receptor with tyrosine kinase activity results in the activation of a series of downstream serine/threonine kinases capable of modulating a host of cell functions from the phosphorylation of transcription factors (Pulverer et al. 1991) to the organization of the cytoskeleton (Gotoh et al. 1991; Tsao et al. 1990). Among the kinases in this cascade are the mitogen-activated protein kinase (MAPK), also known as extracellular signal regulated kinase (ERK), and upstream elements such as MEK, which phosphorylates MAPKs, and MEKK, which phosphorylates MEK, both of which are activated by phosphorylation. Among the known activators of this pathway are growth factors such as epidermal growth factor and nerve growth factor (Cobb et al. 1991). Rapid MAPK activation also occurs via pertussis toxin-sensitive G_i-coupled receptors such as the thrombin receptor (L'Allemain et al. 1991; Nishizawa et al. 1990). These two categories of MAPK activation appear to differ in that growth factor activation is ras-dependent and G protein activation is ras-independent. Recently Gary Johnson has suggested that MEK represents a convergence point in MAPK activation (Lange-Carter et al. 1993). This hypothesis proposes one route leading from tyrosine kinase activation via ras and raf, and a second route leading from G protein-coupled receptors via MEKK. Several MAPKs from vertebrates have been named according to their molecular weights as $p42^{mapk}$, $p44^{erk1}$, $p44^{mapk}$, and p54 (for review see Pelech and Sanghera 1992). MAPKs have a high degree of evolutionary conservation; they are expressed in all cell types, however the distribution of individual isoforms differs among various tissues, with the highest levels found in brain. MAPK substrates are phosphorylated at a serine or a threonine immediately amino to a proline. Several phosphatase genes have been cloned that inactivate MAPK (reviewed in Nebreda 1989).

The Amyloid Precursor Protein

APP is a glycosylated protein containing a short cytoplasmic tail, a once-spanning transmembrane domain, and a long ectodomain. Internalization of the holoprotein or its fragments positions the ectodomain within the lumen of organelles. An approximately 40 amino acid fragment of APP termed $A\beta$assembles extracellularly into the amyloid fibrils found in senile plaques. This fragment, which can be detected in various extracellular compartments normally (Seubert et al. 1992; Haass et al. 1992; Shoji et al. 1992), includes a portion of the membrane spanning domain and a portion of the ectodomain. How this fragment is generated is a central question in Alzheimer's disease reserach. APP_s is found in the medium of several cell types (Schubert et al. 1988, 1989; Ueda et al. 1989; Weidemann et al. 1989); this secreted product arises from cleavage at amino acid 16 within the $A\beta$ peptide, precluding the formation of $A\beta$ and amyloid fibrils (Sisodia et al. 1990; Esch et al. 1990; Anderson et al. 1991).

The normal function of APP is not established, however several studies have attributed trophic activity to this protein. Trophic activity was reported in cultured neurons for a fragment of the $A\beta$ (Whitson et al. 1989). APP_s suppression resulted in growth impairment in fibroblasts that was corrected by the addition of purified APP (Saitoh et al. 1989). The activity for this effect resides in the sequence RERMS (Ninomiya et al. 1993), located in the mid-region of the ectodomain. These data as well as our recent findings that APP_s can activate MAPK (Greenberg et al. 1994) raise the possibility of a receptor for APP_s.

Tau Phosphorylation and the Neuritic Dystrophy

The only known post-translational modification of tau is phosphorylation. A key event in the pathogenesis of Alzheimer neurofibrillary tangles is the hyper-phosphorylation of tau protein. One of the curious features of the Alzheimer-type tau phosphorylation is its similarity to fetal-type tau phosphorylation. At a relatively late time in development when neuronal migration is complete, tau synthesis commences at high levels and the protein is posttranslationally modified by phosphate incorporation. In the juvenile period tau mRNA levels decline, a number of alternatively spliced tau isoforms appear, and much of the phosphate incorporated into tau is lost. In Alzheimer's disease, no clear changes in tau synthesis or splicing have been observed, but tau does become hyper-phosphorylated. The sites which become phosphorylated in Alzheimer's disease, and the relationship of these sites to the phosphorylation sites in fetal tau, are now known (Watanabe et al. 1993; Liu et al. 1994). Therefore, key questions regarding tau phosphorylation are: How is the normal phosphorylation state of tau controlled in adult and fetal life? What enzymes and control mechanisms reduce the degree of tau phosphorylation upon maturation? What are the functions of these different phosphorylation states? How does tau become hyper phosphorylated in Alzheimer's disease?

Neurofibrillary tangles and dystrophic neurites consist of highly insoluble, helically wound pairs of filaments called paried helical filaments (PHF). These filaments consist of the microtubule-associated protein tau in an abnormal phosphorylation state refered to as PHF-tau. It is not known how the tau protein self-assembles into a highly insoluble macromolecular polymer. One of the early steps must be a dissociation of tau from microtubules, leaving the carboxy terminal microtubule-binding domain of tau incompetent to bind microtubules (Wille et al. 1992). Most of the known phosphorylation sites lie on either side of the microtubule-binding domain and, perhaps, conformational effects upon these flanking sequences reduce the binding of tau to microtubules. PHF-tau (the hyperphosphorylated tau of Alzheimer's disease) has a decreased affinity for microtubules, a decreased ability to promote the polymerization of microtubules, reduced electrophoretic mobility (Liu et al. 1991), and reactivity with certain tau antibodies directed against phosphorylation sites. These antibodies include *AT8* (Biernat et al. 1992), *PHF-1* (Greenberg and Davies 1990) and *T3P* (Lee et al. 1991). All of the existing phosphorylation site-directed antibodies reactive with PHF-tau also react with fetal tau.

A number of proline-directed kinases such as MAPK (Drewes et al. 1992; Lu et al. 1993), glycogen synthase kinase 3 (Hanger et al. 1992; Mandelkow et al. 1992) and cdc2 kinase (Ledesma et al. 1992; Vulliet et al. 1992), phosphorylate tau in vitro at many of the same sites (Ledesma et al. 1992). In contrast, other substrates of proline-directed kinases, such as the retinoblastoma gene product, do discriminate among sites phosphorylated by cdc2 and MAPK (personal communication. Ed Harlow). Whether the same tau kinase that is active during fetal life is involved in Alzheimer's disease is an open question.

Phosphorylation at serines or threonines not adjacent to a proline appears to occur within the microtubule-binding domain of PHF-tau (Hasegawa et al. 1992). Phosphorylation at such sites, for example serine 262 in the first repeat of the binding domain, markedly decreases the binding of tau to microtubules (Biernat et al. 1993). One putative kinase for phosphorylation within the microtubule-binding domain is protein kinase C (Correas et al. 1992).

With the onset of maturity the fetal type phosphorylation must be shut off, providing a potential role for the protein phosphatases. Based upon in vitro data, candidate phosphatases that may mediate this fetal to adult transition are protein phosphatase (PP) 2A (Goedert et al. 1992) and 2B (Goto et al. 1985), PP2B is also known as calcineurin, a calmodulin-dependent serine-threonine phosphatase found most abundantly in neurons. A role for phosphatases has been suggested by the induction of hyper-phosphorylated forms of tau protein following okadaic acid treatment of human brain slices (Harris et al. 1993). We recently utilized dissociated cultures from embryonic day 15 rat cerebellum to localize calcineurin (Ferriera et al. 1993). During the outgrowth of neurites, calcineurin is enriched in growth cones where its localization was dependent upon the integrity of both the microtubules and actin filaments. Treatment with cytochalasin shifted the calcineurin to the neurite shaft; with nocadozole, calcineurin shifted to the cell body. Therefore calcineurin is well-positioned to mediate interactions between cytoskeletal systems during neurite elongation. For other phosphatases, subunits have been identified that confer cellular localization. For example, the M subunit of PP-1_c promotes its association with myofibrils where the enzyme can dephosphorylate myosin. When calcineurin in cultured neurons was inhibited with either cyclosporin or the specific calcineurin auto-inhibitory peptide, axonal elongation was prevented (Ferreira et al. 1993). Because of the role that the microtubule-associated protein tau plays in the asymmetric elongation of the single neurite destined to become the axon (Caceres and Kosik 1990), we sought modifications in the phosphorylation state of tau as a result of calcineurin inhibition. In contrast to the normal development of cerebellar macroneurons, in which reactivity with the phosphorylation-dependent antibody, tau-1, progressively increases, when calcineurin was inhibited there was a persistent inhibition of tau-1 reactivity. These data suggest a role for calcineurin in regulating the phosphorylation of tau, and that regulation can control axonal elongation.

Tau Function

When tau protein expression is suppressed in cultured cerebellar macroneurons by antisense oligonucleotides, the cells retain the ability to form minor neurites, but fail to convert one of these neurites to an axon (Caceres and Kosik 1990). By retaining a relatively symmetric array of minor neurites in the presence of tau antisense oligonucleotides, tau protein may have a role in rapid neurite elongation and the acquisition of neuronal polarity. When tau antisense oligonucleotides are administered to more mature cultures after the onset of axonal differentiation, the axons undergo regression, even though the remaining minor neurites continue to increase in complexity with branching patterns suggestive of dendritic differentiation (Caceres et al. 1991). On the other hand, even a prolonged suppression of tau by antisense oligonucleotides from the time of plating leaves the cells locked at a stage with minor neurites only and no dendritic differentiation. One possible speculation from these observations is that cerebellar macroneurons require the antecedent onset of axonal differentiation to develop dendritic morphologies; however, once axonal differentiation begins, these neurons can elaborate dendritic morphologies even as the axons regress.

The interpretation of gene suppression studies requires caution. A myriad of compensatory mechanisms may mask a gene's function; conversely, a secondary effect on downstream elements might suggest a more critical function for the suppressed product than its actual role in the cell. To obtain a more complete understanding of tau function we expressed various neuronal MAPs in *Spodoptera frugiperda* (Sf9) cells using a baculoviral vector (Knops et al. 1991; Baas et al. 1991). In these cells one can observe how a single human brain MAP re-organizes the cell's cytoplasm to create a change in cell shape. The expression of specific neuronal MAPs and MAP isoforms induces distinct phenotypes in the Sf9 host cell (Leclerc et al. 1993). In this capacity the Sf9 cells have served as a cell biological tool for the study of neuronal MAPs and their effect on cell shape.

Tau expression in Sf9 cells is very efficient, with about $20\,\mu g$ of tau protein in 10^6 cells (Knops et al. 1991). At 72 hours after infection, upwards of 75% of the cells react with tau antibodies. Of these reactive cells more than 70% have a single process, a much smaller number have two processes, and very rare cells have three processes. The processes often grow over $100\,\mu m$ in length. After the initial segment they are of uniform caliber and unbranched; there is often a small expansion of cytoplasm at the tip. Processes are elaborated by cells regardless of whether they are attached or in suspension. Within the processes there are discontinuous arrays of tightly packed microtubule bundles that are oriented with their plus ends aligned distally (Baas et al. 1991). Tau expression in Sf9 cells significantly changes the organization of the cytoplasm beyond its effect on microtubules. In the distal cytoplasmic expansion there are collections of organelles, intense reactivity with rhodamine phalloidin, and a relative paucity of microtubules. This actin-rich, microtubule-poor region suggests that the terminus of the tau-induced process resembles a growth cone (Knowles et al. 1994).

Tau expression in Sf9 cells offers the opportunity to study microtubules in cells polymerized with a single MAP. The properties of these microtubules have some interesting features. Specifically they are hyper-stable to nocadozole compared to axonal microtubules, and show essentially no resistance to cold depolymerization, in contrast to the cold-resistant populations of microtubules in axons (Baas et al. 1994). This observation suggests that tau confers specific aspects of microtubule stability.

Before these studies, the MAPs were considered collectively along with taxol as having a role in promoting the assembly of microtubules, stabilizing microtubules, and bundling microtubules. Taxol treatment of Sf9 cells does induce the assembly of microtubules into bundles, but does not induce process formation in Sf9 cells. Therefore the in vitro properties of tau as a molecule involved in microtubule assembly, microtubule stabilization, and microtubule bundling do not sufficiently account for the in vivo ability of tau alone to transduce microtubule assembly into a change in cell shape. By ^{35}S-methionine labeling of the tau-expressing Sf9 cells, we were not able to detect the induction of any additional proteins other than those found following wild-type viral infection (Cheley et al. 1992). Perhaps tau accomplishes its task by creating microtubules with stable proximal domains and more labile distal domains, similar to the distribution of acetylated and non-acetylated domains of microtubules in axons (Baas and Black 1990).

Stimulation of MAPK by Secreted APP

The secreted form of APP, designated APP_s, activates MAPK in PC12 cells (Greenberg et al. 1994). APP_s was purified from medium conditioned by CHO cells stably transfected with either the 695, 751, or 770 amino acid forms of APP. These forms differ by the addition of either one or two exons, one of which encodes a Kunitz protease inhibitor domain (reviewed in Kosik 1992). Purification entailed a column to which two monoclonal antibodies against distinct sites in APP were bound. A single band was eluted from the column with high salt. PC12 cells treated with purified APP_{751} increased the activity of MAPK/ERK-1 in PC12 lysates, as measured in immunocomplexes using myelin basic protein as a substrate. Activity increased from 17- to 25-fold after 10 minutes. Stimulation declined after longer exposures, with a 50% reduction in activity following one hour of treatment. Increased activation was observed at concentrations of APP_{751} as low as 20 pM. APP_{695} appears to have an identical effect on MAPK activity. As an independent assay for MAPK activation, antibodies against phosphotyrosine were used to blot lysates after stimulation. Increased tyrosine phosphorylation of proteins at M_r 44 and 42,000 that comigrate with ERK-1 and ERK-2 was readily detected. MAPK/ERK-1 activity of the medium was reduced by 65% when APP was immunoprecipitated from the CHO conditioned medium. The residual stimulatory activity following immunoprecipitation was probably due to other factors in the CHO medium that have some modest effect on MAPK/ERK-1. Synthetic Aβ peptide at similar concentrations had no effect on MAPK activity.

Other growth factors that activate MAPK, such as nerve growth factor and basic fibroblast growth factor, do so via activation of ras (Thomas et al. 1992; Wood et al. 1992; Kremer et al. 1991; Qui and Green 1992). The requirement for ras following APP_s stimulation was assayed in the PC12-derived cell line GSrasDN6. In this line, dexamethasone induces a dominant inhibitory form of ras (Thomas et al. 1992; Wood et al. 1992; Kremer et al. 1991). Induction of inhibitory ras prevented APP_s-stimulated MAPK activation, as determined by phosphorylation of the MBP substrate and tyrosine phosphorylation.

Linking APP and Hyper-Phosphorylation of Tau

How might APP or its metabolic products alter the phosphorylation of tau? To answer this question, PC12 cells were preincubated with [^{32}P] orthophosphate to saturation followed by the addition of APP_s. Beginning 15 minutes after the addition of APP_s, tau was immunoprecipitated from the PC12 cell lysates. Quantitation of phosphate incorporation by Phosphor imaging revealed a doubling in the radioactive phosphate content of tau following stimulation. Medium from untransfected cells, Aβ peptide, and stimulation in induced GSrasDN6 cells all showed no enhanced tau phosphorylation.

In PC12 cells, as in other cell cultures, tau is present in a fetal-like highly phosphorylated state. This baseline degree of phosphorylation may explain the 17- to 25-fold activation of MAPK/ERK and only a concomittant doubling of the tau phosphorylation. Secondly, because there is similarity between the fetal phosphorylation of tau and its phosphorylation state in Alzheimer's disease, it is important to determine if the enhanced phosphorylation of tau following APP stimulation bears any relationship to the Alzheimer state. To address this question we repeated the experiments in primary neuronal cultures taken from the cerebral cortex and the hippocampus. In these cultures there are many pyramidal cells, one of the principle target cells in the disease process. We found that reactivity with AT8 (Biernat et al. 1992), a monoclonal antibody that reacts with PHF-tau but not with normal adult tau, was enhanced after stimulation of these cells.

As this work proceeds it will be necessary to learn the relationship between the increased tau phosphorylation observed here and the increased tau phosphorylation observed in Alzheimer's disease.

References

Anderson JP, Esch FS, Keim PS, Sambamurti K, Lieberburg I, Robakis NK (1991) Exact cleavage site of Alzheimer amyloid precursor in neuronal PC-12 cells. Neurosci Lett 128: 126–128

Baas PW, Black MM (1990) Individual microtubules in the axon consist of domains that differ in both composition and stability. J Cell Biol 111: 495–509

Baas PW, Pienkowski TP, Kosik KS (1991) Processes induced by tau expression in Sf9 cells have an axon-like microtubule organization. J Cell Biol 115: 1333–1344

Baas PW, Pienkowski TP, Cimbalnik KA, Toyama K, Bakalis S, Ahmad FJ, Kosik KS (1994) Tau confers drug-stability but not cold stability to microtubules in living cells. J Cell Sci 107: 135–143

Biernat J, Mandelkow E-M, Schroter C, Lichtenberg-Kraag B, Steiner B, Berling B, Meyer H, Mercken M, Vandermeeren A, Goedert M, Mandelkow E (1992) The switch of tau protein to an Alzheimer-like state includes the phosphorylation of two serine-proline motifs upstream of the microtubule binding region. EMBO J 11: 1593–1597

Biernat J, Gustke N, Drewes G, Mandelkow EM, Mandelkow E (1993) Phosphorylation of Ser^{262} strongly reduces binding of tau to microtubules: distinction between PHF-like immunoreactivity and microtubule binding. Neuron 11: 153–163

Caceres A, Kosik KS (1990) Inhibition of neurite polarity by tau antisense oligonucleotides in primary cerebellar neurons. Nature 343: 461–463

Caceres A, Potrebic S, Kosik KS (1991) The effect of tau antisense oligonucleotides on neurite formation of cultured cerebellar macroneurons. J Neurosci 11: 1515–1523

Cheley S, Kosik KS, Paskevich P, Bakalis S, Bayley H (1992) Phosphorylated baculovirus p10 is a heat-stable microtubule-associated protein associated with process formation in Sf9 cells. J Cell Science 102: 739–752

Cobb MH, Boulton TG, Robbins DJ (1991) Extracellular signal-regulated kinases: ERKs in progress. Cell Regulation 2: 965–978

Correas I, Diaz-Nido J, Avila J (1992) Microtubule-associated protein tau is phosphorylated by protein kinase C in its tubulin binding domain. J Biol Chem 267: 15721–15728

Drewes G, Lichtenberg-Kraag B, Doring F, Mandelkow EM, Biernat J, Goris J, Doree M, Mandelkow E (1992) Mitogen-activated protein (MAP) kinase transforms tau protein into an Alzheimer-like state. EMBO J 6: 2131–2138

Esch FS, Keim PS, Beattie EC, Blacher RW, Culwell AR, Oltersdorf T, McClure D, Ward PJ (1990) Cleavage of amyloid beta peptide during constitutive processing of its precursor. Science 248: 1122–1124

Ferreira A, Kincaid R, Kosik KS (1993) Calcineurin is associated with the cytoskeleton of cultured neurons and has a role in the acquisition of polarity. Mol Biol Cell 4: 1225–1238

Goedert M, Cohen ES, Jakes R, Cohen P (1992) p42 MAP kinase phosphorylation sites in microtubule-associated protein tau are dephosphorylated by protein phosphatase 2A1. FEBS 312: 95–99

Goto S, Yamamoto H, Fukunaga K, Iwasa T, Matsukado Y, Miyamoto E (1985) Dephosphorylation of microtubule associated protein 2, tau factor and tubulin by calcineurin. J Neurochem 45: 276–283

Gotoh Y, Nishida E, Matsuda S, Shiina N, Kssako H, Shiokawa K, Akiyama T, Ohta K, Sakai H (1991) In vitro effects on microtubule dynamics of purified Xenopus M phase-activated MAP kinase. Nature 349: 251–254

Greenberg SG, Davies P (1990) A preparation of Alzheimer paired helical filaments that displays distinct tau proteins by polyacrylamide gel electrophoresis. Proc Natl Acad Sci USA 87: 5827–5831

Greenberg SM, Koo EH, Selkoe DJ, Qiu WQ, Kosik KS (1994) Stimulation of MAP kinase and tau phosphorylation by secreted β-amyloid precursor protein. Proc Natl Acad Sci USA, in press

Haass C, Schlossmacher MG, Hung AY, Vigo-Pelfrey C, Mellon A, Ostazewski BL, Lieberburg I, Koo EH, Schenk D, Teplow DB, Selkoe DJ (1992) Amyloid β-peptide is produced by cultured cells during normal metabolism. Nature 359: 322–325

Hanger DP, Hughes K, Woodgett JR, Brion JP, Anderton BH (1992) Glycogen synthase kinase-3 induces Alzheimer's disease-like phosphorylation of tau: generation of paired helical filament epitopes and neuronal localisation of the kinase. Neurosci Lett 147: 58–62

Hasegawa M, Morishima-Kawashima M, Takio K, Suzuki M, Titani K, Ihara Y (1992) Protein sequence and mass spectrometric analyses of tau in the Alzheimer's disease brain. J Biol Chem 267: 17047–17054

Harris KA, Oyler GA, Doolittle GM, Vincent I, Lehman RAW, Kincaid RL, Billingsley ML (1993) Okadiac acid induces hyperphosphorylated forms of tau protein in human brain slices. Ann Neurol 33: 77–87

Knops J, Kosik KS, Lee G, Pardee JD, Cohen-Gould L, McConlogue L (1991) Overexpression of tau in a non-neuronal cell induces long cellular processes. J Cell Biol 114: 725–733

Knowles R, Leclerc N, Kosik KS (1994) Organization of actin and microtubules during process formation in tau-expressing Sf9 cells. Cell Motility Cytoskeleton 28: 256–264

Kosik KS (1992) Alzheimer's disease from a cell biological perspective. Science 256: 780–783

Kremer NE, D'Arcangelo G, Thomas SM, DeMarco M, Brugge JS, Halegoua S (1991) Signal transduction by nerve growth factor and fibroblast growth factor in PC12 cells requires a sequence of Src and Ras actions. J Cell Biol 115: 809–819

L'Allemain GL, Pouyssegur J, Weber MJ (1991) p42/Mitogen-activated protein kinase as a converging target for different growth factor signaling pathways: Use of pertussis toxin as a discrimination factor. Cell Regulation 2: 675–684

Lange-Carter CA, Pleiman CM, Gardner AM, Blumer KJ, Johnson GL (1993) A divergence in the MAP kinase regulatory network defined by MEK kinase and raf. Science 260: 315–319

Leclerc N, Kosik KS, Cowan N, Pienkowski TP, Baas PW (1993) Process formation in Sf9 cells induced by the expression of a MAP2C-like construct. Proc Natl Acad Sci USA 70: 6223–6227

Ledesma MD, Correas I, Avila J, Diaz-Nido J (1992) Implication of brain cdc2 and MAP2 kinases in the phosphorylation of tau protein in Alzheimer's disease. FEBS Lett 308: 218–224

Lee VM-Y, Balin BJ, Otvos L, Trojanowski JQ (1991) A68. A major subunit of paired helical filaments and derivatized forms of normal tau. Science 251: 675–678

Liu W-K, Ksiezak-Reding H, Yen S-H (1991) Abnormal proteins from Alzheimer's disease brain. Purification and amino acid analysis. J Biol Chem 266: 21723–21727

Liu W-K, Dickson DW, Yen S-HC (1994) Amino acid residues 226–240 of τ, which encompass the first Lys-Ser-Pro site of τ, are partially phosphorylated in Alzheimer paired helical filament-τ. J Neurochem 62: 1055–1061

Lu Q, Soria JP, Wood JG (1993) p44^{mpk} MAP kinase induces Alzheimer type alterations in tau function and in primary hippocampal neurons. J Neurosci Res 35: 439–444

Mandelkow EM, Drewes G, Biernat J, Gustke N, Lint JV, Vandenheede JR, Mandelkow (1992) Glycogen synthase kinase-3 and the Alzheimer-like state of microtubule-associated protein tau. FEBS Lett 314: 315–321

Nebreda AR (1989) Inactivation of MAP kinases. Trends Biochem Sci 19: 1–2, 1994

Ninomiya H, Roch JM, Sundsmo MP, Otero DAC, Saitoh T (1993) Amino acid sequence RERMS represents the active domain of amyloid β/A4 protein presursor that promotes fibroblast growth. J Cell Biol 121: 879–886

Nishizawa N, Okano Y, Chatani Y, Amano F, Tanaka E, Nomoto H, Nozawa Y, Kohno M (1990) Mitogenic signaling pathways of growth factors can be distinguished by the involvement of pertussis toxin-sensitive guanosine triphosphate-binding protein and of protein kinase C. Cell Regulation 1: 747–761

Pelech SL, Sanghera JS (1992) Mitogen-activated protein kinases: versatile transducers for cell signaling. Trends Biochem Sci 17: 233–238

Pulverer BJ, Kyriakis JM, Avruch J, Nikolakaki E, Woodgett JR (1991) Phosphorylation of c-jun mediated by MAP kinases. Nature 353: 670

Qui MS, Green SH (1992) PC12 cell neuronal differentiation is associated with prolonged p21ras activity and consequent prolonged ERK activity. Neuron 9: 705–717

Saitoh T, Sundsmo M, Roch JM, Kimura N, Cole G, Schubert D, Oltersdorf T, Schenk DB (1989) Secreted form of amyloid β protein precursor is involved in the growth regulation of fibroblasts. Cell 58: 615–622

Schubert D, Schroeder R, Lacorbiere M, Saitoh T, Cole GM (1988) Amyloid β protein precursor is possibly a heparan sulfate proteoglycan core protein. Science 241: 233–226

Schubert D, Jin LW, Saitoh T, Cole GM (1989) The regulation of amyloid β protein precursor secretion and its modulatory role in cell adhesion. Neuron 3: 689–694

Seubert P, Vigo-Pelfrey C, Esch F, Lee M, Dovey H, Davis D, Sinha S, Schlossmacher M, Whaley J, Swindlehurst C, McCormack R, Wolfert R, Selkoe D, Lieberberg I, Schenk D (1992) Isolation and quantification of soluble Alzheimer's β-peptide from biological fluids. Nature 359: 325–327

Shoji M, Golde TE, Ghiso J, Cheung TT, Estus S, Shaffer LM, Cai XD, McKay DM, Tinter R, Frangione B, Younkin SG (1992) Production of the Alzheimer amyloid β protein by normal proteolytic processing. Science 258: 126–129

Sisodia SS, Koo EH, Beyreuther K, Unterbeck AJ, Price DL (1990) Evidence that beta-amyloid protein in Alzheimer's disease is not derived by normal processing. Science 248: 492–495

Tsao H, Aletta JM, Greene LA (1990) Nerve growth factor and fibroblast growth factor selectively activate a protein kinase that phosphorylates high molecular weight microtubule-associated proteins. J Biol Chem 265: 15471–15480

Thomas SM, DeMarco M, D'Arcangelo G, Halegoua S, Brugge JS (1992) Ras is essential for nerve growth factor- and phorbol ester-induced tyrosine phosphorylation of MAP kinases. Cell 68: 1031–1040

Ueda K, Cole GM, Sundsmo MP, Katzman R, Saitoh T (1989) Decreased adhesiveness of Alzheimer's disease fibroblasts: is amyloid beta-protein precursor involved? Ann Neurol 25: 246–251

Vulliet R, Halloran SM, Braun RB, Smith AJ, Lee G (1992) Proline-directed phosphorylation of human tau protein. J Biol Chem 267: 22570–22574

Watanabe A, Hasegawa M, Suzuki M, Takio K, Morishima-Kawashima M, Titani K, Arai T, Kosik KS, Ihara Y (1993) In vivo phosphorylation sites in fetal and adult rat tau. J Biol Chem 268: 25712–25717

Weidemann A, Konig G, Bunke D, Fishcher P, Salbaum JM, Masters CL, Beyreuther K (1989) Identification, biogenesis, and localization of precursor of Alzheimer's disease A4 amyloid protein. Cell 57: 115–126

Whitson JS, Selkoe DJ, Cotman CW (1989) Amyloid β protein enhances the survival of hippocampal neurons in vitro. Science 243: 1488–1490

Wille H, Drewes G, Biernat J, Mandelkow EM, Mandelkow E (1992) Alzheimer-like paired helical filaments and antiparallel dimers formed from microtubule-associated protein tau in vitro. J Cell Biol 118: 573–584

Wood KW, Sarnecki C, Roberts TM, Blenis J (1992) ras mediates nerve growth factor receptor modulation of three signal-transducing protein kinases: MAP kinase, Raf-1, and RSK. Cell 68: 1041–1050

Subject Index

Printing: Saladruck, Berlin
Binding: Buchbinderei Lüderitz & Bauer, Berlin